C0-BYW-876

Stem Cells

ACADEMIC PRESS RAPID MANUSCRIPT REPRODUCTION

Proceedings of a symposium held in October 1975
at McGill University, Montreal in tribute to
C. P. Leblond on the occasion of his
sixty-fifth birthday

Stem Cells

Of Renewing Cell Populations

Edited by

A. B. Cairnie

Department of Biology, Queen's University,
Kingston, Ontario

P.K. Lala

Department of Anatomy, McGill University,
Montreal, Quebec

D. G. Osmond

Department of Anatomy, McGill University,
Montreal, Quebec

ACADEMIC PRESS NEW YORK SAN FRANCISCO LONDON 1976

A Subsidiary of Harcourt Brace Jovanovich, Publishers

ACADEMIC PRESS, INC.
111 Fifth Avenue, New York, New York 10003

United Kingdom Edition published by
ACADEMIC PRESS, INC. (LONDON) LTD.
24/28 Oval Road, London NW1

Library of Congress Cataloging in Publication Data

Main entry under title:

Stem cells of renewing cell populations.

Bibliography: p.
Includes index.
1. Stem cells–Congresses. 2. Cell populations–Congresses. 3. Cell proliferation–Congresses. 4. Cell differentiation–Congresses. 5. Leblond, Charles Philippe, Date I. Leblond, Charles Philippe, Date II. Cairnie, A. B. III. Lala, P. K. IV. Osmond, D. G. V. McGill University, Montreal.
QH587.S73 599′.08′761 76-8003
ISBN 0–12–155050–8

PRINTED IN THE UNITED STATES OF AMERICA

Contents

List of Contributors

Gabriel G. Altmann, Department of Anatomy, Health Science Center, University of Western Ontario, London, Ontario.

N. M. Blackett, Division of Biophysics, Institute of Cancer Research, Sutton, Surrey, England.

H. E. Broxmeyer, Sloan-Kettering Institute for Cancer Research, New York, New York 10021

W. R., Bruce, Ontario Cancer Institute, 500 Sherbourne Street, Toronto, Ontario.

G. Brugal, Laboratoire de Zoologie, Universite Scientifique et Medicale de Grenoble, B.P. no. 53, 38041 Grenoble, Cedex, Grenoble, France.

A. B. Cairnie, Department of Biology, Queen's University, Kingston, Ontario.

Hazel Cheng, Department of Anatomy, McGill University, Montreal, Quebec.

Y. Clermont, Department of Anatomy, McGill University, Montreal, Quebec.

Juliana Denekamp, Cancer Research Campaign, Gray Laboratory, Mount Vernon Hospital, Norwood, England.

N. B. Everett, Department of Biological Structure, University of Washington, School of Medicine, Seattle, Washington 98195.

Jack Fowler, Cancer Research Campaign, Gray Laboratory, Mount Vernon Hospital, Norwood, England.

L. Hermo, Department of Biology, Queen's University, Kingston, Ontario.

Claire Huckins, Biology Division, Oak Ridge National Laboratory, Oak Ridge, Tennessee 37830

A. D. Hugenholtz, Department of Biology, York University, Downsview, Ontario.

G. R. Johnson, Sloan-Kettering Institute for Cancer Research, New York, New York 10021.

J. Kurland, Sloan-Kettering Institute for Cancer Research, New York, New York 10021.

P. K. Lala, Department of Anatomy, McGill University, Montreal, Quebec.

L. F. Lamerton, Division of Biophysics, Institute of Cancer Research, Sutton, Surrey, England.

Edna B. Laurence, Mitosis Research Laboratory, Birkbeck College, University of London, London WClE 7HX, England.

C. P. Leblond, Department of Anatomy, McGill University, Montreal, Quebec.

John M. Lehman, Department of Pathology, University of Colorado Medical Center, Denver, Colorado.

B. I. Lord, Paterson Laboratories, Christie Hospital and Hold Radium Institute, Manchester, England.

Mary A. Maloney, University of California, Laboratory of Radiobiology, San Francisco, California 94143.

H. S. Micklem, Immunobiology Unit, Department of Zoology, University of Edinburgh, Edinburgh EH9 3JT, Scotland.

R. G. Miller, Department of Medical Biophysics, University of Toronto, 500 Sherbourne St reet, Toronto, Ontario, M4X 1K9.

P. B. Moens, Department of Biology, York University, Downsview, Ontario.

M. A. S. Moore, Sloan-Kettering Institute for Cancer Research, New York, New York 10021.

E. F. Oakberg, Biology Division, Oak Ridge National Laboratory, Oak Ridge, Tennessee 37830.

D. A. Ogden, Immunogiology Unit, Department of Zoology, University of Edinburgh, Edinburgh EH9 3JT, Scotland.

D. G. Osmond, Department of Anatomy, McGill University, Montreal, Quebec.

Harvey M. Patt, University of California, Laboratory of Radiobiology, San Francisco, California 94143

W. D. Perkins, Department of Biological Structure, University of Washington, School of Medicine, Seattle, Washington, 98195

Christopher S. Potten, Paterson Laboratories, Christie Hospital and Holt Radium Institute, Manchester, England.

Wendell C. Speers, Department of Pathology, University of Colorado Medical Center, Denver, Colorado.

Alan L. Thornley, Mitosis Research Laboratory, Birkbeck College, University of London, London, WC1E 7HX, England.

J. E. Till, Ontario Cancer Institute, 500 Sherbourne Street, Toronto, Ontario.

John J. Trentin, Division of Experimental Biology, Baylor College of Medicine, Houston, Texas 77025

H. Rodney Withers, Section of Experimental Radiotherapy, University of Texas System Cancer Center, M.D. Anderson Hospital and Tumor Institute, Houston, Texas 77025

Dr. Charles P. Leblond

C. P. Leblond

The dynamic processes taking place within organs, tissues, and cells have always held a special fascination for Dr. C.P. Leblond, in full accord with his own personal dynamism. The vigor of his approach to biology has been and remains a constant inspiration for his many colleagues and students.

An enthusiastic teacher of histology at McGill University since 1941, Dr. Leblond's research productivity is impressive. So far, he has contributed close to three hundred articles to the scientific literature. Those who have worked close to him know the real significance of this figure, since Dr. Leblond's rigorous scientific thinking and perfectionism usually blossom most fully at the time of writing, sometimes to the despair of his students or collaborators!

The technique of radioautography developed by Dr. Leblond in collaboration with Dr. L. Belanger in the mid-1940's was perfectly adapted to view biological phenomena dynamically as a function of time. This technique, now utilized in both light and electron microscopy, provides a most elegant tool to analyze the kinetic processes taking place within either cell populations or the cells themselves. Using radioautography Dr. Leblond and his associates have extensively investigated a wide variety of problems in histology and cell biology, some of which are represented in the present symposium. The diversity of Dr. Leblond's interests may be illustrated by the following examples of areas in which he has made major contributions: the mode of renewal of the intestinal epithelium (studies that started as early as 1947) and of several other cell populations (skin, thymus, esophagus, stomach, testis, etc.), the formation of the thyroid hormone using radioactive iodine as a tracer; the formation and calcification of bone and tooth; the synthesis of proteins in various cells using ^{35}S-labeled methionine as a tracer; the role of the Golgi apparatus in the synthesis of glycoproteins using labeled amino acids as precursors, etc.

It is evident that Dr. Leblond has a particular flair for discerning the significant facts that lead to discoveries. This remarkable quality, combined with his overwhelming enthusiasm in pursuing methodically a research project, has made him eminently capable of training numerous students. Indeed, he is "un maître" in the classical sense of the word and in the best scientific tradition. This facet of Dr. Leblond's personality is happily coupled with warmth and good humor. After long hours of hard work, there is always a place for bursts of joyful relaxation.

Throughout his prolific career Dr. Leblond's scientific endeavors have been acknowledged by a long list of honors, including the Flavelle Medal of the

Royal Society of Canada (1961), the Gairdner Foundation Award (1965), the American College of Physicians Award (1966). A Fellow of the Royal Society of Canada since 1951 and the Royal Society of London since 1965, he also received an Honorary Doctor of Science Degree from Acadia University, Wolfville, Nova Scotia in 1972.

The pioneering studies of Dr. Leblond on the kinetics of cell populations led him to the identification and analysis of the behavior of the "stem cells," which are at the origin of each cell line. Because of his continuous interest in this area of investigation, and also of the importance of this topic in biology, it was felt that an international symposium on stem cells was the best possible way to acknowledge Dr. Leblond's scientific endeavor and honor him as a leading scientist in biology.

Preface

This publication constitutes the proceedings of a symposium entitled, "Stem Cells in Various Tissues" which was organized as a tribute to Charles P. Leblond on the occasion of his sixty-fifth birthday.

A major scientific goal in planning the symposium was that of gathering together investigators working in a wide variety of fields who would not otherwise have the opportunity to interact with one another. For the first time, workers from many disciplines, using several distinct techniques, were able to exchange information on stem cells in a variety of organ systems studied under normal steady state conditions, as well as during growth, aging, regeneration, and neoplasia.

The chairman of the meeting was L.F. Lamerton, widely acknowledged as the *eminence grise* of the field. His opening address and closing summary will indicate to readers why he was considered to be the obvious person to voice both the first and the last word. We are grateful to him for undertaking this arduous task with such skill and grace.

The conference was made possible by generous financial support from the Medical Research Council of Canada, the National Cancer Institute of Canada, le Conseil de la Recherche en Santé du Québec, the Faculty of Medicine and the Faculty of Graduate Studies and Research, McGill University. The organizing committee, consisting of J.E. Till, Y. Clermont, and ourselves, is grateful to those who came so willingly to take part in the program, to those who came to listen and to discuss, and to all those who helped behind the scenes to make the conference a successful and happy scientific occasion. Finally, we thank Academic Press for their assistance in publishing the proceedings.

A.B. Cairnie
P.K. Lala
D.G. Osmond

Chairman's Opening Address at the Symposium on Stem Cells: A Tribute to C. P. Leblond

L. F. LAMERTON

Institute of Cancer Research, Sutton, Surrey, England

This meeting is to honour Professor Charles Leblond. I had wondered why I was given the great privilege of being Chairman and have concluded it was because I am old, old enough to have been in the field during the major biological revolution in which Charles Leblond has played such an important part.

I am, in fact, not quite old enough to have experienced the full impact of the first biological applications of artificial isotopes in the 30's and the early 40's with which the name of George Hevesy will always be associated, and the recognition of the 'dynamic state of body constituents', the title of Schoenheimer's lectures of 1941. The concept of the constituents of the body being in a state of flux, of course, is not a new one. Heraclitus, in the 5th century B.C., believed it. However, the concept was largely forgotten until markers of atoms became available in the form of isotopes, mainly radioactive, though one must not forget that the classic work of Schoenheimer and Rittenberg in 1936 on fatty acid turnover was done with deuterium, which was a stable isotope.

But this was a biochemist's world, and biochemists were too preoccupied with the complexities of their own subject to allow them to pay much attention to the biological problems arising from the heterogeneity of the tissues with which they were dealing, and this led to many difficulties and uncertainties in the interpretation of experimental data.

It was Charles Leblond who recognized the need to study turnover at the cellular level; not only recognized the need but developed the techniques, and over many years has not ceased to produce new and original findings and classic work in the field.

The technique of high resolution radioautography first described by Belanger and Leblond in 1942 has been, and is, of profound importance in biological work because, as Leblond himself has expressed it, it gives histology a fourth dimension - time. The ability to study turnover at the cellular level has opened up a Pandora's box (or as some who are for a quiet life might say, a nest of vipers!) and the fields of cell metabolism and cell population kinetics studied by radioautographic methods now engage a great army of workers.

I am sure Professor Leblond will understand me when I say we should all be grateful he is not a mathematician. This doesn't mean that he has not recognized the place of mathematics in his work and has not collaborated most effectively with very expert mathematical colleagues; but if he had been more mathematically inclined himself, he might have been seduced away from the microscope to indulge in the luxury of devising hypothetical control systems. He did not succumb, and I think he would maintain that before one begins to work out control mechanisms, it is essential to have a thorough understanding of the biology of the system. This remains for us, particularly at the present time, a very important lesson, with computers lurking around every corner demanding employment! I see Professor Leblond's hand in the program of this meeting in that most of the sessions start with a paper on the question of the identification of the appropriate stem cell population.

One of the great debts we owe to Professor Leblond and his colleagues is the classification of the tissues of the body based on their proliferative characteristics. Their application of the principles of the steady state to the renewing tissues of the body has been very fruitful and, increasingly, has concentrated attention on the stem cell populations - comprised of those cells that have the dual property of self-maintenance and of providing cells for the maturation pathway.

The classical theory of control of stem cell population, as presented by Osgood, was based on the assumption of asymmetric division of stem cells and, in a sense, was more a mathematical device than a biological theory. A most important contribution to the subject was made by Professor Leblond when, with Marques-Pereira, he demonstrated that this theory of stem cell control, at least in a simple form, was certainly not applicable to all tissues and possibly to none. This study also raised, in a clear fashion, the fundamental issue of how far the characteristics of stem cells are a consequence of their microenvironment. Is there, as the work suggested, a spatially defined zone in which cells can continue to divide indefinitely without moving on to the maturation pathway? This is a theory with a number of attractions - it shows how a limit could be set to the size of the stem cell population, and could explain the relatively slow turnover of stem cells in most unstimulated tissues as the result of some density-dependent inhibition of division. The problem remains of the particular characteristics of the environment that allow the cells to be shielded from maturation influences; and, in the case of the skin, the problem is made more complex by present work - which we shall hear about - suggesting that the basal layer is by no means entirely composed of stem cells. And then there is the nature of the feed-back message changing the rate of output of cells from the stem cell compartment. Would this primarily be a proliferative stimulus, causing cells to enter the maturation compartment, perhaps as a result of population pressure, or a maturation stimulus leading secondarily to increased proliferation, in other words, *vis a tergo* or *vis a fronte*, or both?

The question of the extent of the influence of the environment in determining stem cell behaviour is, of course, bound up with the ability of cells to proliferate indefinitely and produce cells for the maturation pathway when transplanted into other environments. This requires experiments involving clonogenic assay as, in fact, do many other types of study one would like to do with stem cells. Histological and cell turnover studies can give important clues about stem cell populations, but since the stem cell is defined prospectively, in terms of its future behaviour, the information is bound to be very limited unless we also have functional assays at our disposal.

It is interesting, but perhaps not surprising, that the break-through in providing clonogenic techniques for cells of normal tissues also came from Canada, from the Ontario Cancer Institute. Among their many achievements these workers have demonstrated a remarkable talent for recognizing the significance of some often rather odd biological phenomenon as the basis for quantitative studies of fundamental processes. For us the important example of this was the spleen colony technique of Till and McCulloch for hemopoietic stem cells, and this ushered in the second phase of stem cell investigation which, of course, has now been extended to a number of other tissues, as we shall hear over the next three days.

By means of clonogenic techniques a great deal of information is being gathered about the proliferative behaviour, response to various agents and other characteristics of the repopulating cells, but I believe the real need now is for the integration of the two approaches, histological and clonogenic. One problem here is the relationship between the effective stem cell population *in vivo* and the cell populations measured by various types of clonogenic assay, which has a bearing on the question as to whether the stem cells in various tissues represent a population quite distinct from the maturing cells or whether under stress, for instance under the conditions of clonogenic assay, some cells may revert to stem cell behaviour. However, I suspect that the major problem in the integration of histological and clonogenic approaches is the unequivocal identification of the position and extent of the clonogenic cell population within the tissue architecture. For the gut and the skin, we may be somewhere near the solution. For the bone marrow we need much more information on microarchitecture, and this itself could help in the solution of many outstanding problems.

A great deal of data is going to be presented over the course of the next three days and it is important that the wood should not be lost amongst the trees or, perhaps I should say, the stem amongst the flowers. I think that when there is any doubt, speakers should define what they mean by 'stem cells', and particularly if they use such odd terms as 'committed stem cell'. Let us not forget that the initial stem cell of the body, the fertilized ovum from which all the other stem cell populations are derived, is itself committed to producing *Homo sapiens* and, in fact, still further committed to producing either male or female. Let us relate, whenever we can, our observations to the basic problems, on the one hand, of the nature and mode of operation of the messages to the stem cell compartment that control the rate of output of cells at maturation and, on the other, to the factors that give a cell 'stem' properties. Somehow I feel we shall have enough to talk about.

Now this meeting, which will explore so many facets of dynamic and functional histology, must be very gratifying to Professor Leblond. The voice crying in the wilderness has now become a roar, even if it still has to penetrate some of the fastnesses of classical histology and pathology. Professor Leblond must feel justly proud that the subject which he has been so instrumental in initiating and developing has become one of the major growing points of modern biology.

Session I

Stem Cells in the Intestine

Identification of Stem Cells in the Small Intestine of the Mouse

C. P. LEBLOND and HAZEL CHENG

Department of Anatomy, McGill University, Montreal, Quebec, Canada

Many tissues contain cells which divide throughout life and thus give rise to both differentiating cells and cells similar to themselves. They are the stem cells. In some tissues, especially the seminiferous epithelium, stem cells have been precisely identified. In other locations, such as the bone marrow, they have not been identified with certainty, although some of their properties are known. In the small intestine, the stem cell is believed to be present in the crypts, but the information on the subject is rather vague. In this presentation, an attempt will be made at precise identification, using recently published data (Cheng and Leblond, '74 a,b,c; Cheng '74 a,b). The argument will be preceded by a few comments on the properties of stem cells.

PROPERTIES OF STEM CELLS

Stem cells are found only in those populations of cells which undergo "renewal" throughout life. Such populations are recognized in radioautographs of animals sacrificed one hour after injection of ^{3}H-thymidine. Under these conditions, it is possible to distinguish cell populations which are never labeled by ^{3}H-thymidine; they comprise mainly central and peripheral neurons and are known as *static*. Other cell populations display a few cells labeled by ^{3}H-thymidine, but in smaller and smaller numbers as the animal ages. Such populations are observed in many organs, for instance in liver and kidney, where the few cells arising from mitosis are retained indefinitely (Leblond, '64). These populations are known as *expanding*, although the rate of expansion decreases with age and is very low in the adult. Finally, *renewing* cell populations are those in which many cells are labeled by ^{3}H-thymidine in the adult animal. Hence, many mitoses occur continually. In such cases, the production of new cells must be balanced by the loss of an equivalent number of cells by emigration, degeneration or other processes.

Renewing cell populations include *stem cells*. These have the ability to undergo divisions which give rise to new stem cells and to others known as *differentiating cells*. The latter, with or without further division, evolve into the fully differentiated, non-dividing *end cells*. For instance, in the seminiferous epithelium, the stem cells are the type A spermatogonia; the differentiating cells include type B cells, spermatocytes and spermatids, whereas the end cells are the spermatozoa(Clermont, '72). The restriction of the stem cells to a definite location in the seminiferous tubules, that is, along the limiting membrane, has greatly facilitated their study in this system. In contrast, some of the ambiguity in our knowledge of the bone marrow stem cells arises from the fact that these cells have not yet been clearly identified and their precise localization is not known.

Steady state considerations require that, in the adult animal, the number of stem cells be constant. This constancy - even though not always fully maintained[1] - implies that, on the average, half the daughter cells of stem cell mitoses be new stem cells while the other half evolve toward differentiation. The simplest way for Nature to achieve this goal would be for each division of a stem cell to give rise to one daughter cell that remains a stem cell and one that undergoes differentiation. Such a division may be referred to as differential, since it implies that the two daughter cells receive different amounts of a material leading them to a different fate. Rolshoven ('51) proposed that the divisions of stem cells in renewal systems are differential. However, the only authenticated case of differential mitosis in mammals is that of the oocyte, whose divisions give rise to egg cell and polar bodies. Since in the monkey, rat and human testes, the paired cells arising from division of type A spermatogonia may be identified readily, they were compared and found to always consist of two identical cells, which were either both new type A cells or both differentiating cells (Clermont and Leblond, '59; Clermont, '66; '72). Hence, there was no

---

[1] Qualitative observations indicate a decrease of the number of stem cells with age in at least two renewing cell populations: the subependymal cells (glia producers) and muscle satellite cells (donors of nuclei to striated muscle fibers).

differential mitosis. Somewhat similar, though less compelling evidence, was obtained in the basal layer of the stratified epithelium of the esophagus where, according to recent evidence (Potten, this volume), both stem and differentiating cells are present. The two cells arising from a division in the basal layer are initially identical, but one or the other or both may later undergo differentiation while migrating outward (Marques-Pereira and Leblond '65). Hence, in both testis and esophagus differentiation occurs independently from stem cell mitoses.

The fine structure of stem cells, as seen in the

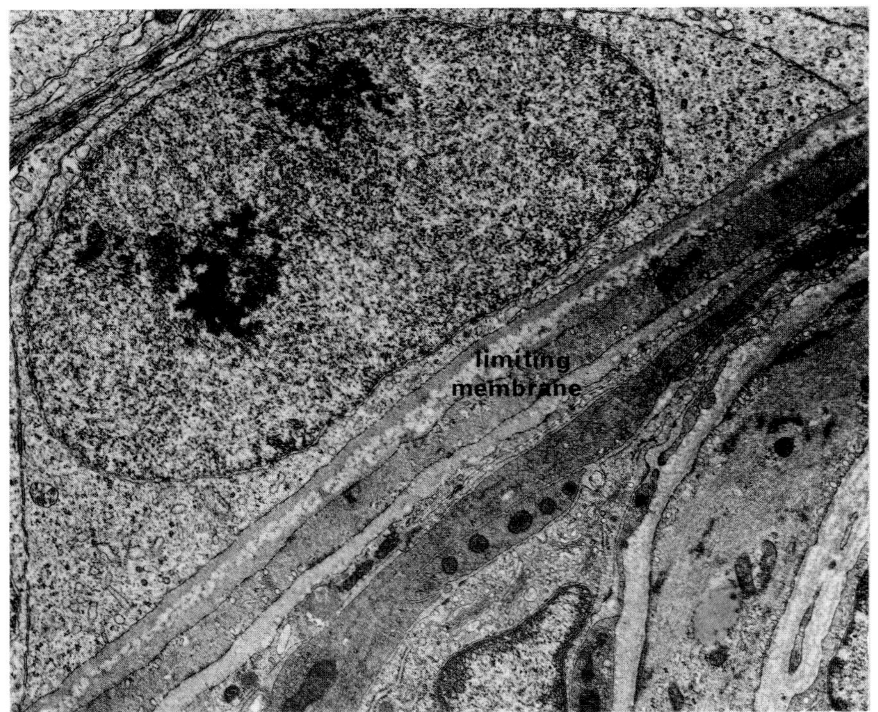

Fig. 1. Type A spermatogonia in a tubule at stage II of the cycle of the seminiferous epithelium. This cell lies adjacent to the limiting membrane of the tubule. The nucleus displays two nucleoli; the chromatin is diffusely distributed. The cytoplasm shows few organelles; small cisternae of endoplasmic reticulum are scattered, but the main component consists of free ribosomes present singly or in groups (preparation of L. Hermo and Y. Clermont). X 15100.

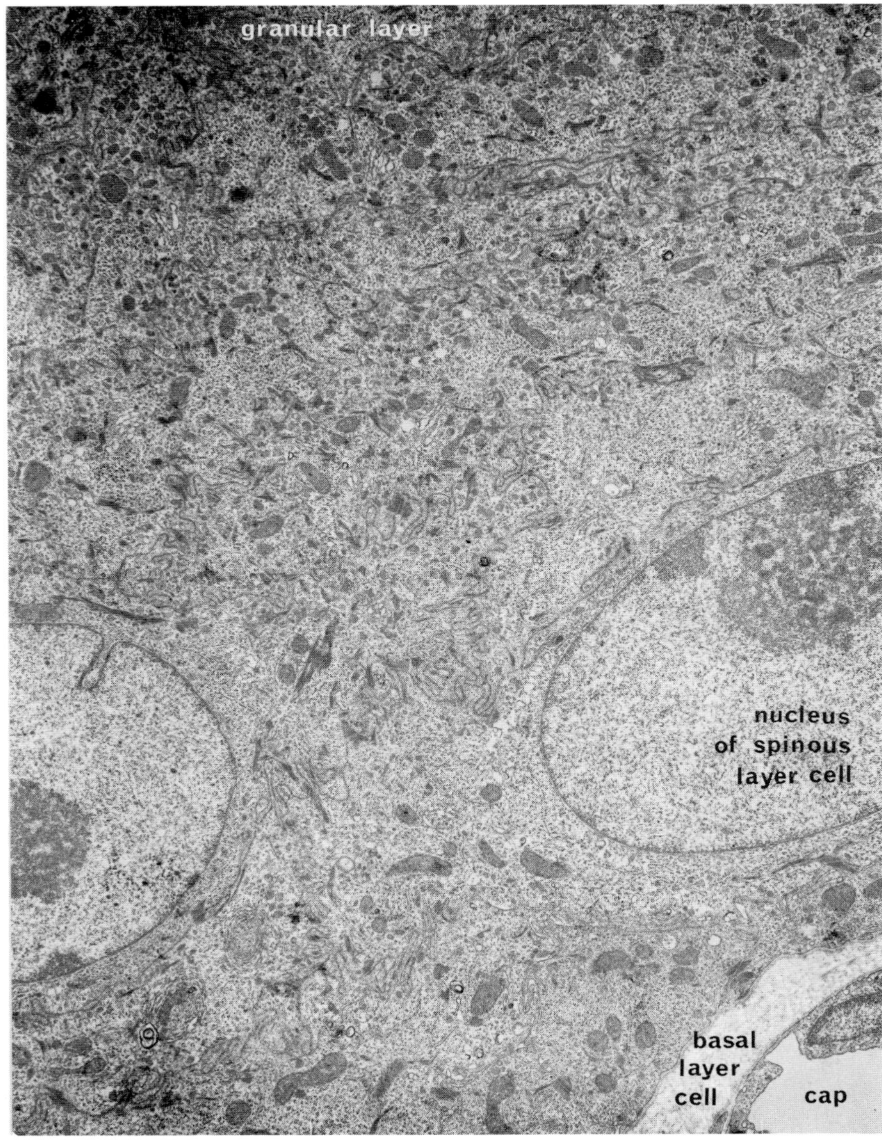

Fig. 2. Malpighian layer of stratified squamous epithelium of mouse forestomach. At lower right, the lamina propria shows a capillary (cap) separated by light staining collagen fibers from the basal layer. The cells of this layer, which include the stem cells, are small and packed

electron microscope has been little investigated. Let us mention a few observations. The clearly identified stem cells of the testis - the type A spermatogonia - has been recently described by Clermont and Hermo in the monkey (this volume). This cell in the rat has a nucleus with diffuse chromatin and a cytoplasm poor in organelles, except for free ribosomes which singly or as polysomes are scattered throughout the cytoplasm (fig. 1). This pattern is reminiscent of embryonic cells. When a dormant type A cell undergoes successive mitoses, the daughter type A cells distribute themselves along the limiting membrane of the seminiferous tubule, a fact suggestive of ameboid properties.

Incidental observations of the stratified squamous epithelium of the mouse forestomach revealed that basal layer cells (which include stem cells) are small and poor in organelles and desmosomes, but rich in free ribosomes; and that as a cell migrates farther from this layer, into the spinous and granular granular, it accumulates more and more organelles and desmosomes (fig. 2).

Similarly, the subependymal cells - the stem cells giving rise to oligodendrocytes and astrocytes - again reveal a pattern of scanty organelles and numerous free ribosomes (Privat and Leblond, '72). Moreover, cytoplasmic processes and nuclear indentations suggest ameboidism (fig. 3), a property which passed on to their daughter cells allows them to migrate into brain tissue while transforming into glial cells.

While only cautious conclusions can be drawn in the absence of systematic observations, the stem cells may be tentatively defined as cells which a) are able to divide in the adult while maintaining their own number and producing differentiating cells and b) seem to retain embryonic features, such as small size, scarcity of organelles and abundance of free ribosomes.

---

with free ribosomes, although mitochondria as well as a few desmosomes and rough endoplasmic reticulum cisternae are also present. Above, two nuclei of spinous layer cells contain nucleoli and mostly diffuse chromatin; there is less packing of free cytoplasmic ribosomes, but more desmosomes along the membrane than in the basal layer. Above, the cytoplasm of transitional and granular layer cells is filled with organelles and granular material; there is still a fair amount of free ribosomes (preparation of J. E. Michaels). X 6000.

STEM CELLS IN SMALL INTESTINE

The epithelium which lines the crypts and covers the villi of the small intestine is composed of cells which undergo renewal (Leblond and Stevens, '48; Leblond and Messier, '58). Columnar cells make up about 90% of the epithelial population. Two other cell types, referred to as mucous and entero-endocrine, are scattered throughout the epithelium in crypts and villi, like the columnar cells themselves. A fourth type, the Paneth cell, is restricted to the base of the crypts. These cells were investigated in the light microscope (fig. 4) and the electron microscope

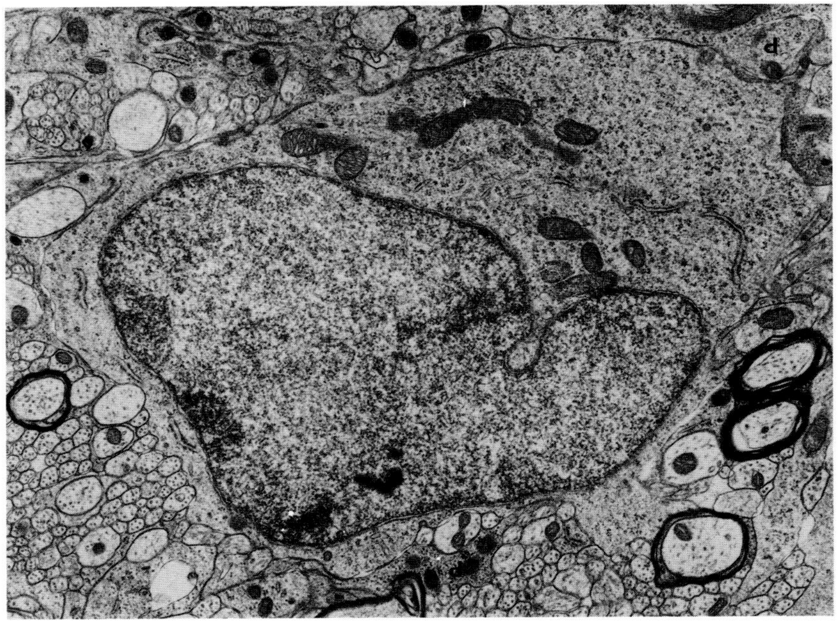

Fig. 3. Subependymal cell at the edge of the subependymal layer in the rat brain. The nucleus is indented. The cytoplasm contains mostly free ribosomes as well as a few mitochondria and cisternae of rough endoplasmic reticulum. The letter P indicates processes with a content similar to that of the cell in center; the processes come from this cell or some other subependymal cells. (C. P. Leblond and J. Paterson). X 4100.

(figs. 5-8) in the hope of finding which one plays the role of stem cell.

The most common cell types are depicted in figures 5-8; they are strikingly different from one another, but none displays the cytological features encountered in stem cells. Furthermore, they are not labeled one hour after the injection of ^{3}H-thymidine and, therefore, do not have the ability to divide. Hence, they are likely to be differentiated end cells.

Let us examine other varieties of the four cell types. In particular, there are different varieties of columnar cells in the various regions of the epithelium. The *crypt base columnar cells,* defined somewhat arbitrarily as those present in the first 9 cell positions counted from the crypt bottom, may appear columnar (fig. 9) or have irregular forms as they adjust to the restricted space available between Paneth cells. The nucleus is oval and contains fairly

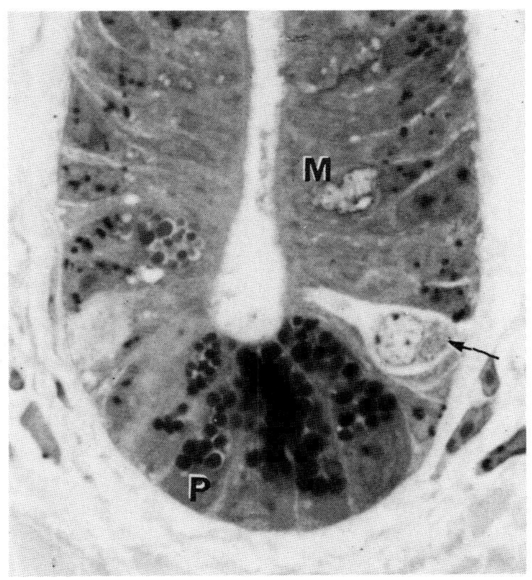

Fig. 4. Radioautograph of a semithin section of mouse crypt of small intestine stained with iron hematoxylin after one day of continuous infusion of ^{2}H-thymidine. At the base of the crypt the Paneth cells (P) are recognized by the large, well stained granules. An entero-endocrine cell at lower right is recognized by the pale cytoplasm, spherical nucleus and a cluster of fine, basal granules (arrow). Higher up, a mucous cell (M) is located between columnar cells. Columnar and mucous cells are labeled, but entero-endocrine and Paneth cells are not labeled at this time (Cheng and Leblond, '74b). X 1300.

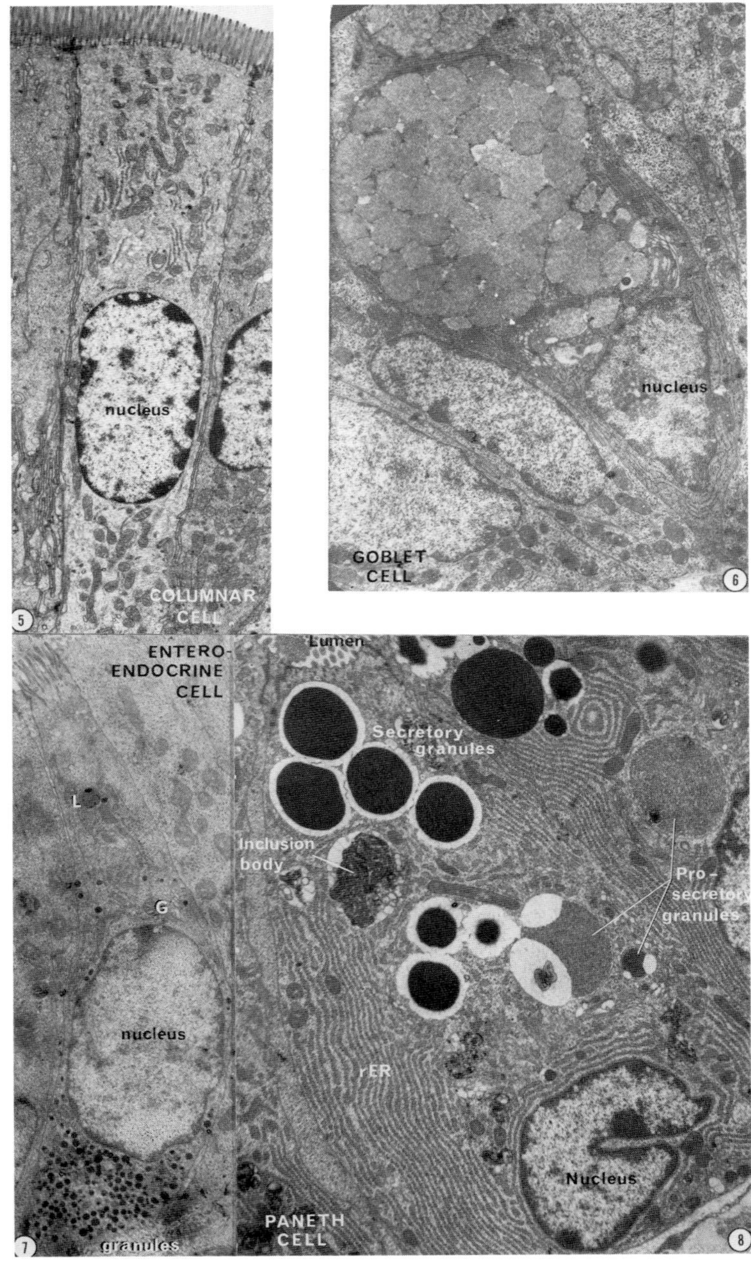
nucleus
COLUMNAR CELL
5
nucleus
GOBLET CELL
6
ENTERO-ENDOCRINE CELL
L
G
nucleus
granules
7
Lumen
Secretory granules
Inclusion body
Pro-secretory granules
rER
Nucleus
PANETH CELL
8

diffuse chromatin. The small cytoplasm contains a small Golgi apparatus, scanty rER cisternae, a few mitochondria, and packed free ribosomes. The microvilli are short. In *mid crypt*, the cells are somewhat taller and organelles more prominent than in the crypt base (fig. 10). The changes are accentuated at the *crypt top* (last five positions of the crypt) where much of the large cytoplasm is occupied by organelles (fig. 11). Finally, the columnar cells found on the *villi* are tall and display numerous organelles, but only few free ribosomes (fig. 5). Thus, cell features gradually change from crypt base to villus, suggesting a progressive evolution.

From this brief survey, it appears that the features described in several types of stem cells: small size, diffuse chromatin, paucity of organelles and abundance of free ribosomes, are also found in crypt base columnar cells. Furthermore, these cells divide frequently (fig. 12). Yet, they are seen in the base of *all* crypts so that their stock is not exhausted by the divisions. Hence, like stem cells, they renew themselves. (We showed that, after

---

Fig. 5. Villus columnar cell. Microvilli are long and lateral membranes interdigitate. The centrally located nucleus contains dense chromatin masses. The cytoplasm contains many organelles, but few free ribosomes (Cheng and Leblond, '74a). X 4200.

Fig. 6. Common mucous cell (goblet cell). The nucleus is irregular and contains dense chromatin. The cytoplasm contains long cisternae of rough endoplasmic reticulum along the lateral membranes, a supranuclear Golgi apparatus and an accumulation of mucous globules (Cheng, '74a). X 4200.

Fig. 7. Typical entero-endocrine cell. This cell is wider at the base than at the apex where microvilli can be seen. In the infranuclear region, there is a cluster of dense granules measuring 200-300nm. The Golgi apparatus (G), dense bodies (L), mitochondria and rough endoplasmic reticulum are situated above the nucleus (Cheng and Leblond, '74b). X 4700.

Fig. 8. Paneth cell. The basal irregular nucleus includes dense masses of chromatin. The cytoplasm contains stacks of rER cisternae, inclusion bodies, prosecretory granules in the Golgi region and secretory granules composed of a dense core and a light halo. (Cheng, '74b). X 3700.

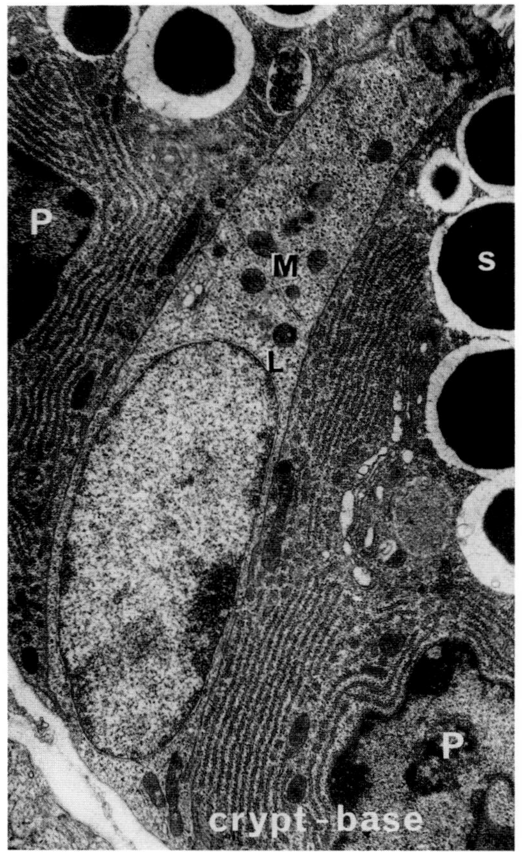

Fig. 9. Crypt base columnar cell between two Paneth cells. This columnar cell has a regularly oval nucleus with a nucleolus at right and mostly diffuse chromatin. The cytoplasm is filled with free ribosomes, but also contains a small Golgi apparatus, a few mitochondria (M), a dense body (L) and rare ER cisternae.

In contrast, the Paneth cells on either side have an irregular nucleus (P) with dense chromatin masses, numerous cisternae, prominent Golgi apparatus and haloed secretory granules (S). Free ribosomes are rare (Cheng and Leblond, '74a). X 6300.

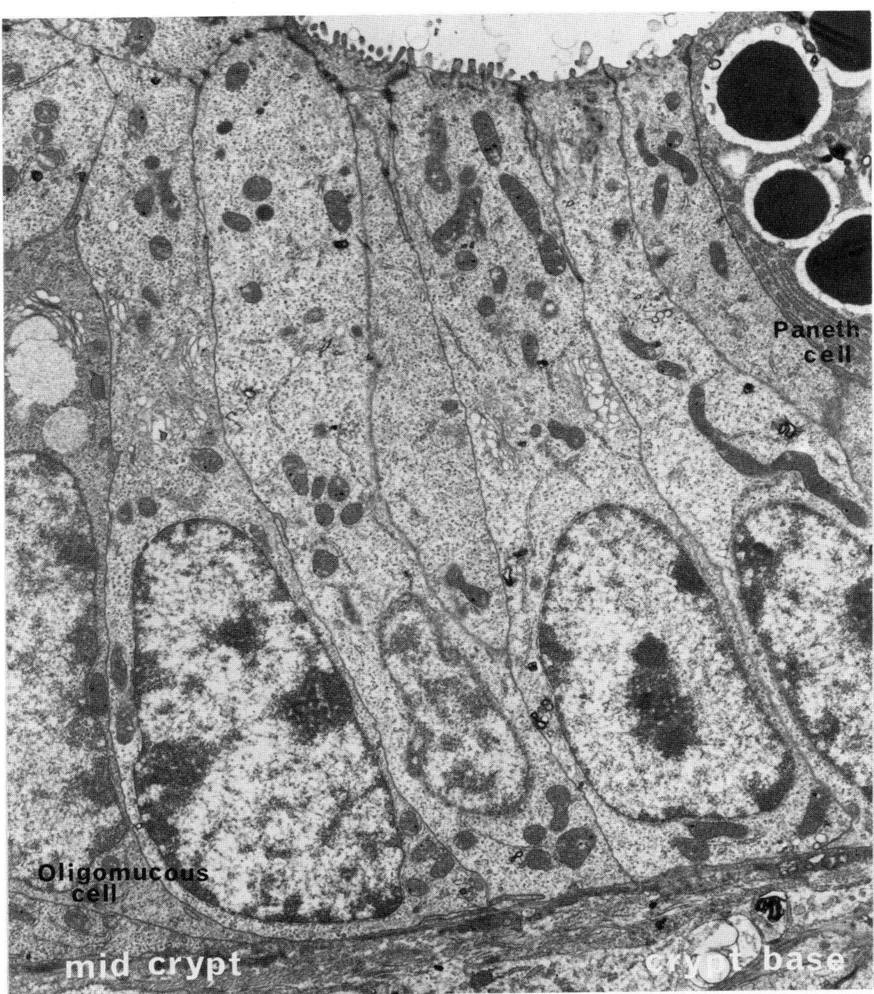

Fig. 10. Junction between crypt base (at extreme right) and mid crypt (in the rest of the figure). While all cells contain free ribosomes, the Golgi apparatus and other organelles are increasingly prominent in the cells at left. X 6700.

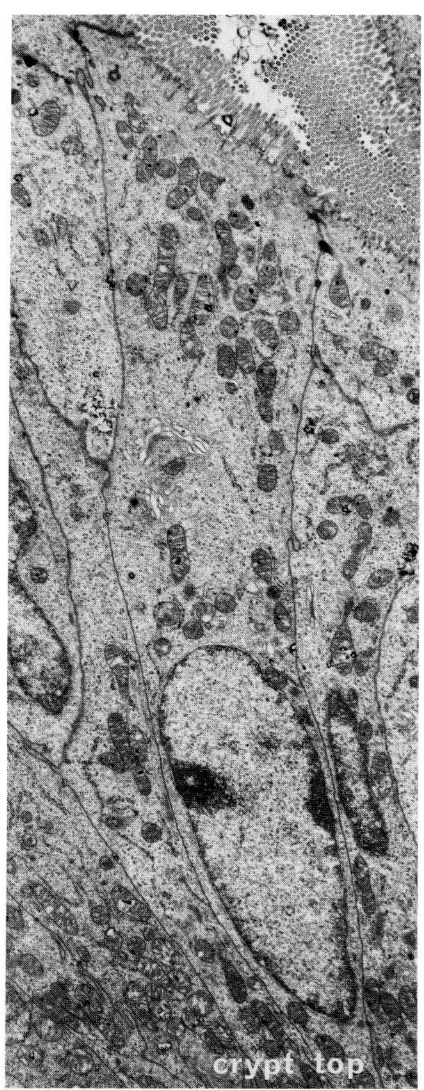

Fig. 11. Crypt top columnar cells. Microvilli, mitochondria and Golgi apparatus are more prominent than in the midcrypt cells shown in figure 10 (Cheng and Leblond, '74a). X 6300.

intravenous injection of labeled blood cells, they may enter the epithelium, but their label is never passed on to crypt base columnar cells; Cheng and Leblond, '74c). Finally, the progressive changes in the structure of columnar cells from crypt base to villus surface (figs. 6-8, 5) suggest that some of the daughters of mitoses of crypt base columnar cells become differentiating cells which in turn transform into villus columnar cells, the end cells of the series. Briefly then, the crypt-base columnar cell is the likely candidate for the stem cell role in the epithelium of the mouse small intestine.

One may well wonder whether the crypt base columnar cells are also related to mucous, entero-endocrine and even Paneth cells. The vast majority of crypt base columnar cells contained no secretory material. However, a few cells which otherwise were indistinguishable from crypt base columnar cells displayed discrete granules, e.g. three mucous globules in figure 12, one entero-endocrine granule in figure 13 and two small Paneth-type granules with a light halo in figure 14. These observations are suggestive of crypt base columnar cells acquiring the ability to elaborate

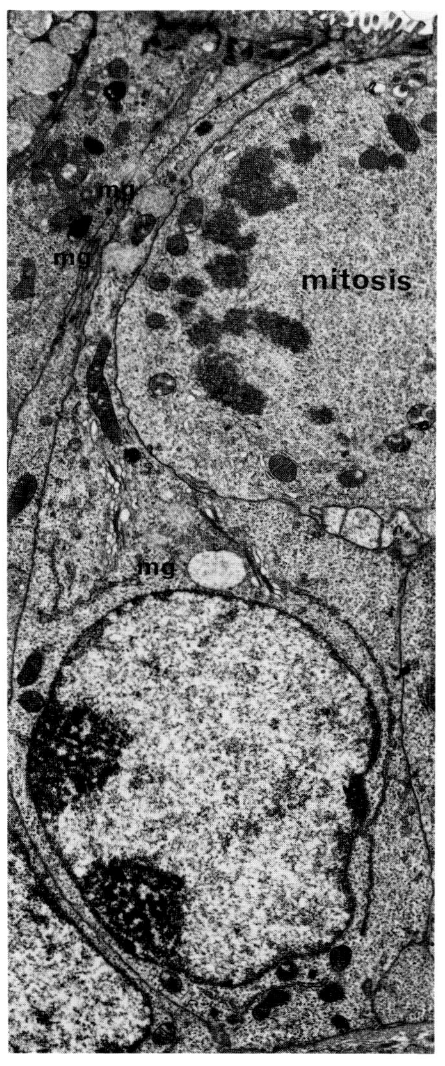

Fig. 12. Crypt base columnar cells with ribosome packed cytoplasm. The one at upper right, which is in metaphase, rounds up next to the lumen and squeezes its neighbors. The cell running from bottom to top contains three mucous globules (mg) as well as a few long cisternae. It is believed to be differentiating into a mucous cell (Cheng, '74a). X 7200.

mucous, entero-endocrine or Paneth granules, thus possibly initiating transformation into mucous, entero-endocrine or Paneth cells respectively.

Mitotic division as well as labeling one hour after ^{3}H-thymidine injection is observed in many crypt base and midcrypt columnar cells and in a few mucous cells with a moderate amount of mucus. Typical entero-endocrine and Paneth cells are not labeled soon after ^{3}H-thymidine injection. However, label may be found in the cells which look like crypt-base columnar cells and contain very few entero-endocrine granules (fig. 13). The division of such cells gives rise to non-dividing cells which become typical entero-endocrine cells. In the case of Paneth cells, crypt base columnar cells with very few Paneth type granules have been observed but not until 12 hours after injection. It is likely that such cells acquired granules after they arose from the stem cell division and then differentiated without further division (fig. 14-16).

That mucous cells come initially from crypt-base columnar cells (Merzel and Leblond, '69) was shown by comparing turnover rates measured in percent cells renewed per day from continuous infusion data or from data obtained one hour after a single injection of ^{3}H-thymidine into mice:

	Columnar cells		Common mucous cells	
	Duodenum	Jejunum	Duodenum	Jejunum
From continuous infusion data	30.0	29.7	31.7	34.8
From data obtained one hour after injection	47.7	41.8	18.5	16.2

The reliable data obtained by continuous infusion demonstrated that the turnover rate of columnar and mucous cells is the same, approximately 30% per day in duodenum and jejunum. Yet, the single injection data yielded much higher figures for columnar than mucous cells. The high labeling of columnar cells one hour after injection indicates that there is more labeling and, therefore, more mitoses of columnar cells than are needed to produce the population of these cells. Presumably then, some of the cells arising from these mitoses evolve into a type other than columnar. Conversely, the low initial labeling of mucous cells indicates that there are not enough mitoses of these cells to produce their own population. In conclusion, the apparent production of too many columnar cells and not enough mucous cells indicates that some of the columnar cells transform into mucous cells.

While the various approaches mentioned so far indicate that crypt base columnar cells may transform into the four mature cell types, direct evidence was derived from the observation that, within 6 hours after mice were given 2μc of ^{3}H-thymidine per gram body weight, a few labeled crypt base columnar cells died, presumably as a result of radiation damage; such cells were then surrounded by narrow processes extending from a neighboring crypt base columnar cell, which eventually engulfed the dead cell. The dead cell appeared as a labeled phagosome within the cytoplasm of a

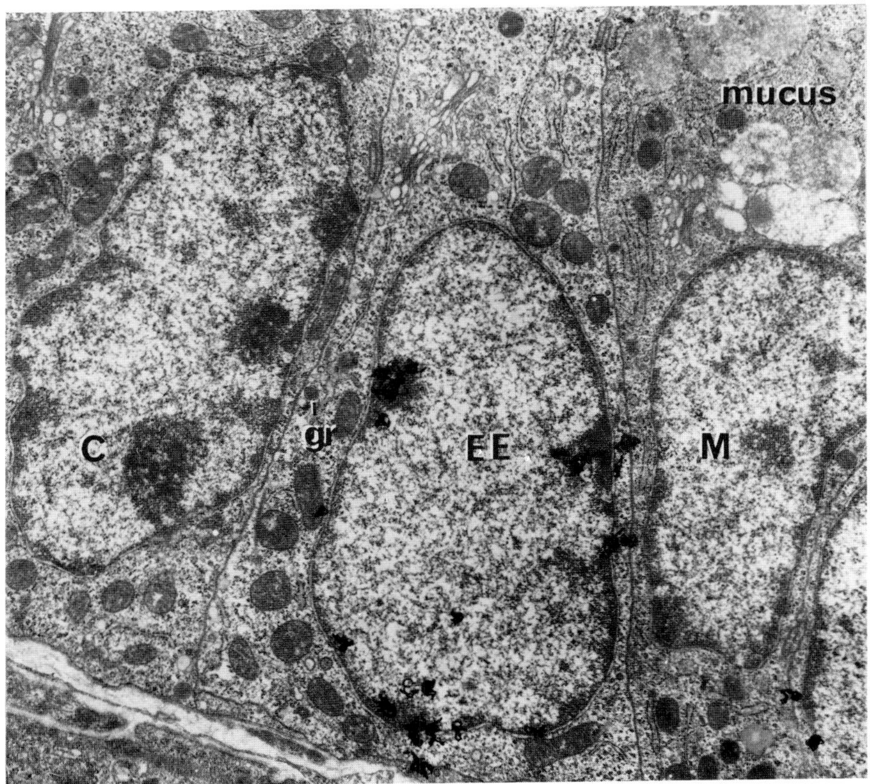

Fig. 13. *Crypt base cells.* *The cell at left (C) is a typical crypt base columnar cell. In the center, a similar cell (EE) contains one entero-endocrine type granule (gr) at center left and is believed to be beginning its differentiation into an entero-endocrine cell. The cell at right (M) is differentiating into a mucous cell (Cheng and Leblond, '74c). The center cell (EE) is labeled. X 9500.*

---

healthy neighbor (Fig. 17) but only in crypt base columnar cells. In such cells, the phagosomes were numerous six hours after injection. They were used as markers to trace the fate of the host cells at later times. By 12 hours after ^{3}H-thymidine injection, numerous labeled phagosomes were observed in midcrypt columnar cells and by 24 hours,

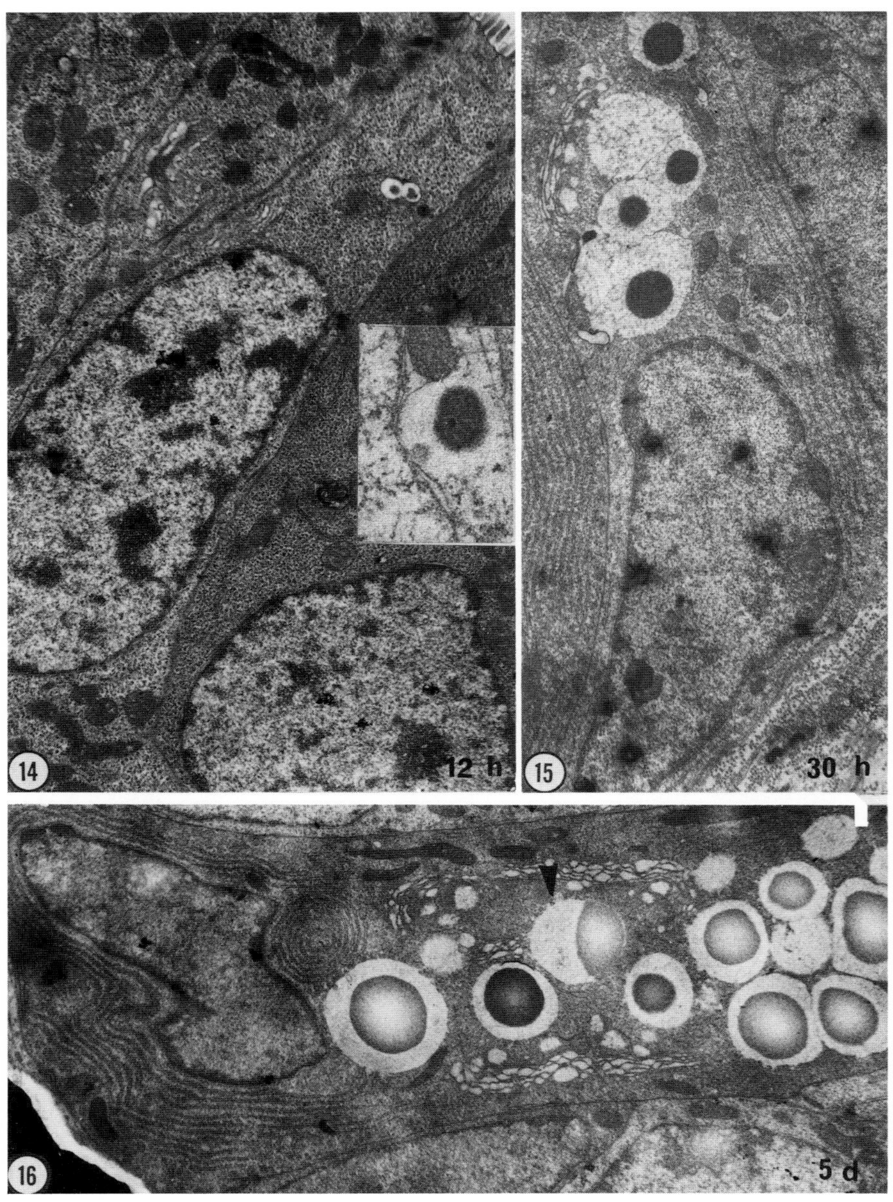
14
12 h
15
30 h
16
5 d

in villus columnar cells. Phagosomes were observed in 9 mucous cells starting from the 12 hour interval (fig. 18). One phagosome was seen in an entero-endocrine cell and two in Paneth cells. There was rough agreement between the number of phagosomes in each cell type and the expected frequency of occurrence of this cell type under normal conditions (Cheng and Leblond, '74c). These results provide direct evidence that the four epithelial cell types come from crypt base columnar cells. These would include the stem cells for all epithelial cells in small intestine (fig. 19).

One may question whether a morphologically homogeneous stem cell pool may not be in fact functionally heterogeneous and be subdivided into distinct stem cell pools for each one

---

Fig. 14. Cell containing Paneth cell granules 12 hours after injection. (The silver grains present are not clearly distinct from chromatin masses). This cell has a basal nucleus and a cytoplasm rich in free ribosomes but poor in other organelles. The lateral cell membrane is smooth while the apical membrane consists of a few short microvilli. Except for the presence of the two small (430nm) Paneth cell granules, this cell is identical to crypt-base columnar cells. X 6700. The inset shows a higher magnification (X 42000) of a small Paneth cell granule taken from a similar cell. Such cells may be considered as Paneth cell precursors.

Fig. 15. Paneth cell 30 hours after injection (silver grains over nucleus). This cell contains more and larger Paneth cell granules (1900nm) than in figure 4. Rough endoplasmic reticulum cisternae begin to be arranged in stacks beside the nucleus. X 6700.

Fig. 16. Paneth cell 5 days after injection (silver grains over nucleus). This cell contains many large Paneth cell granules (2100nm) of the same magnitude as in typical Paneth cells. Golgi apparatus and rough endoplasmic reticulum cisternae are often seen above the nucleus. The nucleus shows an irregular shape. Arrow indicates a forming Paneth cell granule in the Golgi region with dense core and light halo material arranged side by side. X 6700.

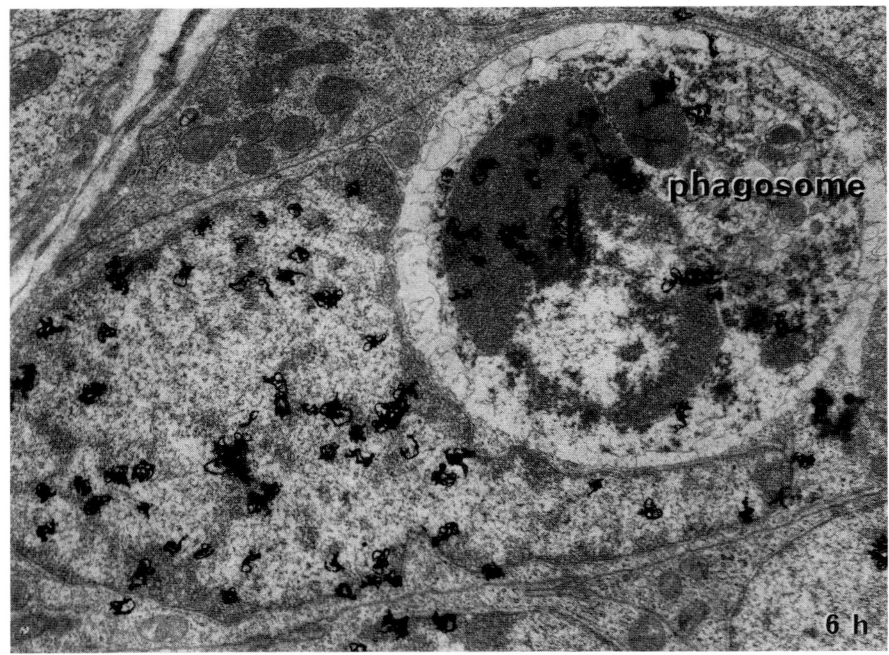

Fig. 17. A labeled crypt-base columnar cell with a labeled phagosome six hours after an injection of 2 µCi/g body weight of ^{3}H-thymidine. The label inside the phagosome is distributed over its nuclear component; the rest of the phagosome consists of cytoplasm. X 10200.

of the epithelial cell types. However, a few epithelial cells with two types of secretory granules, either entero-endocrine and mucous, or Paneth and mucous were occasionally encountered. The precursors of such cells would have both entero-endocrine and mucous cell potentiality in the first case and both Paneth and mucous cell potentiality in the second case. The existence of precursors with more than one potentiality suggests that the stem cell pool was not divided into functionally separate groups for each cell type, but was composed of multipotential stem cells. Hence, all the lines of evidence support a *Unitarian Theory of the Origin of the Epithelial Cell types*.

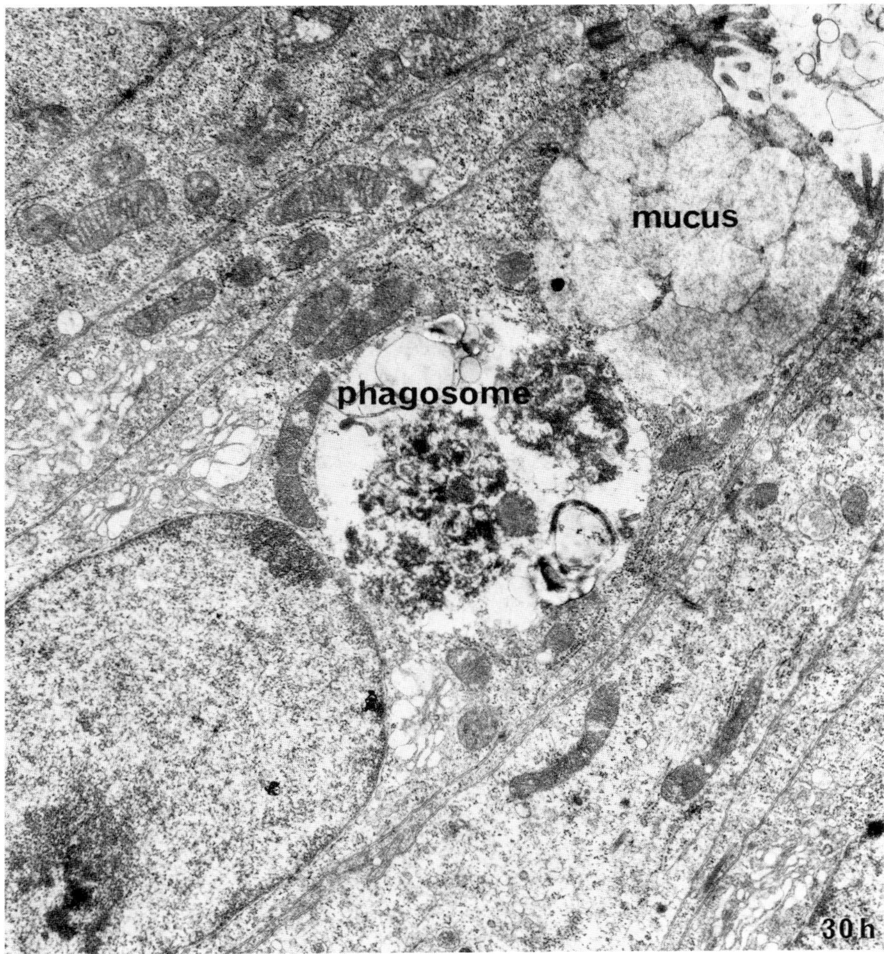

Fig. 18. A mucus-containing cell with a labeled phagosome 30 hours after an injection of ^{3}H-thymidine. The nuclear and cytoplasmic component of the phagosome cannot be recognized because of advanced degeneration. X 12700.

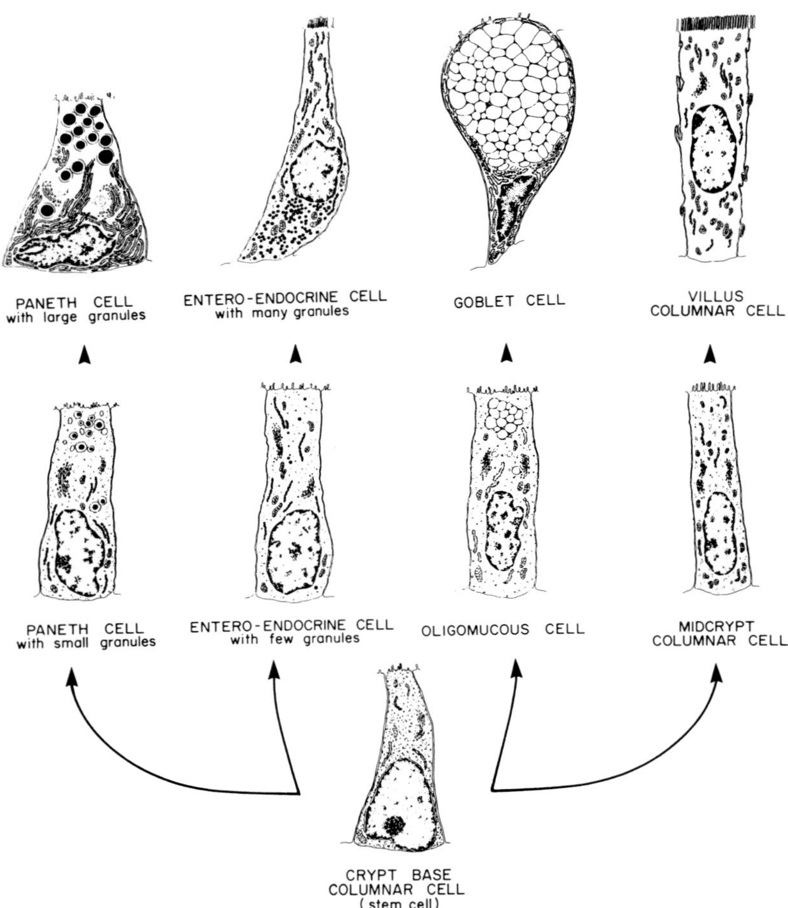

Fig. 19. Unitarian theory of the origin of the four main epithelial cell types in adult mouse small intestine. Some of the crypt-base columnar cells depicted at the base of the diagram function as stem cells and give rise to the four mature cell types. These are from right to left; (1) midcrypt columnar cells which eventually become villus columnar cells; (2) oligomucous cells which become goblet cells; (3) young entero-endocrine cells, which mature by acquiring a full complement of granules; (4) Paneth cells with small haloed granules which differentiate into mature cells with large haloed granules.

PROLIFERATION OF COLUMNAR CELL LINE

	Cell position in crypt	Mean number of progeny generations	%	Sites of mitotic divisions
Crypt top	31 - 33	-		
	28 - 30	-		
	25 - 27	-		
Midcrypt	22 - 24	0.13	13	Fourth division
	19 - 21	0.25	25	
	16 - 18	0.33	33	
	13 - 15	0.41	41	
	10 - 12	0.52	52	Third division
Crypt base	7 - 9	0.83	48	
			35	Second division
	4 - 6	1.15	65	
			50	First division
	1 - 3	0.50	50	

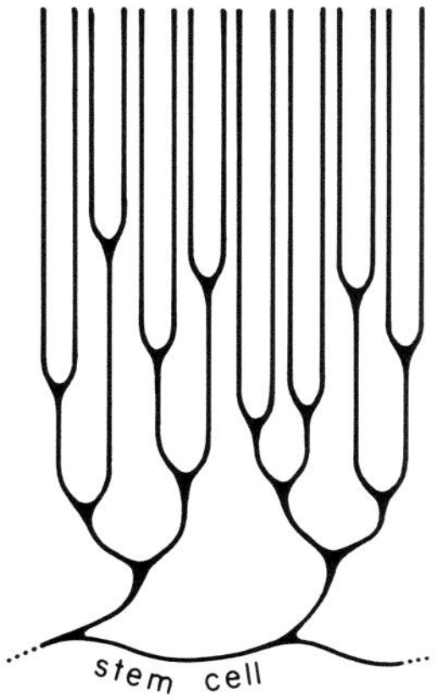

Fig. 20. The number of progeny generations of columnar cells arising from the stem cell was calculated from the labeling index in the positions of jejunal crypts by the method of Chang and Nadler ('75), assuming constant DNA synthesis time and neglecting the small numbers of columnar cells transforming into other cell types. It was pointed out by these authors that the method yields precise values throughout the crypt, except in the initial group of positions (1-3), where an approximation was obtained by assuming a 15 hour cycle time.

Under these conditions, the data indicate that approximately four generations of columnar cells arise from the stem cells.

It has been possible to calculate the number of generations of columnar cells through which the cells go as they arise from stem cell mitoses. This was done by the method of Chang and Nadler ('75). The results depicted in figure 20 for the mouse jejunum suggest that, following the initial stem cell mitosis, three further mitoses of columnar cells take place. Accordingly, the stem cell progeny consists of four successive generations. Similarly, the calculations carried out with the data from duodenum and ileum also yielded four progeny generations for the columnar cell line.

According to this model, only the crypt base columnar cells located in the first 4 or 5 positions would be stem cells. Those in positions 6-9 would be differentiating cells.

Data are not sufficient to make precise calculations for cells other than columnar, but a speculative model is presented in figure 21. It is proposed that the various types of differentiating cells arising from the stem cell divisions in positions 1-5 are "committed". They are referred to as committed progenitor cells in the diagram. They would then undergo three mitoses to produce mature columnar cells, two mitoses to yield goblet cells, one mitosis to give rise to mature entero-endocrine cells and none for Paneth cells. The frequency of the four types of committed progenitor cells would be quite different. (According to speculative calculations on the jejunum, for 66 cells becoming columnar progenitors, 20 become mucous progenitors, one entero-endocrine progenitor, and eight Paneth cells). The mature Paneth cells remain in the crypt base where after two to three weeks they degenerate and die. Columnar, mucous and entero-cells migrate in unison from crypt to villus, in the direction of the villus tip, where after a few days they are extruded into the intestinal lumen.

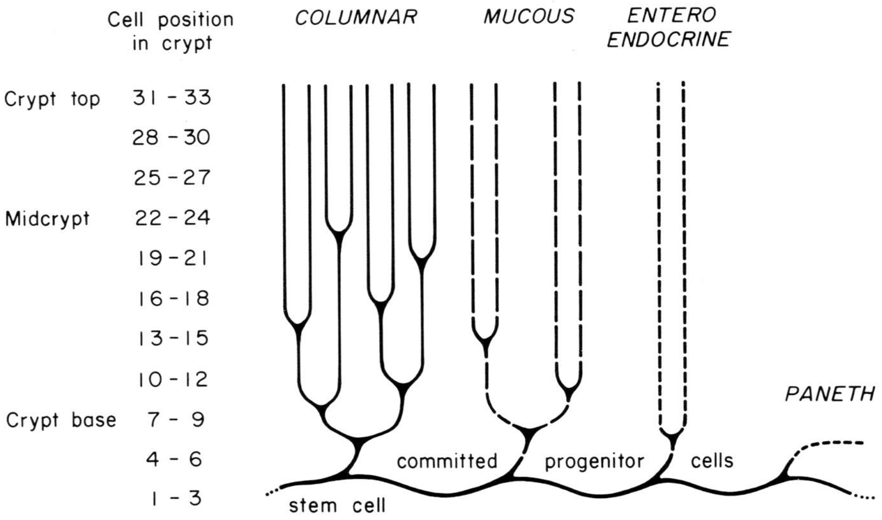

Fig. 21. Model proposed to account for the production of the four main cell lines from the stem cell.

CONCLUSION

The crypt base columnar cells in the epithelium of duodenum, jejunum and ileum have the ability to divide in such a way that they maintain their own stock while giving rise to four main types of differentiating cells. Some of their features (small size, diffuse chromatin, scarcity of organelles, and abundance of free ribosomes) are similar to those of well identified stem cells in other tissues.[1] Their ability to phagocytose nearby dead cells and the possibility of ameboid migration of their daughter cells suggested by Altmann's results (this volume) complete the stem cell picture. It is concluded that the crypt-base columnar cells, or at least some of them, are the stem cells of the epithelium in the mouse small intestine.

ACKNOWLEDGEMENTS

This work was done with the support of a grant from the Medical Research Council of Canada. We are indebted to Dr. Beatrix Kopriwa for assistance in the radioautographic work.

---

[1]In the descending colon of the mouse, the stem cells contain a few secretory granules described by the term "vacuoles". These cells appear less immature than the stem cells in the small intestine (Chang and Leblond, '71; Chang Nadler, '75).

LITERATURE CITED

Chang, W. W. L., and C. P. Leblond. 1971. Renewal of the epithelium in the descending colon of the mouse. I. Presence of three cell populations: vacuolated-columnar, mucous and argentaffin. *Am. J. Anat. 131:*73-100.

Chang, W. W. L., and N. J. Nadler. 1975. Renewal of the epithelium in the descending colon of the mouse. IV. Cell population kinetics of vacuolated-columnar and mucous cells. *Am. J. Anat. 144:*36-95.

Cheng, H. 1974. Origin, differentiation and renewal of the four main epithelial cell types in the mouse small intestine. II. Mucous Cells. *Am. J. Anat. 141:*481-502.

Cheng, H. 1974. Origin, differentiation and renewal of the four main epithelial cell types in the mouse small intestine. IV. Paneth cells. *Am. J. Anat. 141:*521-536.

Cheng, H., and C. P. Leblond. 1974. Origin, differentiation and renewal of the four main epithelial cell types in the mouse small intestine. I. Columnar cell. *Am. J. Anat. 141:* 461-480.

Cheng, H., and C. P. Leblond. 1974. Origin, differentiation and renewal of the four main epithelial cell types in the mouse small intestine. III. Entero-endocrine cells. *Am. J. Anat. 141:*503-520.

Cheng, H., and C. P. Leblond. 1974. Origin, differentiation and renewal of the four main epithelial cell types in the mouse small intestine. V. Unitarian theory of the origin of the four epithelial cell types. *Am. J. Anat. 141:*537-562.

Clermont, Y. 1966. Renewal of spermatogonia in man. *Am. J. Anat. 118:*509-524.

Clermont, Y. 1972. Kinetics of spermatogensis in mammals: seminiferous epithelium cycle and spermatogonial renewal. *Physiol. Rev. 52:*198-236.

Clermont, Y., and C. P. Leblond. 1959. Differentiation and renewal of spermatogonia in the monkey, Macacus Rhesus. *Am J. Anat. 104:*237-274.

Leblond, C. P. 1964. Classification of cell populations on the basis of their proliferative behavior. *Nat. Cancer Inst. Mongr. 14:*119-150.

Leblond, C. P., and B. Messier. 1958. Renewal of chief cells and goblet cells in the small intestine as shown by radioautography after injection of thymidine-^{3}H into mice. *Anat. Rec. 132:*247-259.

Leblond, C. P. and C. E. Stevens. 1948. The constant renewal of the intestinal epithelium of the albino rat. *Anat. Rec. 100:*357-378.

Marques-Pereira, J. P., and C. P. Leblond (with N. J. Nadler). 1965. Mitosis and differentiation in the stratified squamous epithelium of the rat esophagus. *Am. J. Anat. 117:*73-90.

Merzel, J., and C. P. Leblond. 1969. Origin and renewal of goblet cells in the epithelium of the mouse small intestine. *Am. J. Anat. 124:*281-299.

Privat, A., and C. P. Leblond. 1972. The subependymal layer and neighboring region in the brain of the young rat. *J. Comp. Neurol. 146:*227-302.

Rolshoven, E. 1951. Ueber die Reifungsteilungen bei der Spermatogenese mit einer Kritik des bisherigen Begriffes der Zellteilungen. *Verh. Anat. Ges. Jena. 49:*189-197.

Colony-Forming Units in the Intestine

H. RODNEY WITHERS

Section of Experimental Radiotherapy
The University of Texas System Cancer Center
M.D. Anderson Hospital and Tumor Institute
Houston, Texas

The radiation response of jejunal crypt cells provides some insight into the proliferation and differentiation patterns of stem cells and their progeny. The definition of 'stem cells' can be contentious: here it is used to describe a cell capable of reproducing, for all or most of an animal's lifetime, all the cells necessary for the normal function of the tissue of which it is a stem cell. Commonly, the use of the term connotes the existence of daughter cells and their progeny which, although capable of proliferation, have a limited lifetime.

It is difficult to accommodate this definition, and experimental data to be presented, with the recently popular theory advanced by Hayflick (1965), that all cells have a limited lifetime - of the order of 50 divisions. Various assumptions may be made to try to resolve the differences, but I shall consider just one set of assumptions to outline the magnitude of the problem. If it is assumed that stem cells make up 1/1000th of the weight of a 70 Kg man, and that the average cell is 1000 μ^3, the fusion of a sperm and ovum must be followed by at least 25 divisions in which both daughter cells retain their stem cell capacity. This would leave the stem cells of the body with fewer than 25 residual divisions with which to indulge in steady-stage reproduction or to repopulate themselves after injury. Thus, stem cell division would be less than an annual event in people living more than 25 years and even if the S-phase lasted 24 hours, less than 1 in 365 stem cells would be labelled by a single injection of radioactive thymidine. Since we postulated 1/1000 cells were stem cells, one could examine more than 365,000 cells to find one labelled stem cell. Thus, if Hayflick's hypothesis is correct, stem cells would have the following characteristics:

1. Few in number and/or
2. Slowly proliferating (the more stem cells, the slower their cycle)
3. Undetectable by labelling with DNA precursors
4. Essentially non-contributory to the cell renewal process in proliferating steady-state systems

A corollary of these characteristics of stem cells would be that the proliferating cells studied by radioautography are not stem cells and steady-state proliferation is a function of non-stem cells. If these non-stem cells are also limited to 50 divisions they must be renewed frequently: for example, jejunal crypt cells have a mean generation time of less than one day and would need replacing from the stem cell reserve in less than 50 days. Furthermore, if only a small proportion of the proliferative population is stem cells, they must replenish the exhausted non-stem cells more frequently than once every 50 days: for example, if only one proliferative cell in 50 were a stem cell, stem cells would have to divide more than once daily. One division daily in a man living 80 years would involve about 3 x 10^4 divisions.

STEADY-STATE PROLIFERATION OF JEJUNAL CRYPT STEM CELLS

Stem cell is a **retrospective** operational description and I shall use it for cells capable of regenerating a crypt in 3 1/2 days (Fig. 1a) or an easily visible nodule of mucosa in 13 days after irradiation (Fig. 1b).

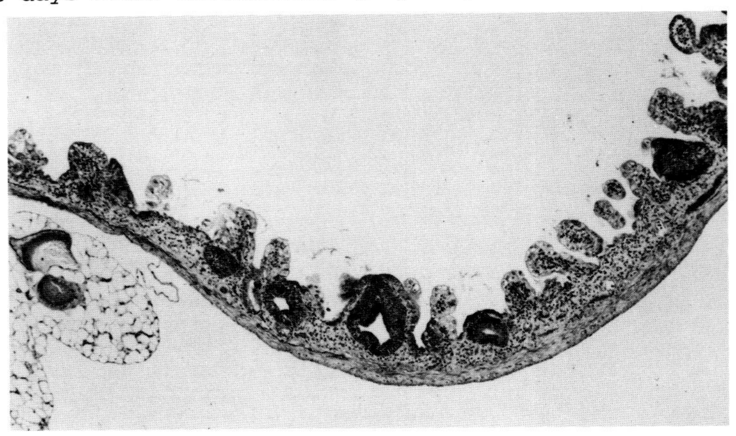

Fig. 1a. Colonies of crypt cells in the jejunal wall 3 1/2 days after 1400 rads of γ-rays.

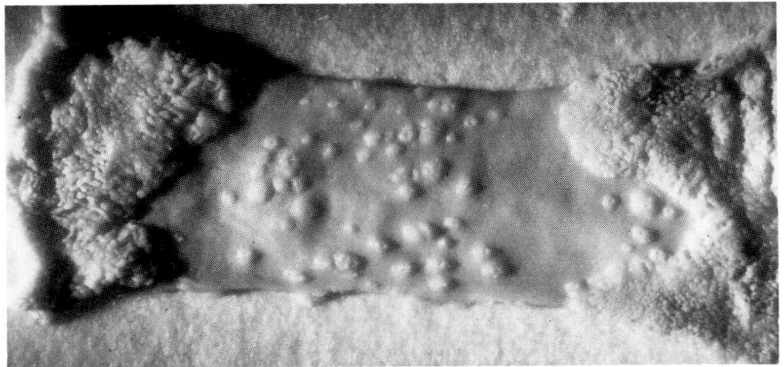

Fig. 1b. Islands of mucosal regeneration viewed from the epithelial surface of the jejunum 13 days after a dose of 1700 rads of x-rays.

At least a proportion of the jejunal crypt stem cells is cycling rapidly. Groups of mice were injected with 5 doses of hydroxyurea at hourly intervals and exposed to 1100 rads of γ-radiation at various times later (Withers *et al.* 1974). The animals were killed 3 1/2 days after irradiation and regenerating crypts per jejunal circumference were scored in histological sections (Fig. 1a). On the reasonable assumption that one or more stem cells could regenerate a crypt, the number of surviving cells was estimated using Poisson statistics (Withers and Elkind, 1970). The results are shown in Figure 2.

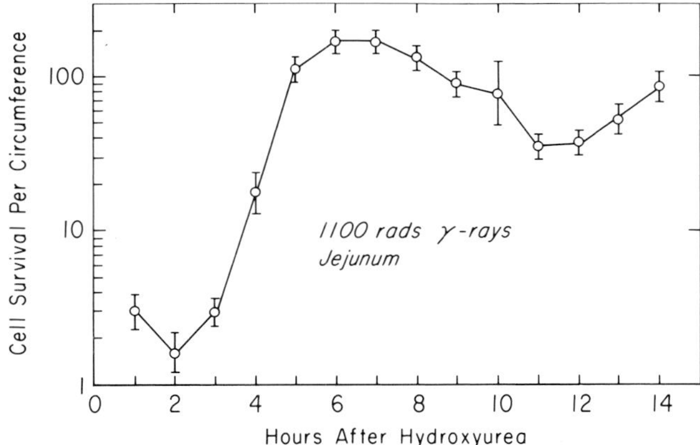

Fig. 2. Variation in survival of mouse jejunal crypt cells exposed to 1100 rads of γ-rays at various times after 5th hourly injection of hydroxyurea. (Withers, Radiology, 1975)

Hydroxyurea kills S-phase cells and blocks progression at the G_1-S boundary. The 100-fold variation in cell survival from 1100 rads reflects the synchronous progression of at least a proportion of the stem cells through the division cycle, varying in radiosensitivity as they go. These data do not exclude the existence of slowly-cycling stem cells but they do establish that some stem cells, as defined by cryptogenic capacity, are cycling rapidly, at least after 4 hours of repeated hydroxyurea injections. Similar fluctuations in radioresponse of bone-marrow stem cells have been demonstrated by Hellman (1972).

REGENERATIVE RESPONSE OF IRRADIATED JEJUNAL MUCOSA

While stem cells of jejunal crypts may cycle rapidly in steady-state, their regenerative response after a dose of 660 rads of x-rays begins relatively slowly. After a first dose of 660 rads, the jejunum was exposed to a second dose of 1415 rads. Survival of stem cells was assayed by counting the number of mucosal nodules regenerated 13 days after the second dose (Fig. 1b). Each nodule was considered to be a clone (Withers and Elkind, 1969). Regeneration occurring between the 2 doses of radiation is reflected in a parallel increase in survival from the second dose. Regeneration begins about 2 days after exposure to 660 rads and proceeds with a doubling time of about 8 hours until the crypt is repopulated (Fig. 3). The horizontal line in Figure 3 defines the level of survival expected had the first dose not been given: when survival from the second dose reaches that level, the initial injury is fully recovered. This happens by about 4 days. There is a brief over-shoot before regeneration stops, followed by a return to normal steady-state proliferation.

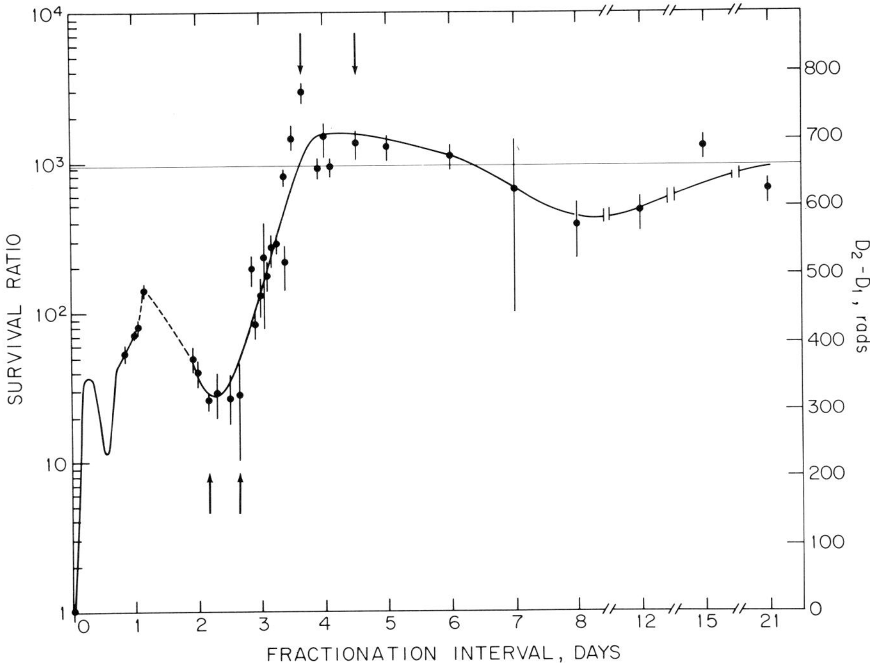

Fig. 3. The ratio of survival from (660 + 1415) rads of x-rays given as 2 doses separated by intervals up to 21 days to survival from a single dose of 2075 rads. The rapid increase in survival ratio between the 2nd and 4th days after the first dose reflects a rapid increase in stem cell number during that interval. After 4 days, increase in stem cell number stops although the crypt cells continue rapid proliferation (in steady-state). (Courtesy Withers and Elkind, 1969).

The delay of 2 days before regeneration begins, together with the evidence for rapid cycling of the stem cells, suggests that, initially, the cells continue their steady-state pattern of cell production and loss. When regeneration begins, loss of cells from the stem cell population must be nearly zero because the doubling time of 8 hours is similar to the maximum cell generation rate measured by radioautography (Lesher and Lesher, 1974).

CONTROL OF REGENERATION IN JEJUNAL CRYPTS

Regeneration of the stem cell population is determined

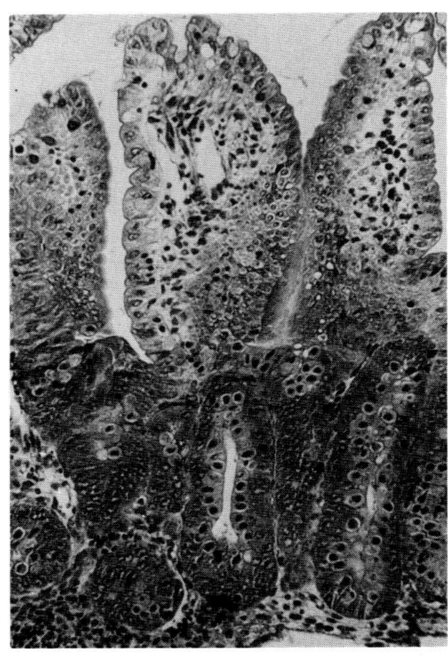

Fig. 4a. Enlarged crypts 4 days after a dose of 1000 rads of γ-rays - a dose insufficient to depopulate crypts completely. The crypt cells are just beginning to move on to the villus.

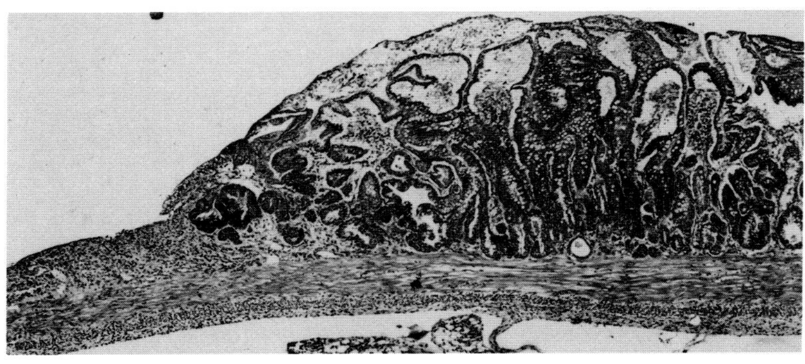

Fig. 4b. Histology of the edge of a nodule of the type shown in Fig. 1b. In the 13 days after irradiation, single surviving cells have repopulated a large number of crypts and have attempted villus formation.

by the pattern of cell loss through terminal differentiation: change in the generation cycle time only affects the rate of regeneration. Histology of irradiated intestine does not answer the important question of what controls cell differentiation and loss but it allows correlation of the kinetics of regeneration with changes in the crypt cell population, the mature differentiated cell population on the villus, and the mesenchymal microenvironment.

When regeneration begins, 2 days after irradiation (Fig. 3), the crypts are depleted of cells whereas the villi show only some shortening and reduced epithelial cell density. Regeneration of stem cells ends on the fourth day with the movement of cells from enlarged crypts on to shortened depopulated villi (Fig. 4a). When larger doses are given and cells survive in only a small proportion of crypts, the movement of cells out of a crypt about 4 days after irradiation may allow them to repopulate adjacent, depopulated, crypts. By 13 days after irradiation, when a surviving stem cell has produced an easily visible nodule of mucosa, there has been extensive repopulation of crypts and some attempt at villus formation (Fig. 4b). At lower doses, when all crypts contain surviving stem cells, and there are no empty crypts to repopulate, the cells that leave the crypt move on to the villus and lose their stem cell capacity (Fig. 4a and Fig. 3). These observations suggest that the mesenchymal microenvironment, that is, the nonepithelial structure of the depopulated crypt, plays an important role in determining the differentiation or clonogenicity of stem cells, whereas a feedback mechanism from the mature villous population seems a less likely form of control.

LITERATURE CITED

Hayflick, L. 1965. The limited *in vitro* lifetime of human diploid strains. *Exper. Cell Res. 37:* 614-636.

Hellman, S. 1972. X-irradiation of the hematopoietic stem cell compartment. *Front. Radiat. Ther. and Onc. 6:* 415-427.

Lesher, J. and S. Lesher. 1974. Effects of single-dose partial-body x-irradiation on cell proliferation in the mouse small intestinal epithelium. *Radiat. Res. 57:* 148-157.

Withers, H. R. and M. M. Elkind. 1969. Radiosensitivity and fractionation response of crypt cells of mouse jejunum. *Radiat. Res. 38:* 598-613.

Withers, H. R. and M. M. Elkind. 1970. Microcolony survival assay for cells of mouse intestinal mucosa exposed to radiation. *Int. J. Radiat. Biol. 17:* 261-267.

Withers, H. R., K. Mason, B. O. Reid, N. Dubravsky, H. T. Barkley, Jr., B. W. Brown and J. B. Smathers. 1974. Response of mouse intestine to neutrons and gamma-rays in relation to dose fractionation and division cycle. *Cancer 34:* 39-47.

Withers, H. R. 1975. Cell cycle redistribution as a factor in multifraction irradiation. *Radiology 114:* 199-202.

Presence of Intestinal Chalones

G. BRUGAL[1,2,3]

Laboratoire de Zoologie, Universite Scientifique et Medicale de Grenoble, France

The concept of autoregulation in cell proliferation homeostasis by means of a chalone mechanism has been successfully applied to several tissues (for review articles, see Nat. Cancer Inst. Monogr., 38, '73). Nevertheless, there is a lack of experimental data with which to elucidate the control mechanisms of intestinal epithelium cell proliferation although a considerable amount of information has been obtained concerning the cell kinetics of this tissue (Leblond and Messier, '58; Lipkin, '73).

Partial resection of the rat intestine induces a depletion of the intestinal epithelium and an increase in the mitotic activity in the remaining intestine which were both ascribed to a humoral stimulating proliferation factor (Loran et al., '64). However, this hypothesis has not been corroborated by other resection experiments, and X-irradiation of the rat intestine yielded evidence for a local inhibitory feedback control of the intestinal mitotic activity (Galjaard et al., '72; Meer-Fieggen, '73). Furthermore, possible chalone-like substances seemed to be involved in *in vivo* and *in vitro* inhibition of the intestinal proliferation in young chick embryos treated with duodenal extract from old chick

---

[1]Mailing address: Laboratoire de Zoologie, Université Scientifique et Médicale, B.P. 53, 38041 Grenoble-Cedex, France.

[2]I am grateful to Professors P. Chibon, J. Pelmont and J.P. Bertrandias for their advice, support and encouragement.

[3]This work was supported by grants from Centre National de la Recherche Scientifique (ATP N° A655 1799) and Institut National de la Santé et de la Recherche Médicale (N° 74142036) of France.

embryos (Bischoff, '64). Moreover, convincing evidence for the occurence of a tissue-specific G_2 inhibitor in the gastric mucosa of old chick embryos has been provided by Philpott ('71). We have shown more recently that crude intestinal extract from the adult newt *Pleurodeles waltlii* inhibited cell proliferation in the intestinal epithelium of the newt embryos. The inhibiting factor(s) was suspected to be tissue-specific as intestinal extract from adult did not prevent cell proliferation in embryonic liver or encephalon and since liver extract from adult did not inhibit cell proliferation in embryonic intestine (Brugal, '73). The present paper deals with a preliminary attempt to purify and characterize the intestinal factor(s) accounting for mitotic inhibition and to specify its biological effect on embryonic cell kinetics.

INTESTINAL CELL PROLIFERATION AND DIFFERENTIATION IN PLEURODELES EMBRYOS

Functional differentiation of the intestinal epithelium of the pleurodele occurs during the last period of embryogenesis (stage 34) and completes during the first steps of larval development, before the onset of feeding (stage 37). In the course of this 4-day period the intestinal epithelium, which is a closed endodermal cylinder at stage 34, opens at stage 35 while goblet cells and columnar absorbing cells differentiate. At stage 36 the cell-nests appear beneath the epithelium and are assumed to be analogous to the intestinal crypts of higher vertebrates (Patten, '61).

Between stages 34 and 37, the intestinal mitotic index (MI) and the growth fraction (GF) decrease about three-fold while cell number (N) increases (Fig. 1). Since the product GF.N remains constant during this period, it is obvious that the number of cycling cells remains constant while the number of non-dividing cells is increasing. The cell cycle durations are: T = 29 hr, G_1 = 2 hr, S = 24 hr, G_2 = 1 hr and M = 2 hr at stages 34 and 35, and T = 37 hr, G_1 = 8 hr, S = 25 hr, G_2 = 2 hr and M = 2 hr at stage 36. Therefore, about 3 cell cycles take place between stages 34 and 37 and the increase in non-dividing cell number, after every cell cycle time interval, is equal to the number of proliferating cells. Since the cells prevented from dividing are arrested in G_1 phase, as shown by cytophotometry (Brugal and Pelmont, '75), it is likely that each mitosis gives rise to another cycling cell and a resting one which is blocked in G_0 phase and possibly differentiates (Brugal, '75).

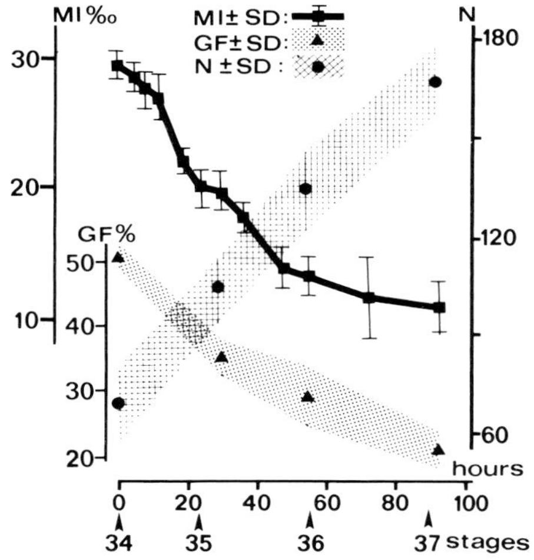

Fig. 1. Variations of the cell proliferation parameters in the intestine of newt embryos between stages 34 and 37. MI: mitotic index; GF: growth fraction; N: number of cells in a transverse section of embryonic intestine.

PRESENCE OF MITOTIC INHIBITORS IN INTESTINAL EXTRACT FROM THE ADULT NEWT

Owing to the decrease of cell proliferation in the differentiating intestinal epithelium of pleurodele embryos, the question has arisen whether such a depression of mitotic activity could also be mediated by a chalone mechanism similar to that involved in the steady state cell populations of adult animals. As reported above, preliminary experiments have shown that crude extract from fully differentiated intestinal epithelium of adult pleurodele caused a drastic mitotic inhibition of immature intestinal cells when injected in embryos at stage 34 (Brugal, '73). In further experiments, crude intestinal extracts were fractionated by G-200 Sephadex chromatography and the effects of the fractions on the intestinal cell proliferation were tested *in vivo* by both mitotic index and ^{3}H-thymidine incorporation methods. Figure 2 shows that among embryos sacrificed 5 hr after injection of fractions, those treated with fractions 36-38 exhibited a significant depression of the intestinal MI. On the contrary, among embryos sacrificed 26 hr after injections, only those treated with fractions 8-10 exhibited a significant depression of the intestinal MI. Furthermore, animals treated with fractions 36-38 exhibited a significantly higher intestinal MI attributed to the division of those cells that previously were inhibited by these same fractions. It was subsequently demonstrated that fractions 8-10 did not only decrease the intestinal MI 26 hr after their injection, but also depressed the ^{3}H-thy-

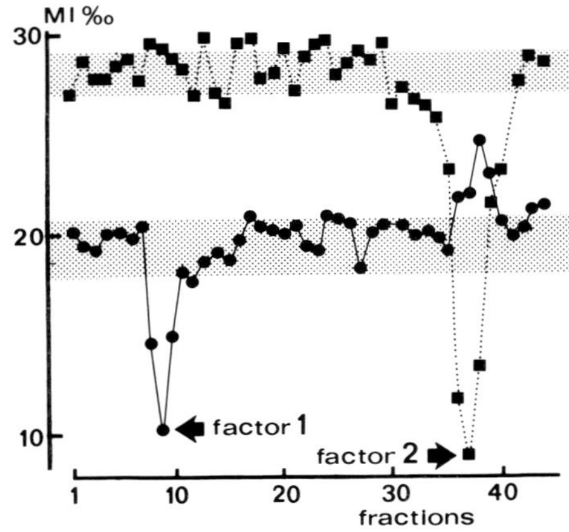

Fig. 2. Mitotic index in the intestine of the newt embryos sacrificed 5 hr (■) or 26 hr (●) after injection, at stage 34, of adult newt intestinal extract fractionated by G-200 Sephadex chromatography.

midine incorporation (Brugal and Pelmont, '74; '75). Since the observed inhibitions appeared as tissue-specific among investigated tissues, it was suggested from these results that two mitotic inhibitors were present in the intestinal extracts from the adult pleurodele: one (factor 1) was eluted with fractions 8-10 and blocked the cells at the end of the G_1 phase and the other (factor 2) was eluted with fractions 36-38 and blocked the cells in the G_2 phase. A cytophotometric study of the inhibited intestinal cell populations was performed by measuring the nuclear DNA content of cells stained with the Feulgen reagent. Age distributions of cells within the cycle were thus computed and undoubtly demonstrated the accumulation of cells either in G_1 phase after injection of factor 1 or in G_2 phase following injection of factor 2. Moreover, cytophotometric study showed that inhibitory effects of both factor 1 and factor 2 lasted for about 12 hr (Brugal and Pelmont, '75).

BIOCHEMICAL PROPERTIES OF THE INTESTINAL MITOTIC INHIBITORS

As a first step to characterize the intestinal mitotic inhibitors, the effect of factor 1 and factor 2 containing fractions on the intestinal MI was tested after heating at various temperatures. Factor 1 was inactivated at 80°C (pH 7; 5 min) but not at 60°C. On the contrary, the inhibitory activity of factor 2 was not altered by heating at 100°C (pH 7; 5 min). These results are opposite to those concerning the epidermal chalones as G_2 epidermal inhibitor is heat labile (Hondius Boldingh and Laurence, '68) whereas the G_1 epidermal inhibitor is heat stable at 100°C (pH 6) (Marks, '73).

In order to evaluate their molecular weight, the factors 1 and 2 were chromatographed through a G-200 Sephadex gel column previously standardized by proteins of known molecular weights (MW). The results indicated that factor 1 is characterized by a MW ranging from 120,000 to 150,000 and that factor 2 has a MW less than 2,000 (Brugal and Pelmont, '74; '75). Two similar mitotic inhibitors, characterized by the same sizes and inhibiting specifically the G_1 and the G_2 intestinal cells of the newt embryos, were extracted from the intestine of adult quail and mouse. From these results, we conclude that the intestinal mitotic inhibitors are probably not species-specific among vertebrates.

The chalones already extracted from various tissues have a MW ranging from 100,000 to less than 2,000. Most of them seem to be proteins or glycoproteins with a MW of about 30,000-50,000 or peptides of one tenth of that size (Houck and Hennings, '73). The diversity of chalone size appears more likely to be related to the variety of extraction and purification procedures used than to the tissues of origin. Supporting this assumption, Marks ('73) has shown that epidermal G_1 inhibitor had an apparent MW somewhere between 1.10^5 and 3.10^5 as determined by gel filtration whereas it had an apparent MW of about 1.10^4 to 2.10^4 when the epidermal extract was pronase-digested prior to Sephadex chromatography. It was thus concluded that the G_1 chalone is constituted of a small active group linked to a macromolecular entity. Such a structure has also been suggested for the intestinal G_1 inhibitor (Brugal and Pelmont, '75).

In order to get some information on the chemical and physical properties of the intestinal G_1 inhibitor, the factor 1 containing fractions obtained from gel filtration were submitted together to electrofocusing with Ampholine pH 3-10. The collected fractions were injected in stage 34 embryos and the intestinal and telencephalonic MI were calculated 26 hr later and plotted against fractions number (Fig. 3). These results demonstrated a specific inhibition of the intestinal MI in animals treated with fractions 34-35 thus providing evidence that factor 1 is a protein with an isoelectric point at pH 6.7-7.0. The chemical nature of factor 2 is still unknown.

EFFECTS OF REPEATED INJECTIONS OF INTESTINAL MITOTIC INHIBITORS ON EMBRYONIC CELL PROLIFERATION

As our preliminary hypothesis was that the intestinal cell proliferation decrease, correlated to differentiation, may be triggered by an increasing concentration of intestinal mito-

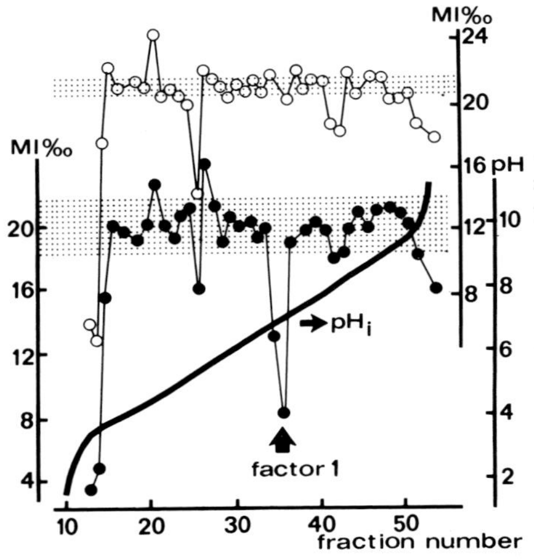

Fig. 3. Mitotic index in the telencephalon (O) and in the intestine (●) of newt embryos sacrificed 26 hr after injection, at stage 34, of fractions obtained by electrofocusing of factor 1 containing fractions from G-200 Sephadex chromatography.

tic inhibitors, we have performed repeated in ections (one injection every 4 hr) of factor 1 or 2 in stage 34 embryos and sacrificed after various time intervals up to 93hr following the first injection.

Figure 4 shows that the intestinal MI of embryos treated

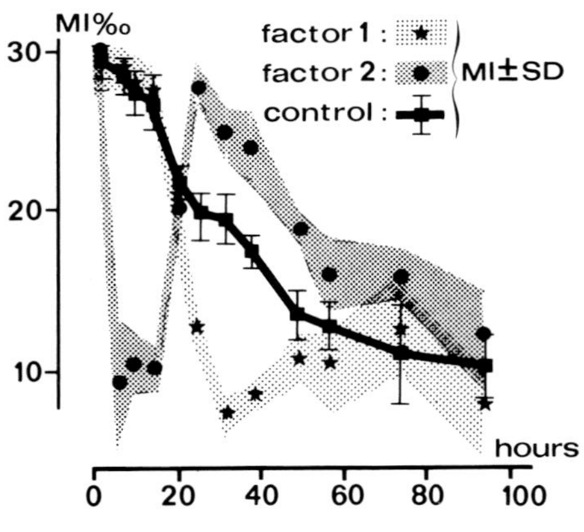

Fig. 4. Effect of repeated injections (one injection every 4 hr from stage 34 to 37) of factor 1 or factor 2 containing fractions, on the intestinal MI in the newt embryos sacrificed at different time intervals after the beginning of the treatment.

with factor 2 collapsed transiently, increased afterwards above the control values and finally decreased. This later diminution correlated to differentiation was delayed for about 12 hr in comparison to that of the controls. Therefore, both single and repeated injections of factor 2 could not delay the intestinal embryonic G_2 cells for more than about

12 hr. Owing to the inability of factor 2 to maintain G_2 inhibition in spite of repeated injections, and as cytophotometric study of age distribution of cells failed to demonstrate any resting G_2 cell sub-population within the embryonic intestine of the newt, it is very likely that this factor cannot be considered as one of the regulators of embryonic cell proliferation. It may have a role in adult animals since Pederson and Gelfant ('70) have provided evidence for a resting G_2 cell sub-population in the mouse duodenum.

As expected, the intestinal MI of embryos treated with factor 1 decreased drastically between 24 hr and 30 hr after the first injection (Fig. 4). Nevertheless, the increase of this MI up to control values between 36 hr and 48 hr showed some of the cells initially delayed in G_1 phase later divided. On the other hand, since the intestinal MI in treated animals never increased significantly above control values, other cells inhibited in G_1 phase did not divide again in the course of the experiment. It is likely that these cells entered G_o phase and possibly differentiated (Brugal, '75). Owing to this result and since normally differentiating cells are arrested in the G_1 phase, it is likely that factor 1 may be considered as a potential regulator of cell proliferation in the intestine of the newt embryo. The adult newt intestinal extracts, however, did not induce any alteration of the cell proliferation pattern in entirely proliferating intestine of young newt embryos at stages 28 and 30 (unpublished data). It was thus suggested that responsiveness to chalone indicates some commitment of the cells to the path toward functionalization. In support of this conclusion, it has been shown that the responsiveness to G_1 chalone in the epidermis of newborn mice is also developed with increasing age (Bertsch and Marks, '74).

CONCLUSIONS

Taking into account the present experimental data, we have shown that the intestinal mitotic inhibitory factors extracted from the adult newt intestine are tissue-specific and species-unspecific; therefore, according to Bullough ('62), they should be called intestinal G_1 and G_2 chalones.

A hypothesis on the relations between cell proliferation and differentiation may be elaborated by focusing on the following assumptions: (i) during normal embryonic development, the progression of cell differentiation in a cell lineage is thought to be correlated to a definite number of cell generations, so that every developmental stage is reached after a given tissue-specific number of cell cycles (Dettlaff,

'64; Brugal, '71); (ii) two kinds of cell cycles subserving two distinct functions have been postulated: one is leading to a simple duplication of the mother cell's phenotype ("proliferative" cell cycles), the other is leading to a differentiation step of this phenotype ("quantal" cell cycles) (Holtzer et al., '72); (iii) the cell responsiveness to chalones was expected to require an advanced state of differentiation (Bertsch and Marks, '74; Brugal, '75) . We thus assume that at least two "switches" are involved when an embryonic G_1 cell has to "decide" between entering a new cell cycle or taking the path toward functionalization. The first switch is turned on when the required number of quantal cell cycles has occured, leading to a cell that is sufficiently differentiated to respond to chalone. The second switch is turned on when the chalone concentration within the tissue is high enough, i.e. when the required number of cells is obtained in this tissue. When both switches are turned off or when a single one is turned on, the G_1 cell enters a new quantal or proliferative cell cycle according to its differentiation state. On the contrary, when both switches are turned on, the G_1 cell leaves the mitotic cycle to enter the G_0 phase. It may then complete its differentiation to become functional. When the tissue specific function begins (feeding in the newt larvae for example) additional cell proliferation regulating mechanisms, such as G_2 chalone feedback, may happen in response to the occurence of cell death and cell loss.

LITERATURE CITED

Bertsch, S. and Marks, F. 1974. Lack of an effect of tumor-promoting phorbol esters and of epidermal G_1 chalone on DNA synthesis in the epidermis of newborn mice. *Cancer Res. 34:* 3283-3288.

Bischoff, R. 1964. Inhibition of mitosis by homologous tissue extracts. *J. Cell Biol. 23:* 10A-11A.

Bullough, W. S. 1962. The control of mitotic activity in adult mammalian tissues. *Biol. Rev. 37:* 307-342.

Brugal, G. 1971. Etude autoradiographique de l'influence de la temperature sur la proliferation cellulaire chez les embryons ages de *Pleurodeles waltlii* Michah. (Amphibien, Urodele). *Wilhelm Roux' Archiv 168:* 205-225.

Brugal, G. 1973. Effects of adult intestine and liver extracts on the mitotic activity of corresponding embryonic tissues of *Pleurodeles waltlii* Michah. (Amphibia, Urodela). *Cell Tissue Kinet. 6:* 519-524.

Brugal, G. and Pelmont, J. 1974. Presence, dans l'intestin du Triton adulte *Pleurodeles waltlii* Michah., de deux facteurs antimitotiques naturels (chalones) actifs sur la proliferation cellulaire de l'intestin embryonnaire. *C. R. Acad. Sc. Paris, Ser. D 278:* 2831-2834.
Brugal, G. and Pelmont, J. 1975. Existence of two chalone-like substances in intestinal extract from the adult newt, inhibiting embryonic intestinal cell proliferation. *Cell Tissue Kinet. 8:* 171-187.
Brugal, G. 1975. Effets des chalones intestinaux sur la proliferation et la differenciation cellulaires de l'epithelium intestinal embryonnaire du Triton *Pleurodeles waltlii* Michah. (in preparation).
Dettlaff, T. A. 1964. Cell divisions, duration of interkinetic states and differentiation in early stages of embryonic development. In: *Advances in Morphogenesis 3:* 323-362.
Galjaard, H., van der Meer-Fieggen, W., and Giesen, J. 1972. Feedback control by functional villus cells on cell proliferation and maturation in intestinal epithelium. *Exp. Cell Res. 73:* 197-207.
Holtzer, H., Weintraub, H., Mayne, R. and Mochan, B. 1972. The cell cycle, cell lineages, and cell differentiation. *Current Topics in Developmental Biology 7:* 229-256.
Hondius Boldingh, W. and Laurence, E. B. 1968. Extraction, purification and preliminary characterisation of the epidermal chalone: a tissue specific mitotic inhibitor obtained from vertebrate skin. *European J. Biochem. 5:* 191-198.
Leblond, C. P. and Messier, B. 1958. Renewal of chief cells and goblet cells in the small intestine as shown by radioautography after injection of thymidine-^{3}H into mice. *Anat. Record 132:* 247-259.
Lipkin, M. 1973. Proliferation and differentiation of gastrointestinal cells. *Physiol. Rev. 53:* 891-915.
Loran, M. R., Crocker, T. T. and Carbone, J. V. 1964. The humoral effect of intestinal resection on cellular proliferation and maturation in parabiotic rats. *Fed. Proc. 23:* 407.
Marks, F. 1973. A tissue-specific factor inhibiting DNA synthesis in mouse epidermis. *Natl. Cancer Inst. Monogr. 38:* 79-90.
Meer-Fieggen, W. van der. 1973. Regulation of cell proliferation and differentiation in intestinal epithelium. Thesis; Rotterdam.
Patten, S. F. Jr. 1961. Renewal of the intestinal epithelium of the Urodele. *Exp. Cell Res. 20:* 638-641.

Pederson, T. and Gelfant, S. 1970. G_2-population cells in mouse kidney and duodenum, and their behavior during the cell division cycle. *Exp. Cell Res. 59:* 32-36.
Philpott, G. W. 1971. Tissue-specific inhibition of cell proliferation in embryonic stomach epithelium *in vitro*. *Gastroenterology 61:* 25-34.

Factors involved in the differentiation of the epithelial cells in the adult rat small intestine.

GABRIEL G. ALTMANN

Department of Anatomy, Health Science Centre
The University of Western Ontario
London, Ontario, Canada

According to calculations based on cell number estimations along the small intestine, there may be as many as two billion cells released daily from the epithelium into the lumen of the rat small intestine (Altmann and Enesco, '67). This cell loss is compensated by the production of new cells arising from mitosis in the crypts. In 1948, Leblond and Stevens had indeed proposed that the intestinal epithelium is continually being renewed, the sequence of events being cell production in crypts, migration to villi and along the villi, and cell loss from villus tips. A balance between production and loss ensures a steady state of the epithelium (Leblond and Walker, '56), so that, even though the size of the epithelium varies in different regions of the small intestine, it seems to remain constant in any one of these regions (Altmann and Enesco, '67). It follows that regulatory factors must continually exert their influence on the renewal process. During the past years, our research was oriented toward exploring these factors. In this presentation, our progress in this field will be briefly reviewed.

LIFE SPAN OF THE VILLUS EPITHELIAL CELLS

DNA content estimations, the colchicine technique, and histometric measurements were combined (Altmann and Enesco, '67) to show that in the various regions of the small intestine the number of crypt cells, mitotic as well as nonmitotic, per unit intestinal length was not different whereas the number of villus epithelial cells decreased gradually from duodenum to ileum (Fig. 1). This decrease

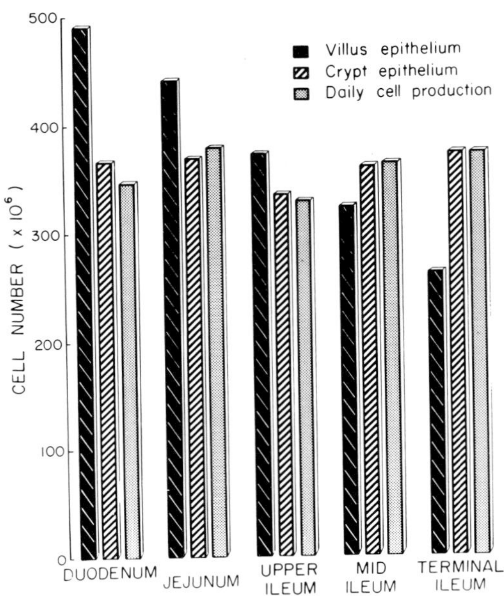

Fig. 1. Calculated total number of cells per intestinal segment in villus epithelium and crypt epithelium. These data were taken from Altmann and Enesco, '67;they were calculated by combining results of DNA determinations and histometry. The daily cell production means the total number of crypt cells entering mitosis daily per segment; it was measured by the colchicine technique. A segment measured 1/5th of the length of the small intestine. A segment was taken from each of the five intestinal regions shown.

was associated with a similar decrease in the size of the villi; this phenomenon was collectively called "the villus size gradient". It was deduced from the cell number data that the rate of cell production was not significantly different in the various intestinal regions and therefore was not directly related to the differences in villus size. It appeared that the time spent by the epithelial cells on the villus between their emergence and exfoliation was related to villus size. This time may be referred to as the life span of the villus epithelial cells. Since these cells presumably are migrating in a cohort, the turnover time of villus epithelium may serve as an estimate of the average life span. Using the cell number data, this turnover time was calculated to be 41 hours in the duodenum, 27 hours in the jejunum, 21 hours in the upper ileum and 17 hours in the lower ileum. Thus, this study has shown that while the crypts supply new cells for renewal at about the same rate in every intestinal region, the time these cells spend in the villus epithelium displays regional differences which in turn seem to be proportional to villus size.

PARAMETERS OF RENEWAL

Since 1970, our estimations of renewal rate are made in histological sections and involve the measurement of essentially three parameters: 1) the number of epithelial cells per representative villus section (V), 2) the number of cells per representative crypt section (C) and 3) the number of mitotic figures per representative crypt section (M). Assuming steady state conditions in adult animals, V and C remain constant and the rate of renewal is then determined by M. If M cells are produced and renewed per mitotic time (T_m), V cells will be renewed per VT_m/M time, which is the turnover time of the villus epithelium, and C cells will be renewed within CT_m/M time, which is the turnover time of the crypt epithelium. These values are to be multiplied by a correction factor (K) to correct for the three-dimensional arrangement of the crypts and the villi and thereby to obtain actual turnover times (Altmann, '72).

This method gave very consistent results as long as rigid criteria were followed for selecting representative sections of the villi and the crypts (Altmann and Leblond, '70; Altmann, '72). V and C have usually been referred to as villus size index and crypt size index, respectively, as they are proportional to other size parameters such as height, area, etc. K appears to be linearly related to the distance of the samples from pylorus (D) and may hence be approximated as follows:

$$K_{villus} = 1.47 - 9.9 \times 10^{-3} D$$

$$K_{crypt} = 1.82 - 7.8 \times 10^{-3} D$$

D is measured as the percent distance taking the whole length of the intestine from pylorus to caecium as 100%. When V, C and M are determined by other methods, for example as the numbers of cells per unit intestinal length, the above formulae for turnover time will still apply but the value of K will be different.

VILLUS ENLARGING AND REDUCING FACTORS

To examine the factors behind the villus size gradient, various surgical transpositions of intestinal segments were carried out and the resulting effects on villus size were observed (Altmann and Leblond, '70; Altmann, '71). It was shown that factors in the chyme cause the gradient and without such factors, a medium villus size would only be maintained. In the chyme of the upper intestine, villus enlarging factors were shown to predominate whereas the ileal chyme was shown to contain villus reducing factors. By connecting the pyloric stomach, the duodenum or the pancreas to the ileum, it was shown that the secretions of these organs contained villus enlarging factors. The villus reducing factors were recently examined (Altmann, '74a). Consideration was given to the fact that the transit of chyme in the ileum is relatively slow so that there is a relative stagnation which may give rise to chemical or biological alterations. Various components of the ileal chyme were injected into the lumen of ligated jejunal loops and the morphological effects were examined 24 hours later. It was found so far that bile after stagnation causes villus reduction; probably the bile salts are involved.

By exposing the ileum to the villus enlarging factors, it is possible to convert ileal areas into duodenum-like areas, at least as far as villus size is concerned. To examine how this change in villus size was related to other parameters of renewal, we have recently examined samples from six animals (unpublished) in which the duodenal papilla was transplanted to the mid-ileum as described by Altmann('71). In these animals, the pancreatic secretions, which are transmitted through the duodenal papilla, caused the ileal villi to enlarge markedly. The most affected samples were taken and V, C and M were measured and compared to control samples of the mid-ileum. V was increased from 137 $\pm$ 4 (control) to 279 $\pm$ 4 (experimental), C remained the same as in controls, that is 88 $\pm$ 2, M was 5.4 $\pm$ 0.3 in controls and was somewhat higher, 6.5 $\pm$ 0.4, in the experimental animals. Thus, the villi were enlarged by more than 100%, mitotic rate was increased only by 20% and crypt size was unchanged. Clearly, villus size was specifically affected without comparable permanent increase in mitotic rate. The turnover time of the villus epithelium was increased by about 70% indicating a marked lengthening of the life span of the villus epithelial cells.

STARVATION AND REFEEDING

It is becoming increasingly recognized that the intestinal chyme contains factors which have morphological effects on the intestinal tissue (Altmann, '74a; Dowling, '74; Clarke, '74). Aside from containing villus enlarging and reducing factors, the chyme also appears to have an effect on the epithelial renewal rate according to our study on the effects of starvation and refeeding (Altmann, '72).

In this study, groups of rats were starved for 3, 5 and 7 days, and one group was refed for a day after 6 days of starvation. Briefly, V, C and M progressively decreased during starvation, but turnover time became stabilized at a relatively high level (fig. 2). After refeeding, no

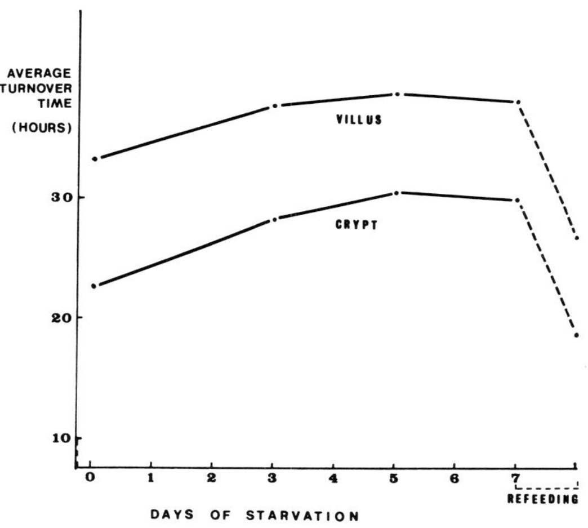

Fig. 2. Plot of average turnover time of villus epithelium and crypt epithelium against the days of starvation. Turnover time was estimated as described in the text. The turnover time estimates were obtained from five intestinal regions and the averages from these five measurements are shown. The data were taken from Altmann, '72.

significant change was noticed in V and C but M increased by about 50% indicating an upsurge in the rates of renewal and cell migration; correspondingly, turnover time decreased below control values (fig. 2). When refeeding was combined with the administration of methotrexate, a mitotic inhibitor, the upsurge of M was blocked but instead now the values of C showed a decrease of about 40%. Histological observations revealed that the upper crypt region was partially or totally emptied. As intestinal epithelial cells continue to migrate during methotrexate treatment (Altmann, '74b), it appears that the crypt decrease was caused by migration of crypt cells to the villi with concomitant transformation of these cells into absorptive cells. This migration was not compensated by production of new cells as there was no significant mitotic activity. Thus, after refeeding a pull is presumably exerted by the villi on the crypts eliciting increased rate of migration of crypt cells to the villi. This migration may have been a primary event which triggered the increase in mitotic activity after simple refeeding when mitotic inhibitor was not used. It may be calculated that the number of cells provided for the villi per unit time and per crypt was about the same in our two types of refeeding which respectively resulted in the increased mitosis and in the crypt-emptying.

Recently these studies were extended (unpublished) in order to find out what caused the increased renewal rate after refeeding. In one experiment, a defunctionalized isolated segment of the jejunum was created (as in Jeynes and Altmann, '75). This segment was excluded from the stream of chyme. A month after operation, the animals were starved and refed. M did not show increase in the defunctionalized segments. In another experiment, the distal ileum was ligated before refeeding so that the chyme accumulated above the ligature after the animals ate. Mitotic activity showed the expected increase above the ligature but not below. These experiments showed that the contact with the chyme elicited the increase in renewal rate after refeeding. Similar conclusions were reached by Clarke ('74) after studying mitotic activity in various intestinal sacs. In the next experiments, refeeding was carried out with specific diets containing only certain components of the normal diet, e.g., bulk only, sugar only or protein hydrolsate only. Only the diet with the protein hydrolsate elicited the expected increase in M. It may be concluded that the intestinal chyme and the absorption of proteins may have a role in regulating the rate of epithelial renewal.

ROLE OF DNA AND RNA SYNTHESIS

An inhibitor of nucleic acid synthesis and mitosis, methotrexate, was used in most of this study (Altmann, '74b). An initial injection of 5 mg was given to normally fed untreated adult male rats. This was followed by methotrexate mixed in the food so that a high blood level of this drug was continuously maintained. An hour before the initial injection, a dose of tritiated thymidine was administered to many of the animals. Tritiated thymidine labeled the DNA synthesizing cells. These cells could not progress into mitosis because of the methotrexate-block. Groups were killed at half day intervals up to three days when the toxic effects of methotrexate became apparent. During these three days, villus size and crypt size gradually decreased and mitotic activity was absent. Tritiated thymidine labeled a large proportion of the crypt cells initially. Within 24 hours, the labelled cells moved up the villi at near normal rate while the crypts became partially emptied. Thus, cell migration and differentiation of crypt cells into villus epithelial cells continued in spite of the block of nucleic acid synthesis and mitosis. After 48 hours, the entire villus epithelium contained mostly labelled cells indicating that this epithelium was formed during the methotrexate treatment from crypt cells which were in their DNA synthesis phase at the beginning of the treatment. Since there was no RNA synthesis, the cells surviving must have had stored RNA necessary for migration and differentiation and for associated protein synthesis. It appeared that the length of survival of the individual cells was linked to the amount of stored RNA.

This study has thus indicated that the differentiating and differentiated epithelial cells may not need the synthesis of new RNA and they may therefore function by using stored RNA. It was indicated that the acquisition of this RNA takes place among the cycling cells and is completed by the time of the DNA synthesis phase or during this phase. It may be speculated that in normal animals the signal for differentiation is received by some of these cells. This signal may be exogenous or endogenous to the cells and may stop any further RNA synthesis. There is evidence in the literature that significant RNA synthesis is present only in the crypts (Amano _et al._, '65), and also that the RNA-caused staining decreases along the villi (Padykula _et al._, '61; Padykula, '62; Pearse and Rieched, '67) probably indicating a decrease in the amount of stored RNA as the epithelial cells approach the end of their life span.

ROLE OF PROTEIN SYNTHESIS

Thus far, we have used two approaches to study the role of protein synthesis. In the first one, we have measured triatiated leucine uptake (unpublished); in the second one, we have used cycloheximide, an inhibitor of protein synthesis (Altmann,'75).

Adult male rats received a single dose of tritiated leucine. The animals were killed an hour later and samples of the intestine were processed for radioautography. In the radioautographs, the number of grains over representative cell sections was estimated. Those sections of epithelial cells were selected which were cut along the central longitudinal plane of the cells; the method of selecting such sections has been presented elsewhere (Jeynes and Altmann, '75). The grain counts were carried out at seven levels along the epithelium and in five intestinal regions (Table 1). The results (Table 1, Fig. 3) have

TABLE 1

H^3-Leucine Incorporation per Columnar Epithelial Cell (average grain count per representative cell section)

	Crypt Base	Mid Crypt	Upper Crypt	Villus Base	Mid Villus	Upper Villus	Villus Tip
DUODENUM	17.6 ±1.0	26.3 ±1.7	33.0 ±2.3	52.0 ±1.9	31.0 ±1.9	17.1 ±1.9	7.3 ±0.9
JEJUNUM	17.5 ±1.7	25.7 ±1.1	31.7 ±1.2	41.6 ±1.8	25.4 ±1.9	15.4 ±1.2	6.4 ±0.7
UPPER ILEUM	17.4 ±1.2	23.7 ±1.7	30.4 ±2.3	34.1 ±2.5	22.3 ±1.7	14.4 ±1.3	5.3 ±0.5
MID ILEUM	16.6 ±1.4	22.3 ±2.2	24.3 ±2.4	26.0 ±3.2	18.0 ±1.8	11.6 ±0.8	5.1 ±0.6
TERMINAL ILEUM	16.9 ±1.4	19.0 ±2.2	22.1 ±2.6	25.1 ±2.0	17.1 ±2.2	11.1 ±1.4	4.0 ±0.4

shown that the overall rate of protein synthesis per cell increases along the crypts and reaches a maximum at the villus base. Thereafter this rate gradually declines and

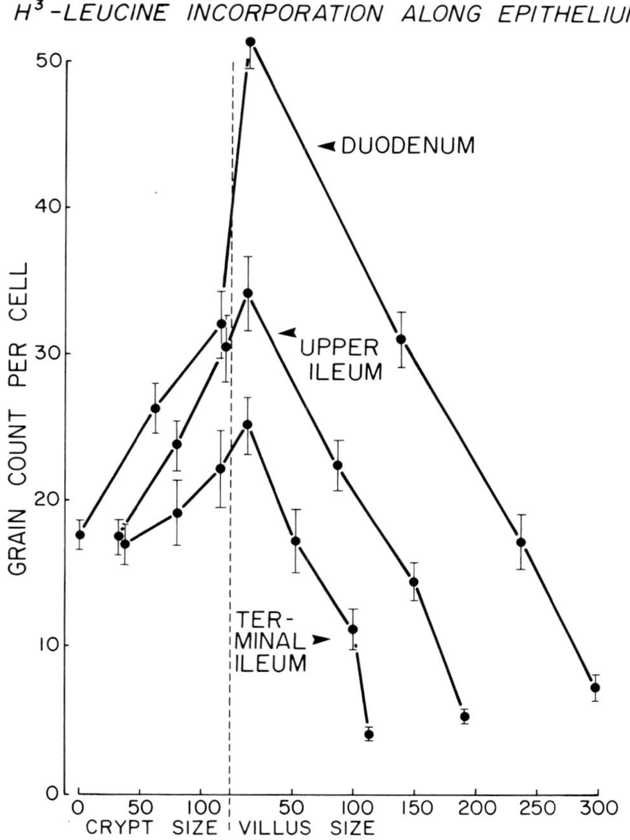

Fig. 3. Graphic representation of some of the data from Table 1 to show the gradual increase in the rate of protein synthesis along the crypts, and the gradual decline along the villi. Crypt size is given as the number of epithelial cells per representative crypt section. Villus size is given as the number of epithelial cells per representative villus section. The vertical bars represent the standard error of the mean.

reaches a minimum at the villus tip. This pattern is consistent with other reports in the literature (Lipkin and Quastler, '62; Bennet et al., '74). Thus, the process of differentiation of crypt cells into villus epithelial cells is associated with an upsurge in protein synthesis. In the differentiated cells however, the protein synthesizing activity gradually subsides presumably because of degradation of the stored RNA.

As pointed out earlier, villus size did not appear to be directly related to the mitotic activity in the crypts. Protein synthesis on the other hand appeared to be related directly to villus size as shown in Figure 4. Here the grain counts at the villus base are plotted against distance of the samples from the pylorus and a similar plot of villus size is simultaneously presented. The parallelism of the two curves indicates that villus size is directly related to the capacity for protein synthesis. Since this capicity differs along the length of the intestine, it is probably under the influence of the villus enlarging and reducing factors.

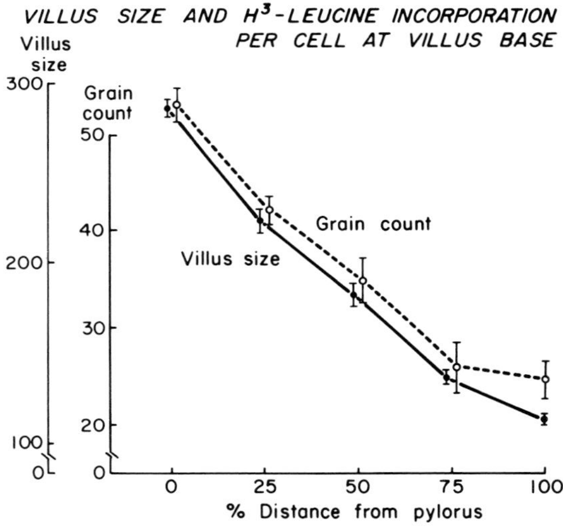

Fig. 4. Simultaneous plot of villus size and the rate of protein synthesis at the villus base against the distance of the samples from the pylorus. Villus size is measured as the number of epithelial cells per representative villus section. The rate of protein synthesis is measured as the number of grains in radioautographs over representative sections of epithelial cells.

In our other study of protein synthesis a large single dose of cycloheximide was administered to adult male rats. Within hours, the villi markedly decreased. According to histological observations, this decrease was caused by an accelerated exfoliation of the epithelial cells at the villus tip. About 10% of the villus epithelium was lost per hour. It appears from these results that the synthesis of certain proteins, presumably surface ones, is responsible for epithelial cohesion. Cycloheximide presumably stopped the synthesis of these proteins but their degradation was continued resulting in massive exfoliation of epithelial cells. Since the loss of cells was gradual, cells toward the villus base were likely to have increasing amounts of these proteins stored. The life time of these proteins would be about 10 hours.

Assuming the existence of these proteins, a tentative model of the normal epithelial cell exfoliation may be proposed. The degradation of these proteins would be balanced by synthesis in the lower villus where the cells have marked capacity for protein synthesis. As this capacity subsides along the villus, there would be less and less synthesis. There would be no synthesis in the upper villus during the last 10 hours of the life span of the epithelial cells. The lack of synthesis and the progressing degradation would eventually lead to the exfoliation of the cells.

PRELIMINARY OBSERVATIONS

A few preliminary observations have been made recently. They contribute to the conclusions of this presentation and will therefore be briefly described.

Adult male rats received actinomycin D in varying doses in order to suppress the synthesis of messenger RNA. Final conclusions from this investigation are not yet available but the observations suggest so far that a large proportion of the intestinal epithelial cells survive and function for at least a day after the administration of this drug. These cells apparently possess messenger RNA in stored form. There seems to be an early damage however, within about six hours. This consists of lysis of many of the stem and mitotic cells, indicating that these cells need the active production of messenger RNA. Within hours after the administration of methotrexate, degeneration of some of the stem cells was also observed. The stem cells may be identified by their position in the crypt-bottom (Cheng and Leblond, '75).

Another study is under way in which the recovery of the intestinal epithelium is being studied after damage by large sublethal doses of methotrexate. The damage consists mainly of diminution of villi and crypts, and flattening of the columnar epithelial cells. Recovery starts in some of the crypts by the recovery of the columnar shape of the cells, mitosis, and uptake of tritiated thymidine. Cells from these crypts eventually migrate up the villi and gradually replace the flat villus epithelium with normal looking columnar cells. In the few animals examined so far, when such recovered crypts were associated with damaged and diminished villi, DNA synthesis and mitosis were present all along these crypts including the upper third or maturation zone where cycling cells are normally rare or absent. When the crypts were associated with recovered villi, the maturation zone was clearly present. These observations substantiate the view that a negative feedback on cell proliferation originates from the intestinal villi. In the maturation zone, there is a decreasing probability of DNA synthesis and mitosis toward the villi (Cairnie et al., '65; Cheng and Leblond, '75). The existence of negative feedback from the villi has been suggested on the basis of studies using irradiation (Galjaard et al., '72; Rijke et al., '74). Also the existence of an intestinal chalone was demonstrated in the newt (Brugal, '73; Brugal and Pelmont, '75), and a substance supressing cell proliferation was isolated from the rabbit intestinal mucosa (Sassier and Bergeron, unpublished).

CONCLUSIONS

Leblond and co-workers have emphasized the importance of a dynamic view of the various histological structures (Leblond, '65a). They have introduced the concepts of cell renewal, and continuous protein synthesis in cells (Leblond and Walker, '56; Leblond, '65b). In view of these concepts and the several lines of results presented here, it may now be possible to introduce a few new, as yet tentative, concepts regarding the differentiation of the epithelial cells within the renewal system of the intestinal epithelium.

It appears that active transcription takes place mainly among the stem and mitotic cells. In the mitotic zone, some of the cycling cells acquire stored RNA which enables these cells to differentiate and to function without the synthesis of new RNA. These cells probably receive a signal, perhaps through a chalone, and further cycling and perhaps further RNA synthesis stop as a result and the process of differentiation into absorptive cell begins. This process initially involves an upsurge of protein synthesis until the villus base is reached. After this, protein synthesis gradually declines presumably in association with the degradation of the stored RNA. A morphological sign of this degradation may be the gradual decline of RNA staining along the villi. The decrease of protein synthesis along the villi apparently leads to cell exfoliation at the villus tip when the proteins responsible for epithelial cohesion are degraded and are no longer synthesized. It appears that the life span of the villus epithelial cells between emergence on villus and exfoliation is dependent on the capacity for protein synthesis and may therefore be linked to the amount and the degradation rate of the stored RNA. The capacity for protein synthesis seems to be also related to villus size and may therefore be subject to modifications by the villus enlarging and reducing factors present in the intestinal chyme.

It may be speculated that the various states of protein synthesis along the villi reflect differing absorptive function of the epithelial cells. In the lower villus, where protein turnover is rapid, enzymes, transport proteins, receptor sites, and perhaps a chalone, are probably rapidly produced as well as used up. It is a possibility that the proteins of the chyme are absorbed here and this absorption somehow triggers cell migration and mitotic activity in crypts. In the upper villus, where protein turnover is less, active transport across the epithelium is probably less important or it is connected with the less frequent materials of the chyme.

LITERATURE CITED

Altmann, G. G. 1971. Influence of bile and pancreatic secretions on the size of the intestinal villi in the rat. *Am. J. Anat. 132:* 167-178.

Altmann, G. G. 1972. Influence of starvation and refeeding on mucosal size and epithelial renewal in the rat small intestine. *Am. J. Anat. 133:* 391-400.

Altmann, G. G. 1974a. Demonstration of a morphological control mechanism in the small intestine. Role of pancreatic secretions and bile. In: *Intestinal Adaptation.* Ed. Dowling and Riecken, F. K. Schattauer Verlag, pp. 75-86.

Altmann, G. G. 1974b. Changes in the mucosa of the small intestine following methotrexate administration or abdominal X-irradiation. *Am. J. Anat. 140:* 263-280.

Altmann, G. G. 1975. Morphological effects of a large single dose of cycloheximide on the intestinal epithelium of the rat. *Am. J. Anat. 143:* 219-240.

Altmann, G. G. and M. Enesco. 1967. Cell number as a measure of distribution and renewal of epithelial cells in the small intestine of growing and adult rats. *Am. J. Anat. 121:* 319-336.

Altmann, G. G. and C. P. Leblond. 1970. Factors influencing villus size in the small intestine of adult rats as revealed by transposition of intestinal segments. *Am. J. Anat. 127:* 15-36.

Amano, M., C. P. Leblond and N. J. Nadler. 1965. Radioautographic analysis of nuclear RNA in mouse cells. *Exptl. Cell Res. 38:* 314-340.

Bennett, G., C. P. Leblond and A. Haddad. 1974. Migration of glycoprotein from the Golgi apparatus to the surface of various cell types as shown by radioautography after labeled fucose injection into rats. *J. Cell Biol. 60:* 258-284.

Brugal, G. 1973. Effects of adult intestine and liver extracts on the mitotic activity of corresponding embryonic tissues of *Pleurodeles waltlii* Michah. (Amphibia, Urodela). *Cell Tissue Kinet. 6:* 519-524.

Brugal, G. and J. Pelmont. 1975. Existence of two chalone-like substances in intestinal extract from the adult newt, inhibiting embryonic intestinal cell proliferation. *Cell and Tissue Kinet. 8:* 171-187.

Cairnie, A. B., L. F. Lamerton and G. G. Steel. 1965. Cell proliferation studies in the intestinal epithelium of the rat. I. Determination of the kinetic parameters. *Exp. Cell Res. 39:* 528-538.

Cheng, H. and C. P. Leblond. 1975. Origin, differentiation and renewal of the four main epithelial cell types in the mouse small intestine. I. Columnar cell. *Am. J. Anat. 141:* 461-480.

Clarke, R. M. 1974. Control of intestinal epithelial replacement - lack of evidence for a tissue specific blood-borne factor. *Cell Tissue Kinet. 7:* 241-250.

Dowling, R. H. 1974. The influence of luminal nutrition on intestinal adaptation after small bowel resection and by-pass. In: *Intestinal Adaptation.* Ed. Dowling and Riecken, F. K. Schattauer Verlag, pp. 35-46.

Galjaard, H., W. Van Der Meer-Fieggen and J. Giesen. 1972. Feedback control by functional villus cells on cell proliferation and maturation in intestinal epithelium. *Exptl. Cell Res. 73:* 197.

Jeynes, B. J. and G. G. Altmann. 1975. A region of mitochondrial division in the epithelium of the small intestine of the rat. *Anat. Rec. 182:* 289-296.

Leblond, C. P. 1965a. The time dimension in histology. *Am. J. Anat. 116:* 1-28.

Leblond, C. P. 1965b. What radioautography has added to protein lore. In: *Use of Radioautography in Investigation of Protein Synthesis,* pp. 321-339. Academic Press Inc., New York.

Leblond, C. P. and Stevens, C. E. 1948. The constant renewal of the intestinal epithelium in the albino rat. *Anat. Rec. 100:* 357-378.

Leblond, C. P. and B. E. Walker. 1956. Renewal of cell populations. *Physiol. Rev. 36:* 255-275.

Lipkin, M. and H. Quastler. 1962. Studies of protein metabolism in intestinal epithelial cells. *J. Clin. Invest. 41:* 646.

Padykula, H. A. 1962. Recent functional interpretations of intestinal morphology. *Fed. Proc. 21:* 873-879.

Padykula, H. A., E. W. Strauss, A. J. Ladman and F. H. Gardner. 1961. A morphologic and histochemical analysis of the human jejunal epithelium in nontropical sprue. *Gastroenterology 40:* 735-765.

Pearse, A.G.E. and E. O. Riecken. 1967. Histology and cytochemistry of the cells of the small intestine, in relation to absorption. *Brit. med. Bull. 23:* 217-222.

Rijke, R.P.C., W. Van der Meer-Fieggen and H. Galjaard. 1974. Effect of villus length on cell proliferation and migration in small intestinal epithelium. *Cell Tissue Kinet. 7:* 577-586.

Sassier, P. and M. Bergeron, University of Montreal. Unpublished data.

Homeostasis in the Small Intestine

A. B. CAIRNIE

Department of Biology, Queen's University, Kingston, Ontario, Canada

Last year I submitted a paper to Cell and Tissue Kinetics and one referee said, "A referee rarely has to suggest that a paper be lengthened, but this one is almost written in shorthand. .... the model for cell population control in the small intestine should be elaborated". I was unable to resist this invitation to pontificate, even though, or perhaps because, I know so little about cell population control in the intestine. The revised version of the paper contains the model - which I suppose might easily have been rejected by the referee as unsubstantiated if it had been in the original version (Cairnie and Millen, '75).

There are five postulates and I shall say a few words about each in turn.

(1) Each crypt is a closed self-perpetuating clone. It originated from a single cell and cells neither enter nor leave, except of course as post-mitotic cells going to the villus. There are no G_0 cells but some cells in the base of crypt have a longer G_1.

A few years ago there was some interest in the idea that cells *in vivo* have a limited capacity for division which would be a manifestation of the aging process. Such a limitation on capacity for division has been claimed for tissue culture cells (Hayflick, '65), though this is not now so clear-cut as it once seemed. Harrison ('75) most recently, and others present at this meeting, have shown that if bone marrow stem cells have an intrinsic decline in function, manifested in transplantation assays, it is not observed within the normal lifespan of the organism. I presume the same is true for intestinal stem cells and there is therefore no *need* to postulate non-cycling G_0 reserve stem cells either inside or outside the crypt which would take over sequentially as the animal ages. I have claimed that Paneth cells come from a precursor outside the crypt, though I am not aware that anyone has supported my view (Cairnie, '70).

The studies of Cairnie, Lamerton and Steel ('65) showed that labelling data were consistent with the view that all cells in the lower part of the crypt are proliferating and that at the base the cell cycle is longer. This has been confirmed several times.

I have more trouble in dealing with the steady state achieved under continuous irradiation where G_1 is much shortened, and yet the percentage labelling of cells in the lower part of the crypt is reduced (Cairnie, '67). I was driven to invent an R-cell, a non-cycling cell which persists in the crypt and I suppose moves to the villus eventually. This is model-building at its worst, and no one else has ever referred to R-cells though our P and Q cells are well ensconced in the literature. This meeting gives me an opportunity to pull out an old unpublished experiment (Fig. 1). Rats were injected with ^{3}H-thymidine at 5-hour intervals and killed 1 hour

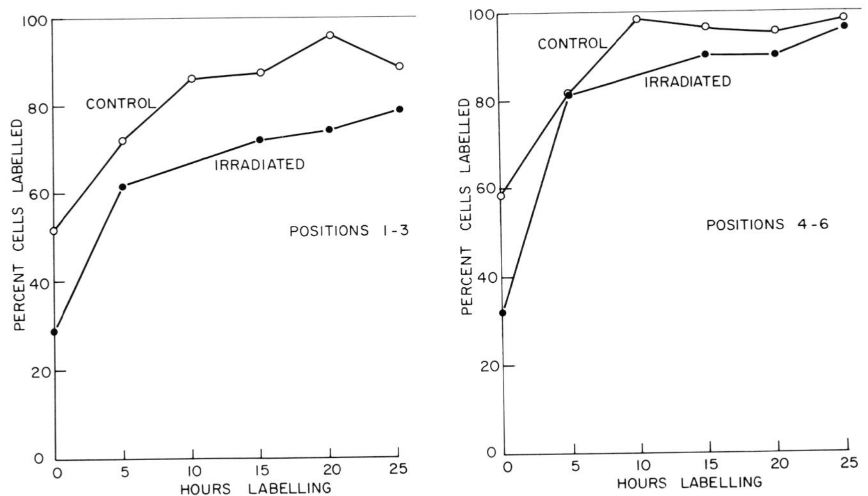

Fig. 1. Repeated labelling with ^{3}H-thymidine of cells in positions 1-3 (left) and 4-6 (right) in the crypt.

after 1 to 6 injections. There was a control series of animals and a series which received 350 rads/day for 5 days before the injections and throughout the time to sacrifice. In positions 4-6 in the crypt, counting from the base, labelling in both series approached 100 percent. In positions 1-3, having excluded Paneth cells, I saw somewhat slower labelling in the control intestine, as one would expect, and very much slower labelling in the continuously irradiated cells. This never got above 80 percent. PLM measurements of cell cycle duration in the bottom of the crypt did not explain these results at all. What are these non-labelling cells found in continuously irradiated animals? This unsolved problem hardly affects the first postulate as a generalisation.

(2) All proliferating cells are potential stem cells. This is a slight modification of the premise adopted by Hagemann, Sigdestad and Lesher ('71) and Withers and Elkind ('69).

In the bone marrow there is 1 CFU/10^4 nucleated cells and in the past I have argued that by analogy only a small percentage of intestinal proliferating cells would be stem cells. But Hagemann *et al.* and Withers and Elkind have claimed that all the intestinal proliferating cells are stem cells, and Roti Roti and Dethlefsen ('75) have concluded that following hydroxyurea treatment post-mitotic Q cells are recruited to the P cell proliferating compartment. It is necessary to have about 160 stem cells per crypt, as Hagemann *et al.* and Withers and Elkind point out, in order to have a significant proportion of crypts able to regenerate after doses in excess of 1000 rads. They both assume that the D_o is about 100 rads. Dr. Potten, who I hope will speak in the discussion, has carried out an analysis which has led him to the conclusion that there are between 20 and 80 stem cells per crypt (Potten and Hendry, '75). Their work is based on the Withers and Elkind microcolony assay. Dr. Leblond's elegant studies point to the cells in the base of the crypt in intimate contact with the Paneth cells as the stem cells. There are only a few such cells. I have therefore proposed a distinction between functional stem cells, those Dr. Leblond talks about, and the much larger group of potential stem cells which includes the other proliferating cells and possibly even, in the light of Dethlefsen's work, some post-mitotic cells. This much larger group of potential stem cells may form the pool from which survivors of radiation are diverted to be functional stem cells. However, I also have an alternative hypothesis to which I shall return.

One problem which has been with us in this field for several years is the question of an assay for intestinal stem cells. Withers and Elkind ('70) developed both the microcolony and macrocolony assay which Dr. Withers has discussed in this symposium. The D_o values they give are 109 and 107 rads respectively and extrapolation numbers 58 and 300. Hagemann, Sigdestad and Lesher ('71) developed a different assay in which crypt numbers in the intestine are estimated at 3 to 4 days following irradiation. The D_0 value is 330R and extrapolation number is 10. The difference between these two results is reproducible and is too big to explain in terms of different experimental conditions; it has therefore been conveniently ignored. All three assays have been used for RBE and radioprotection studies by different laboratories and the results obtained have been consistent with expectations from other cellular systems.

I have sought an explanation for this difference in terms of a model proposed by Gilbert ('69) for relating whole animal mortality to the survival curve of single stem cells. He proposed an empirical relation

$$\text{probit}\ (P_m) = -\ \beta\ \ln\ (\alpha S)$$

where P_m is the mortality probability of an organism whose sensitive cells have been reduced to S viable cells,
α is the average probability of a viable cell causing the organism to survive,
and β is a parameter describing the sensitivity of organism survival to changes of S near the critical value.

Figure 2 shows on the left the probit mortality curve for mice and on the right the dose survival curve for bone marrow

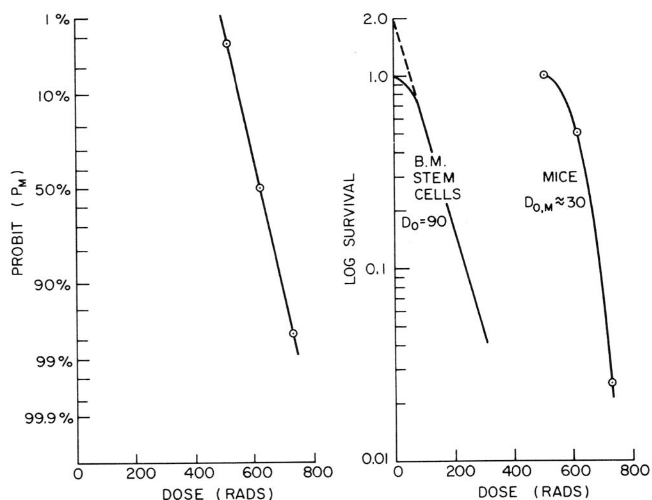

Fig. 2. Comparison of mouse survival, derived from probit mortality curve on left, and mouse bone marrow stem cell survival. Open circles represent the $LD_{50,30}$ and 2 standard deviations on either side.

stem cells taken from Gilbert's paper. The second curve on the right is his mortality data replotted by me on a log scale. Gilbert derives a relation

$$K = \frac{D_o}{\beta}$$

where K is the probit width given by

$$\text{probit } (P_m) = \frac{D - LD_{50}}{K}$$

and D_o is the slope for stem cells. The exponential slope of the mortality curve on the right, $D_{O,M}$, is a function of both D_o and β and has a value of about 30 rads. The extrapolation number n_M is very large. In this case we all realise that the $D_{O,M}$ and n_M values obtained do not pertain directly to the cells which are critical for survival.

A closely related problem was treated by Elkind and Whitmore ('67) who discussed the shape of survival curves for groups of cells. "If only M < N cells in a group population of N cells per group must be inactivated to suppress colony formation - i.e. if these cells survive dependently - the D_o value will be smaller than if all N cells must be inactivated".

Returning now to the problem of the difference between Withers and Elkind's assay and Hagemann *et al.*'s, I would like to suggest as a hypothesis that Hagemann *et al.*'s assay represents intestinal stem cell survival, or at least stands one step closer, and that Withers and Elkind's macrocolonies are structures, derived from surviving stem cells at low levels of survival, which have a greater chance of formation if there is a cluster of surviving stem cells. The macrocolonies are analogous to the animal in Gilbert's analysis or to the group of cells in Whitmore and Elkind's analysis. Yet another way of expressing this is that the plating efficiency of intestinal stem cells is dependent on the local density of survivors. The essential point is that I am suggesting that cells survive dependently after high radiation doses. In Figure 3 I show Hagemann *et al.*'s results, modified to correct for weight change, and Withers and Elkind's data. In the latter case we have only a D_o value obtained between 1600 and 2200 rads and I have arbitrarily chosen to draw it over the same range of survival though we do not know where the line in fact lies relative to 100 percent on the survival axis.

If I am correct in postulating that Hagemann *et al.*'s crypt assay represents intestinal stem cell survival, and if a crypt survives if one stem cell survives, the extrapolation number obtained (10) represents the product of the stem cell

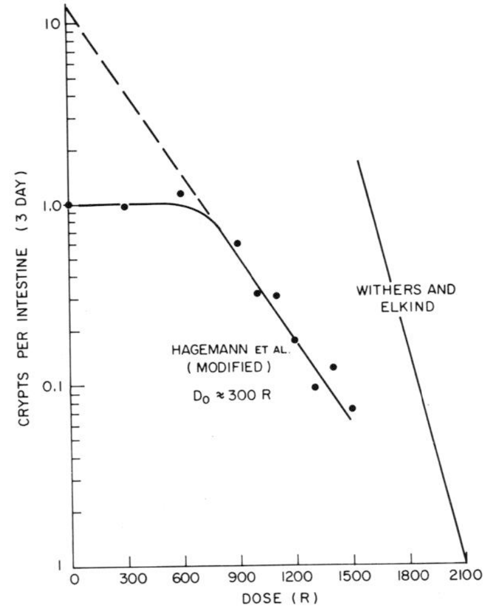

Fig. 3. Hagemann et al. ('71) obtained a dose survival curve for crypts/mg. intestine. I have modified this to crypts per intestine and obtain a D_o of 300 rads and n of 10. The line on the right shows Withers and Elkind's findings. The ordinate scale in this case is in arbitrary units.

number per crypt and the extrapolation number for crypt stem cells. It is awkward to suggest this when Hagemann *et al.* appear to have preferred the Withers and Elkind results to their own. One final point is that while a D_o value of 300R is at the high end of the normal range, Till ('61) reported that several mouse L cell lines have D_o values between 270 and 290 rads.

(3) Only those cells in contact with Paneth cells in the base of the crypt are currently functioning as stem cells. Cell migration in the crypt is orderly and sister cells tend to migrate together. Therefore only those at the bottom of the crypt are self-perpetuating and can be considered functional stem cells (as distinct from potential - see postulate 2). This has most recently been explored by Cheng ('72) but it has been implicit in earlier discussions.

This is a summary of the ideas developed over the years by Dr. Leblond and his co-workers and presented to us this morning. I have nothing whatever to add.

(4) The number of crypts present in the intestine increases with age (Cairnie, unpublished results). There is negative feedback control which regulates the number at the preset level.

The number of crypts was measured in our experiments by the method derived from Hagemann, Sigdestad and Lesher ('70) as described by Cairnie and Millen ('75) (Fig. 4). The young

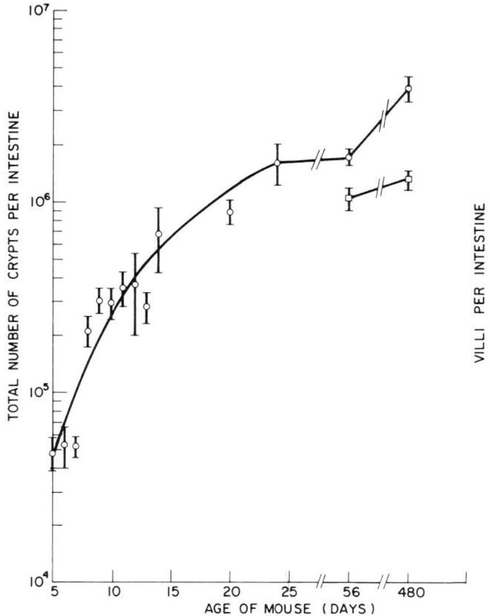

Fig. 4. Growth of the small intestine in terms of numbers of crypts per intestine. Squares refer to villi, counted only in older animals.

mice were Swiss Webster obtained from Biobreeding of Ottawa and the second part of the experiment was done on CBA/CAJ from Jackson Laboratory, Bar Harbor, Maine. As shown in the figure there is a striking increase throughout the first 4 weeks of life which is extended to late adult at a slower rate.

Feedback control was demonstrated in the case of the intestine recovering from 1600 rads X-rays where the number of crypts was initially reduced to one-third and then returned to control levels (Cairnie and Millen, '75). Another aspect of feedback control is seen in the response to continuous irradiation (Cairnie, '67 and '69) and in response to the physiological demand of lactation (Harding and Cairnie, '75). There are changes in cell cycle time, and the cut-off position where cells stop entering new P-cell cycles and produce post-mitotic Q-cells. Surprisingly, the functional lifespan of villus cells does not seem ever to change but the rate of migration through the maturation zone does.

I have found a very useful way of thinking about feedback control is in terms of the Lotka-Volterra equations used by population biologists in analysing predator-prey relationships. This can be represented graphically as in Figure 5a where the villus cell population is the "predator" and the crypt cell population the "prey". Points within the central shaded area represent the region of homeostatic control where negative feedback will restore the situation to the set point represented by the intersection of the equilibrium and utilisation lines. This central area is like a saucer in which a marble

Fig. 5. Lotka-Volterra plots of homeostatic responses in the small intestine. 5a (top left) The shaded area represents the co-ordinates of points from which the system can return to stability at the centre. 5b (lower left) There is a minimal level of villus population consistent with survival. Arrows show responses to various doses of acute radiation. 5c (lower right) Responses to continuous irradiation and lactation. A and B represent new set points.

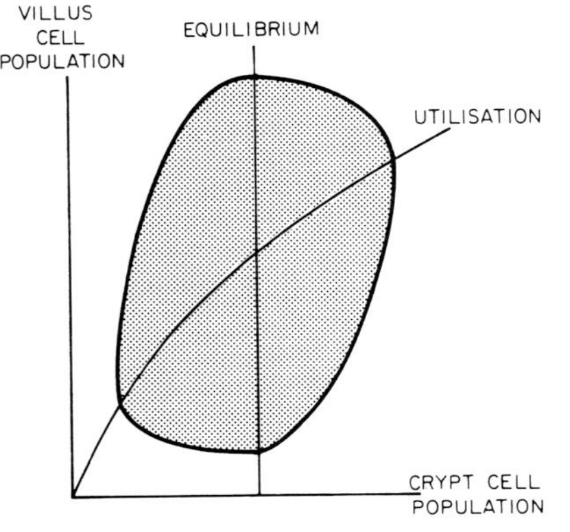

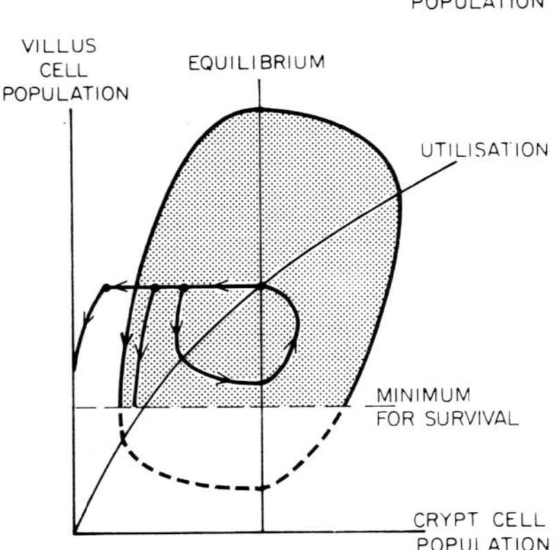

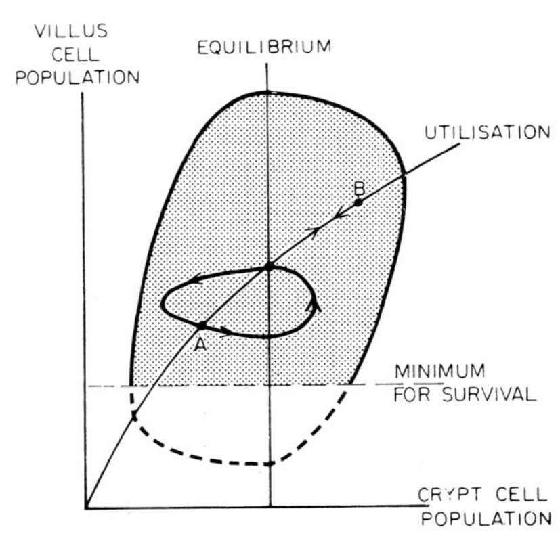

will eventually come to rest in the base.

Imagine now that a dose of acute irradiation is given to the intestine (Fig. 5b). The crypt population shrinks immediately and then the villus population shrinks. There is a minimum level of villus cell population which is consistent with survival. If the animal is to survive, homeostasis must bring the system round without reaching this minimal level. Predator-prey relationships are well known for wild oscillations. The predator increases rapidly when the prey is abundant and does not limit his population voluntarily when he is increasing fast. The result is that the prey becomes scarce and both populations crash. The saucer is quite shallow and the populations spiral in slowly. In the crypt and villus we obviously have a sophisticated control system superimposed on simple competition where cyclic oscillations are damped out and the system is restored to the set point in the minimum time. Nothing is known about either the detection of "error" signals or the feedback pathways in these control systems but the chalones described by Dr. Brugal offer an exciting challenge.

Figure 5c represents the establishment of new steady states under continuous irradiation at 350 rads/day or during lactation. The equilibrium line is shifted to the right or the left to give a new set point, either A or B. If the functional lifespan of villus cells had increased in either situation, this would have been represented by a counter-clockwise rotation of the utilisation line. When either condition ends the set point is returned to its normal co-ordinates. It is striking that within 6 hours of stopping or starting irradiation at 350 rads/day, changes in cell proliferation patterns were seen (Cairnie, '67).

(5) Crypts are replaced at long intervals by fission of existing crypts. This is also the mechanism of increase in number with age or after irradiation. While the initial bud is not necessarily at the base of the crypt, it extends to this level in the mucosa and acquires Paneth cells before fission is completed.

I have little to add to what I have already published this year. I have slides of crypts in fission in normal animals, in 13-day mice, and in animals recovering from irradiation with 1000 rads.

In conclusion, it is always fun to do experiments which could upset the whole applecart. Drs. Cheng and Leblond have recently attempted unsuccessfully to detect incorporation of ^{3}H-uridine-labelled rat bone marrow cells in rat intestine (Cheng and Leblond, '74). The hypothesis under test by them was that the stem cell for one or more cell types in the intestine might be a bone marrow cell. I have carried out a stem cell transfer experiment in which either bone marrow or intestinal cells, genetically marked, were injected in either unirradiated mice or mice which the previous day received 1200 rads. The genetically marked cells came from C57 Bl/6J-sla/sla mice which are distinguished by the accumulation of iron in the villus cells because of failure of the transport mechanism (Pinkerton, '68). I am indebted to Dr. E. S. Russell for drawing this marker to my attention. The recipients were killed 6 or 8 days after irradiation. No cells exhibiting the sla phenotype were detected in the intestinal villi of the recipient mice. There was thus no transplantation of either intestinal or bone marrow cells to the intestine detected in this experiment.

This research was supported by National Research Council Grant A4298 and Defence Research Board Grant 1675-15.

LITERATURE CITED

Cairnie, A. B. 1967. Cell proliferation studies in the intestinal epithelium of the rat: response to continuous irradiation. *Radiation Res. 32:* 240-264.

Cairnie, A. B. 1969. Further studies on the response of the small intestine to continuous irradiation. *Radiation Res. 38:* 82-94.

Cairnie, A. B. 1970. Renewal of goblet and Paneth cells in the small intestine. *Cell Tissue Kinet. 3:* 35-45.

Cairnie, A. B., L. F. Lamerton and G. G. Steel. 1965. Cell proliferation studies in the intestinal epithelium of the rat. II Theoretical aspects. *Exper. Cell Res. 39:* 539-553.

Cairnie, A. B. and B. H. Millen. 1975. Fission of crypts in the small intestine of the irradiated mouse. *Cell Tissue Kinet. 8:* 189-196.

Cheng, H. and C. P. Leblond. 1974. Origin, differentiation and renewal of the four main epithelial cell types in the mouse small intestine. V Unitarian theory of the origin of the four epithelial cell types. *Amer. J. Anat. 141:* 537-562.

Elkind, M. M. and G. F. Whitmore. 1967. The Radiobiology of Cultured Mammalian Cells. Gordon and Breach (publishers), New York.

Gilbert, C. W. 1969. The relationship between the mortality of whole animals and the survival curve for single cells. *Int. J. Radiat. Biol. 16:* 287-290.

Hagemann, R. F., C. P. Sigdestad and S. Lesher. 1970. A method for quantitation of proliferative intestinal mucosal cells on a weight basis: some values for C57Bl/6. *Cell Tissue Kinet. 3:* 21-26.

Hagemann, R. F., C. P. Sigdestad, and S. Lesher. 1971. Intestinal crypt survival and total and per crypt levels of proliferative cellularity following irradiation: Single X-ray exposures. *Radiat. Res. 46:* 533-546.

Harding, J. D. and A. B. Cairnie. 1975. Changes in intestinal cell kinetics in the small intestine of lactating mice. *Cell Tissue Kinet. 8:* 135-144.

Harrison, D. E. 1975. Normal function of transplanted marrow cell lines from aged mice. *J. Gerontol. 30:* 279-285.

Hayflick, L. 1965. The limited *in vitro* lifetime of human diploid strains. *Exper. Cell Res. 37:* 614-636.

Pinkerton, P. H. 1968. Histological evidence of disordered iron transport in the X-linked hypochromic anaemia of mice. *J. Path. Bact. 95:* 155-165.

Potten, C. S. and J. H. Hendry. 1975. Differential regeneration of intestinal proliferative cells and cryptogenic cells after irradiation. *Int. J. Radiat. Biol. 27:* 413-424.

Roti Roti, J. L. and L. A. Dethlefsen. 1975. Matrix simulation of duodenal crypt cell kinetics. II Cell kinetics following hydroxyurea. *Cell Tissue Kinet. 8:* 335-353.

Till, J. E. 1961. Radiosensitivity and chromosome numbers in strain L mouse cells in tissue culture. *Radiation Res. 15:* 400-409.

Withers, H. R. and M. M. Elkind. 1969. Radiosensitivity and fractionation response of crypt cells of mouse jejunum. *Radiat. Res. 38:* 598-613.

Withers, H. R. and M. M. Elkind. 1970. Microcolony survival assay for cells of mouse intestinal mucosa exposed to radiation. *Int. J. Radiat. Biol. 17:* 261-267.

Small Intestinal Crypt Stem Cells

CHRISTOPHER S. POTTEN

Paterson Laboratories
Christie Hospital and Holt Radium Institute
Manchester, England

Cell kinetics in mouse intestinal epithelia have been extensively studied for many years (Quastler and Sherman, '59; Lesher et al., '61; Cairnie et al., '65; Leblond et al., '67). Cell proliferation is restricted to the crypts from where the cells migrate on to the villus from the tip of which they are lost 1·5-2·0 days after production. The average cell cycle time (Tc), taking the crypt as a whole, is between 9 and 15 hours with the cells at the top of the crypt having shorter cell cycles than those at the base (Cairnie et al., '65; Al-Dewachi et al., '75).

The crypts contain several distinctive cell types which may all be derived from a small population of pluripotent stem cells situated towards the base of the crypt Fig. 1 (Cheng and Leblond, '74a).

The data I should like to present support this concept of a relatively small group of crypt base cells acting as precursors for the remaining crypt cells. The evidence is largely a consequence of a simple modification of the crypt squash technique (Wimber et al., '60) allowing the position of labelled cells within the crypt to be estimated (Kovacs and Potten, '73). This technique showed that of the 250-260 crypt cells, 135-164 are proliferating rapidly (i.e. having a high labelling index:- LI) while at the base of the crypt there are 40-50 relatively dormant cells (i.e. not labelled) Fig. 2 (Kovacs and Potten, '73; Hendry and Potten, '74; Potten and Hendry, '75; Potten, '75). This lowest crypt zone contains 10-20 functional and committed precursor Paneth cells (Hampton, '68; Cheng and Leblond, '74b; Al-Dewachi et al., '75), leaving 20-30 other undefined undifferentiated cells. If there is a subpopulation of stem cells in the crypt they would be expected to be passing through the cell cycle more

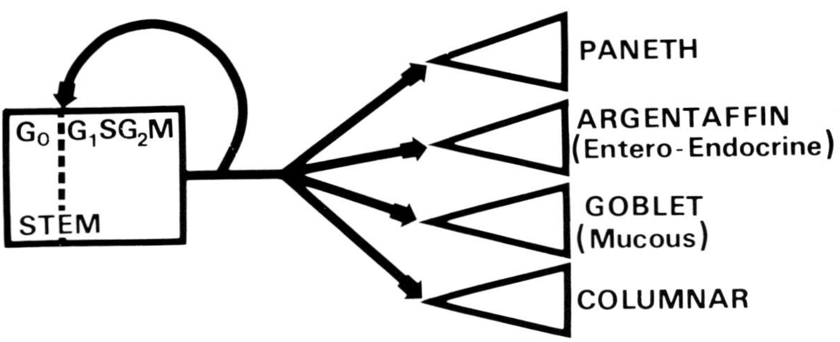

Fig. 1. A schematic representation showing a self-maintaining stem cell population producing cells that enter a second proliferative compartment where there is commitment to a given differentiation pathway (from Potten, '75, courtesy of Masson et Cie).

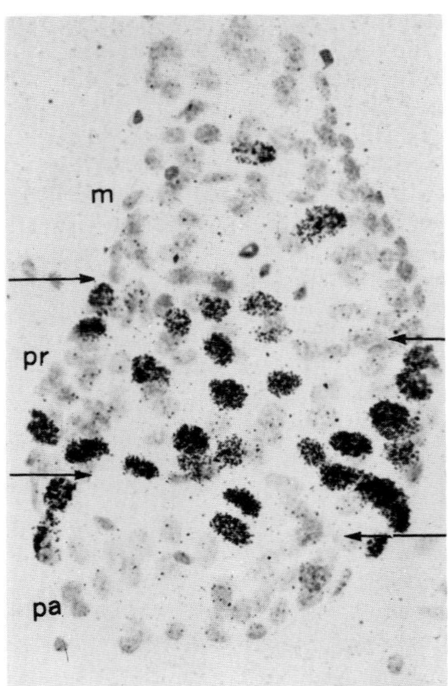

Fig. 2. An isolated under-squashed small intestinal crypt autoradiograph, sampled during the daylight hours, showing the band of proliferative (pr) cells defined by the majority of the labelled cells. At the base of the crypt are about 40 cells in the Paneth (pa) cell zone. At the top of the crypt is the maturation (m) zone. Feulgen stained; total magnification X 400 (^{3}HTdR) (from Potten, '75, courtesy of Masson et Cie).

slowly than the majority of the proliferative cells and to be located in this zone. Continuous labelling and PLM studies have indicated the presence of 20-25 cells with Tc values of more than 20 hours at the border between the proliferative and Paneth zones (compared to 9-10 hours in the proliferation zone) (Potten et al., '74; Al-Dewachi et al., '75).

Assuming that the length of S (Ts) does not vary as greatly as the variations in Tc then a flash exposure to labelled thymidine (^{14}CTdR) will leave a greater proportion of the slowly cycling cells unlabelled. If the proliferative cells are derived from the crypt base cells then with time after ^{14}CTdR an increasing number of the lower proliferative cells would be unlabelled as the unlabelled stem cells feed new unlabelled cells into the proliferative zone. This can be observed by an increase in the size of the crypt base zone (Paneth zone) as defined by the position of the band of high LI (Table 1).

TABLE 1

Increase in the lowest crypt compartment (Paneth zone) with time after flash labelling.

Time after ^{14}CTdR	Total cells	Unlabelled cells
40 mins (control)	42·4 ± 1·0	39·1
15 hrs	35·3 ± 2·0	30·7
24 hrs	67·0 ± 2·1	61·3
48 hrs	81·6 ± 7·2	73·8

15 crypts scored per mouse; mean and standard error of 3 mice per group. The autoradiograph exposure time was increased from 2 to 6 weeks to counteract dilution of label by division (data from Potten, '75).

These figures illustrate the fact that the arbitrary lower border of the proliferative zone, defined by the presence of many labelled cells, moves up the crypt with time after flash labelling because new, and predominantly unlabelled, cells are being produced beneath this border i.e. the proliferative compartment is gradually replaced by cells from a small group of crypt base stem cells.

The presence of a distinctive subpopulation of stem cells is also suggested by radiobiological studies where estimates for the number of cells capable of crypt regeneration (cryptogenic cells) are consistently a factor of two less in steady state than the observed number of cells within the zone defined by the labelled cells (Hendry and Potten, '74;

Potten and Hendry, '75). This difference can be as high as a factor of 18 during regeneration (Table 2).

TABLE 2

Crypt changes after irradiation

	Cells per crypt		
	Crytogenic	Proliferative	Paneth
Unirradiated control	86	164	40•5
Time after 900 rads			
1 day	-	63	17•8
2	7	70	20•4
3	15	277	27•8
4	35	426	58•2
5	-	272	56•1

Data from Potten and Hendry, '75, and Potten, '75.

These data suggest that there are two functionally distinct cell populations in the crypt; those capable of crypt regeneration and those lacking this ability but capable of cell division. Some degree of overlap between these two might be expected. The estimates for cryptogenic cells are upper limits. Since one would expect to find the cryptogenic stem cells in the upper Paneth zone the cryptogenic and Paneth cell data in Table 2 produce a mutually consistent picture. During early regeneration all stem cells are called into rapid cycle and the empirical lower proliferative border is displaced down the crypt: day 1 Paneth cell values in Table 2. These cells cycle rapidly for several days (and the Paneth zone remains at about 20 cells) until the stem cell compartment is re-established (cryptogenic numbers approaching normal and proliferative compartments restored). Then the stem cells switch off, no longer label, and the border moves back up the crypt (the Paneth compartment size is back to normal, or overshot on day 4; Table 2).

These data all suggest that the crypt contains 20-40 (cell kinetic studies) or no more than 80 (radiobiological studies) stem cells situated in the crypt base. The crypt also contains a number of cells (probably in excess of 100) capable of division but not of self-maintenance or crypt regeneration.

The work has been supported by grants from the Cancer Research Campaign and the Medical Research Council.

LITERATURE CITED

Al-Dewachi, H. S., N. A. Wright, D. R. Appleton and A. J. Watson. 1975. Cell population kinetics in the mouse jejunal crypt. *Virchows Arch. B. Cell Path. 18:* 225-243.

Cairnie, A. B., L. F. Lamerton and G. G. Steel. 1965. Cell proliferation studies in the intestinal epithelium of the rat. I. Determination of the kinetic parameters. *Exp. Cell Res. 39:* 528-538.

Cheng, H. and C. P. Leblond. 1974a. Origin, differentiation and renewal of the four main epithelial cell types in the mouse small intestine. V. Unitarian theory of the origin of the four epithelial cell types. *Am. J. Anat. 141:* 537-561.

Cheng, H. and C. P. Leblond. 1974b. Origin, differentiation and renewal of the four main epithelial cell types in the mouse small intestine. IV. Paneth cells. *Am. J. Anat. 141:* 521-535.

Hampton, J. C. 1968. Further evidence for the presence of a Paneth cell progenitor in mouse intestine. *Cell Tissue Kinet. 1:* 309-317.

Hendry, J. H. and C. S. Potten. 1974. Cryptogenic cells and proliferative cells in intestinal epithelium. *Int. J. Radiat. Biol. 25:* 583-588.

Kovacs, L. and C. S. Potten. 1973. An estimation of proliferative population size in stomach, jejunum and colon of DBA-2 mice. *Cell Tissue Kinet. 6:* 125-134.

Leblond, C. P., Y. Clermont and N. J. Nadler. 1967. The pattern of stem cell renewal in three epithelia (oesophagus, intestine and testis). In: *Canadian Cancer Conf.*, Pergamon Press, Oxford, pp. 3-30.

Lesher, S., R.J.M. Fry and H. I. Kohn. 1961. Age and the generation time of the mouse duodenal epithelial cell. *Exp. Cell Res. 24:* 334-343.

Potten, C. S. 1975. Kinetics and possible regulation of crypt cell populations under normal and stress conditions. *Bull. Cancer.* In press.

Potten, C. S. and J. H. Hendry. 1975. Differential regeneration of intestinal proliferative cells and cryptogenic cells after irradiation. *Int. J. Radiat. Biol. 27:* 413-424.

Quastler, H. and F. G. Sherman. 1959. Cell population kinetics in intestinal epithelium of the mouse. *Exp. Cell Res. 17:* 420-438.

Wimber, D. E., H. Quastler, O. L. Stein and D. L. Wimber. 1960. Analysis of tritium incorporated into individual cells by autoradiography of squash preparations. *J. Biophys. Biochem. Cytol. 8:* 327-331.

Discussion, Session I - Stem Cells in the Intestine

Dr. Leblond's presentation

Dr. Cairnie asked if any of the columnar cells in the mid crypt died after the injection of a dose of ^{3}H-thymidine which induced death in some crypt-base columnar cells. Dr. Hazel Cheng replied that on examination of the crypts, 60 phagosome-containing cells were enumerated six hours after injection of 2 μCi of ^{3}H-thymidine/gram body weight into mice. Of the 60 cells, 59 were crypt-base columnar cells. The one exception was a mid-crypt cell close to the border line of the crypt base. Neither did the cells of the crypt-base undergoing differentiation into mucous or other cell type die. Hence, only crypt-base columnar cells are easily killed by radiation. And only the survivors amongst these cells have the ability to phagocytose nearby dead cells. Incidentally, the phagocytosed crypt-base columnar cells were lysed, and eventually disappeared, so that no phagosome was observed later than 30 hours after the ^{3}H-thymidine injection.

Dr. Altmann mentioned that this type of phagosome could be produced by other means than the beta radiation of ^{3}H-thymidine since he observed them after injection of large doses of methotrexate, a drug which inhibits RNA and DNA synthesis. Hence, crypt-base columnar cells are sensitive to various types of damage.

Dr. Potten stated that some of his results support the eight-fold amplification of the number of columnar cells arising from stem cells, as proposed in Dr. Leblond's model. Furthermore, his results indicate that about 20 stem cells are at the bottom of each crypt.

Dr. Wither's presentation

Much of the discussion centered on the manner in which the little colonies believed to arise from single cells surviving radiation damage grow. Dr. Withers proposed that they extend from the crypt where they arise into neighboring ones. Otherwise, the colonies could not grow as large as those seen at 13 days. He even thinks that cells may migrate from the top of a crypt (where columnar cells are normally unable to divide) and, as they reach the next crypt, regain stem cell capacity and divide.

Dr. Sigdestad suggested another possible explanation for Dr. Withers' results, that is, the crypts initially formed by division of surviving cells would split by fission from the base up and thus produce new crypts, as initially proposed by Dr. Cairnie. Dr. Withers answered that he had occasionally seen patterns of fission but these were not frequent enough to account for his results. Dr. Sigdestad also commented that the micro-environment of the crypts was not necessarily the main factor controlling the expansion of the colonies and that events taking place in the villi must also be considered. There is evidence that conditions such as thyrotoxicosis or germ-free living produce an enlargement of villi without any apparent change in the crypt.

Dr. Brugal's presentation

In reply to a question, Dr. Brugal emphasized that the chalone properties of the extract he worked on were tested only in embryonic amphibian tissue. To Dr. Laurence who expressed surprise at the long duration of the S-phase in the cell cycle, Dr. Brugal answered that this was the case in embryonic as well as adult frogs; perhaps because they are homeotherms and, therefore, their cells are usually at a cooler temperature than those of mammals.

Dr. Bergeron of the University of Montreal investigated the chalonic properties of an ammonium sulfate extract of mucosa from rabbit small intestine. This extract decreased the rate of DNA synthesis, as measured after an injection of labeled thymidine by biochemical methods. The effect lasts 2-4 hours after an injection of the extract and disappears by 7 hours. This extract has a potent inhibitory effect on cell divisions in the small intestine but has no influence on cell division in the regenerating kidney. Dr. Bergeron also confirmed his results by another method which consists in injecting ^{3}H-thymidine every 4 hours into mice. While in controls the label reaches the top of jejunal villi in about 45 hours, it takes about 17 hours longer in the animals receiving the extract.

Dr. Cairnie's presentation

Dr. Withers commented on the possibility that survival of cells in the irradiated intestine may be related to two, three or more cells being in contact with each other. Such a hypothesis, even though fascinating, does not account for his observations.

Dr. Withers repeated his previous statement that the fission of crypts may occur but does not account for the expansion of the colonies. Dr. Cairnie replied that, while forking crypts are rare in the normal intestine, they are common 6 or 8 days after a large dose of radiation.

Dr. Leblond stated that, after continuous infusion of ^{3}H-thymidine for several days, the columnar, mucous and entero-endocrine cells in the crypt are labeled; only some Paneth cells are unlabeled. When serial 1-μ thick plastic sections are examined, any unlabeled nucleus which seems to be present in the crypt-base at that time, turns out to belong to a Paneth cell whose granules are in an adjacent section. Dr. Cairnie said that while the number of labeled cells may reach 100% in the normal, there may be 20% of the nuclei that are not labeled (R cells) in continuously irradiated animals.

With regard to Hayflick's view that stem cells cannot divide more than about 50 times - or possibly 100 to 200 times in some cases - the participants were skeptical. In the intestine the fact that after a few days of continuous ^{3}H-thymidine injection all crypt cells are labeled is difficult to reconcile with any kind of Hayflick's limit, even a high one. On the other hand, Dr. Micklem pointed out that in the case of bone marrow, one can calculate a potential mitotic burden for the stem cell population in the region of 100 to 150 mitoses, a figure which is compatible with Hayflick's views. Dr. Cairnie agreed, but added that there is no need to introduce the idea of a reserve stem cell in the small intestine.

Dr. Altmann's presentation

Dr. Withers asked whether the observations in animals treated with cycloheximide gave information on feedback from villus to crypt. Dr. Altmann answered that only the experiments with methotrexate gave such information. He added that the cells undergoing DNA synthesis are the most resistant to methotrexate.

Dr. Leblond commented on the decrease in protein synthesis by epithelial cells from the base to the tip of the villus reported by Dr. Altmann and the hypothesis that the extrusion of cells from villus tip into lumen was due to the absence of a short-lived protein required for cell adhesion. In work done with Dr. Gary Bennett it was found by radio-autography that ^{3}H-fucose is taken up into glycoproteins in the Golgi region of villus columnar cells and, within 20 minutes, migrates to the cell surfaces. These phenomena, like the protein synthesis described by Dr. Altmann, are intense in the cells at the base of the villi and decrease gradually in the cells up to the villus tip, where they are lacking. Since the rate of Golgi synthesis and surface migration of glycoprotein decreases from villus base to tip and since glycoproteins are believed by some to be effective in cell adhesion, those findings support Dr. Altmann's hypothesis that newly-synthesized proteins (presumably including glycoproteins) keep cells sticking to one another until they reach the villus tip.

In answer to a question Dr. Altmann mentioned that the epithelial cells of small intestine do not migrate individually but in a coherent sheet and that there is no evidence that the basement membrane also moves.

Session II

Stem Cells in the Epidermis

Identification of Clonogenic Cells in the Epidermis and the Structural Arrangement of the Epidermal Proliferative Unit (EPU)

CHRISTOPHER S. POTTEN

Paterson Laboratories,
Christie Hospital and Holt Radium Institute
Manchester, England

EPIDERMAL CELL POPULATIONS

The basal layer of the epidermis contains several discrete cell populations with the majority of the cells being related to the keratinocytes. Melanocytes and Merkel cells would generally be regarded as having little to do with the daily production of new keratinizing cells, though their precise functional role in epidermis is uncertain. Langerhans cells should perhaps be considered in the same way but even less is known about their function, though these cells may play a more important role in keratinocyte production or in its control (Potten and Allen, '75a). These three cell types make up about 15% of the basal cells in normal mouse ear or dorsal epidermis, with the Langerhans cells constituting about 10%, the melanocytes about 3-5% with the Merkel cells being rare in interfollicular epidermis (Allen and Potten, '74; Potten, '75a; MacKenzie, '72).

The remaining 85% of the basal cells are keratinocytes, some of which are actively involved in the division cycle while others will have completed the proliferative process (post-mitotic) and will be maturing (i.e. beginning keratin synthesis). The number of these post-mitotic maturing (Potten, '75b; Potten and Allen, '75b; Christophers et al., '74) or basal differentiating (Christophers, '71a, b; Iversen et al., '68) cells is not reliably established and probably varies according to site.

These cells can be seen as weakly fluorescein-isothiocyanate positive cells (Christophers, '71a, b)or be inferred from transit experiments (Potten, '75b), or experiments where the time of appearance of the first suprabasal labelled cell is studied (Potten, '75b; Christophers et al., '74; Iversen et al., '68; Etoh et al., '75) and also from the study of migrating cells after mechanical or radiation wounding (Potten and Allen, '75b; Etoh et al., '73, '75).

A discrepancy is observed when one compares the overall time taken for label to reach the surface after flash labelling basal S phase cells and the estimates for the rate of transit through the known numbers of cell layers. This discrepancy would be explicable if cells were to spend 2-3 days on the basal layer maturing prior to migration. Knowing the cell production rate (see below) 2-3 days supply is equivalent to 20-30% of the basal cells. Wounding experiments indicated that the basal layer normally contains cells capable of rapid migration. This rapid migration after mechanical wounding was one of cells with sufficient maturity (or size) to enable them to stack normally into columns and resulted in a basal cell depletion which may be the stimulus for the burst in DNA synthesis and mitosis.

Fifteen per cent non-keratinocytes and 20-30% basal maturing cells result in a figure of about 0•6 for the growth fraction (GF) for the basal layer, a value which can also be derived in other ways (Iversen et al., '68; Hamilton and Potten, '72). A further consideration is whether all of these cells have the same proliferative or regenerative potential or whether there might be a special minority class of cells from which the majority are derived, i.e. a subpopulation of stem cells. Some of the radiobiological data seem to suggest that only 10% or less of the basal cells have the ability to regenerate the epidermis after severe radiation depletion (Potten and Hendry, '73) (Table 1).

TABLE 1

Clonogenic Cells in Mouse Dorsal Epidermis (per mm^2)

Clonogenic cells	300-1100
Total basal cells	15000
Cornified cell colums (EPUs)	1400
Basal cells per column	10

Data derived from several sources (Potten and Hendry, '73; Potten, '74, '75a).

EPIDERMAL ORGANIZATION

The organization of the epidermis can be considered at two levels; the level of the cells within groups or columns and the relationship between these groups.

The heterogeneity of the epidermis has recently been supported by the findings that the superficial keratinizing and keratinized cells are arranged in precise columns in epidermis from a variety of sites and species (MacKenzie, '69, '72; Christophers, 70, '71a; Menton and Eisen, '71). These observations indicated that when cells migrate from the basal layer they align themselves precisely beneath the previously migrated cells and flatten to cover a roughly hexagonal surface area approximately ten times that occupied by the basal cells (Fig. 1). The lack of many intermediates in this migration, flattening and alignment process suggests that the process is rapid, and since the volume as well as surface area appears large it seems likely that only the largest basal cells can undergo this process. This suggests that the basal cells might need a growing or maturing time before migration (see above). If the cell production rate is raised then the cells cannot spend as long maturing, the columnar pattern breaks down, the keratinizing cells being smaller and less flattened (Potten and Allen, '75b; Potten, '75a).

The columns seem to have an even greater degree of organization if the small regions of overlap between neighbouring columns are studied (Fig. 2). The interdigitation appears precise and since each column has an average of six neighbours this implies a considerable degree of inter-column control to ensure that one column does not add another cell before all six neighbours have done so. This is also suggested by the fact that, cornified cells appear to become detached from the surface only when all the edges are free of the neighbouring cornified cells and also by the constancy in epidermal thickness at a given site (Potten and Allen, '75c; Allen and Potten, '74,'75). It seems unlikely that migrating cells can jostle around to find the right column since all these cells are rich in desmosomes and cytoplasmic interlocking. Also mistakes seem to be rare (i.e. cells do not get caught between columns or enter columns at the same time as other cells. This precision in organization suggests that the cell production per column is precisely controlled and also that the cells of the column can be produced only by the 10 cells that lie beneath it. This is further supported by the fact that the edges of the column can be traced in electron micrographs to the basal layer (Allen and Potten, '74) and that in sheets of epidermis separated at the

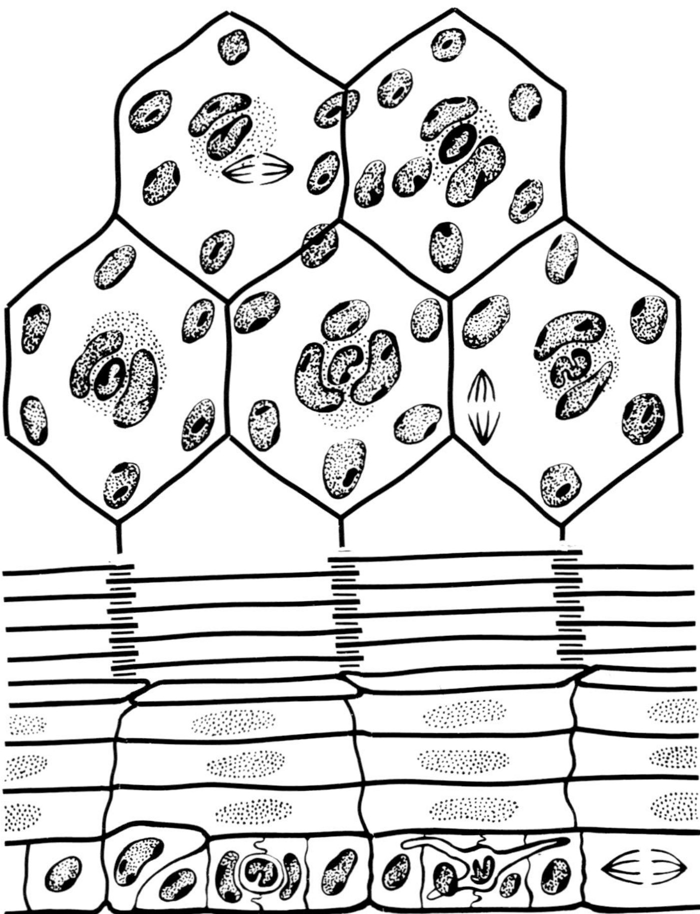

Fig. 1. Schematic representation of mouse dorsal or ear epidermis showing the columns of hexagonal cornified cells with precise regular overlap of their edges above 3 nucleated keratinizing cells which are also aligned in columns. Beneath each column there is a group of about 10 basal cells. Towards the centre of this group is a morphologically distinctive cell with an indented nucleus with extensive marginal chromatin condensation. This central cell may be rounded or dendritic. Migration from the basal layer tends to occur from the periphery of this group of cells. Each group of basal cells with its column of superficial cells has been regarded as a discrete Epidermal Proliferative Unit (EPU). The epidermis is represented both in section and surface view (based on data presented in Potten, '74; Allen and Potten, '74).

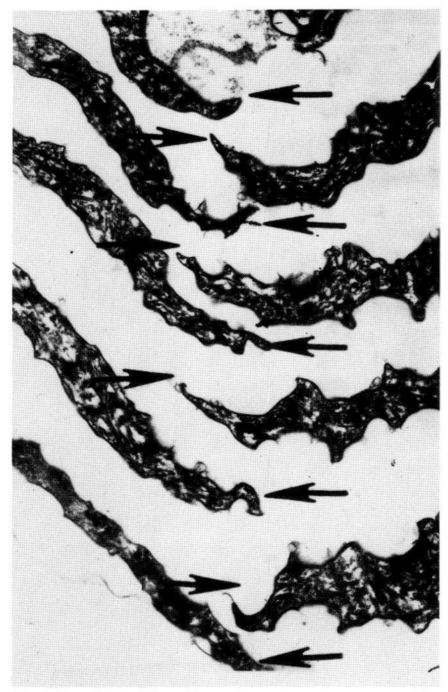

Fig. 2. Electron micrograph of the region of overlap between 2 adjacent columns of cornified cells in mouse ear. Sample swollen by 7 days exposure to organ culture conditions. See also unswollen figures in: Allen and Potten, '74 and Potten and Allen, '75c. The regular alternating pattern can be easily seen (arrows). Magnification X 9300.

basement membrane some patterns of basal cells can often be seen in association with the columns (Potten, '74; Goerttler et al., '73) (Fig. 1). These 10 basal cells are arranged in a rather characteristic fashion with a ring of about 6-7 peripheral cells and 3-4 cells forming a central sub-unit (Allen and Potten, '74). This has led to the suggestion that the epidermis is divided into a series of discrete epidermal proliferative units (EPUs) which are defined by the limits of the hexagonal columns (Potten, '74). The cells on the basal layer beneath the columns can feed cells only into that column immediately above themselves.

Consideration of the regular overlap between columns, the stability of epidermal thickness and the appearance of the surface cells in scanning electron micrographs has indicated a considerable degree of control at the inter-column level. Using hexagonal grids as a model for skin it was concluded that the organization can only be perpetuated by an arrangement whereby groups of EPUs are in some form of synchrony (Potten and Allen, '75c). The only arrangement which maintains an organized epidermis without inter-cell spaces is one where about one-third of the units are in synchrony at any given time (Fig. 3). This produces a regular pattern with a trio arrangement about a given column.

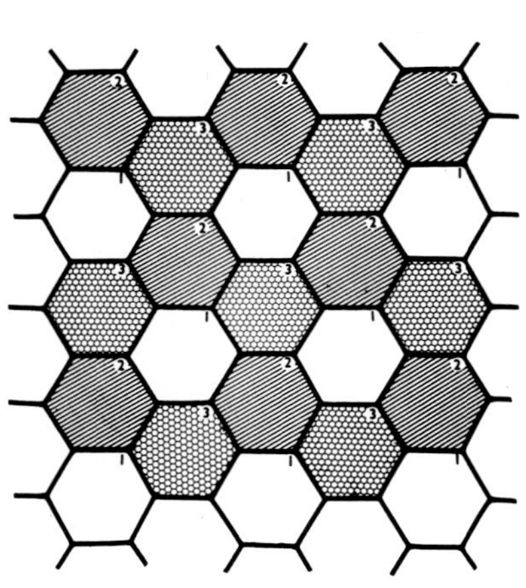

Fig. 3. Scheme based on model hexagonal columns of cells showing the probable pattern of synchrony between groups of columns. This scheme would ensure the regular alternating overlap of neighbouring columns. The scheme suggests that approximately a third of the columns are synchronous, that is they have a limited period of time within which they must add a cell to the base of the column. Those columns marked 1 will add a cell (or lose one) followed in turn by those marked 2 and 3. Groups of 3 columns (trios) about any given column will be synchronous (from Potten and Allen, '75c, courtesy of Springer-Verlag).

EPIDERMAL CELL PRODUCTION RATES

The number of cells per column is easy to determine but varies with site, animal species and possibly strain, being about 10 layers in the dorsum and 15 in the ear of our strain of mouse (Potten, '75a). Knowing any two of the following; the numbers of cells per column, the rate of production of cells, and the transit rate or time, the third parameter can be deduced since all three are inter-related. Taking into consideration, the organization, the cell populations, the probable growth fraction and the marked diurnal rhythms, the literature can be surveyed for data on transit times, cell cycles, and labelling and mitotic indices (LI and MI) to derive some values for the average cell production rates (Table 2). Each value is based on between 1 and 17 reports. The different techniques used and the lack of attention to the full diurnal variations results in some discrepancies. The LI and MI data also often show some differences (not apparent in the Table). However, the overall picture is one where each column in both ear and dorsum receives about one new cell per day.

TABLE 2

Approximate epidermal cell production rates per day per column.

	Dorsum	Ear
Average from transit experiments	1·0	1·0
Average from 24 hours LI or MI	1·5 - 1·8	1·8
Average of all other LI or MI data	1·4	0·5
Average from published T_C assuming GF=0·6	1·2	0·4
Overall average	1·3	0·9

(Data summarized from a review of the literature in Potten, '75a)

EPIDERMAL PROLIFERATIVE MODELS

This high degree of organization seems inconsistent with any scheme of proliferation depending on random processes. One is therefore led to seek some scheme in which cells are produced in a precisely programmed sequence. Any scheme must incorporate the following points: 1) the presence of post-mitotic maturing basal cells and the topographical patterns of basal cells; 2) the observations that migration (Allen and Potten, '74) and much of the normal cell division (Potten, '74; MacKenzie, '72, '75) occurs in the cells at the periphery of the area defined by the column; 3) the probability that the cells towards the centre which are capable of responding dramatically to a wounding stimulus (Potten, '74) might be less mature and possibly contain a stem cell for the EPU (Potten and Hendry, '73). Any proposed model must also accommodate the existing cell kinetic observations, namely: 1) the observed cell production rate figures (Table 2) and the minor variations in the number of basal cells per column (78% of the columns have between 8 and 12 basal cells, Allen and Potten, '74); 2) the continuous labelling data which indicate that the outer ring of cells accumulate label more rapidly than the central cells (Potten, '74) and that eventually all basal cells become labelled (Iversen et al., '68; Potten et al., '74; Hegazy and Fowler, '73); 3) the per cent labelled mitosis (PLM) data (Hegazy and Fowler, '73).

To conclude I should like to present a model (Fig. 4) for

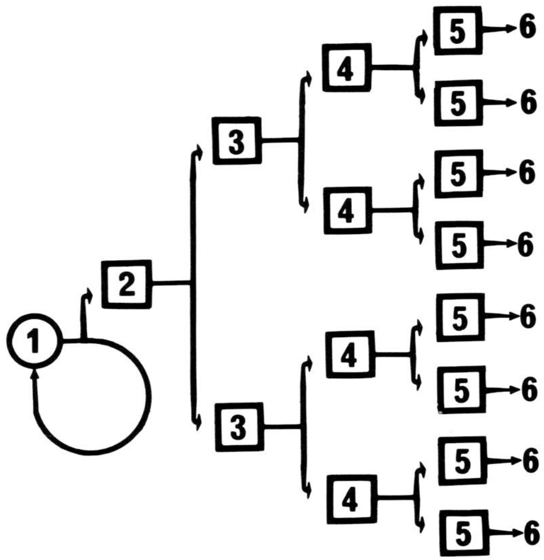

Fig. 4. Schematic representation of the sequential division processes in epidermis following the initial division of a stem cell; which is amplified eight-fold in this diagram. (Eight new keratinizing cells are produced for each stem cell division.) Each circle or box represents a cell with the maturity (differentiation) being represented numerically. A total of 4 division stages is presented beginning with the division of an immature cell (numbered 1). With each further division the cells move irreversibly towards a postmitotic maturation stage (numbered 5) before migrating from the basal layer (numbered 6), i.e. at each stage the cells become more committed to differentiation as their proliferative potential is progressively restricted. Since the number of cells per EPU does not vary very greatly it suggests that the basal cells within an EPU must be at all stages within this heirarchy, with probably the greatest proportion undergoing their terminal division.

consideration which I believe fulfils all these requirements and implies a precise functional and topographical relationship between the basal cells beneath a column. The model suggests a possible mechanism for the control of cell proliferation in epidermis based on the columnar hexagonal organization, the biological variability of which exerts a distorting influence on the schematic representations.

It has been assumed that the basal layer contains a

sub-population of stem cells present at a frequency of 10% of the basal cells, i.e. one per EPU, but the principle could apply to any small percentage. With one stem cell per EPU and 9 other cells a system with three sequential divisions following an initial stem cell division seems likely (Fig. 4).

Finally, a model can be suggested for the operation of this scheme within the EPU (Fig. 5) which will maintain a stable model skin. This scheme is based on a number of simple rules: 1) It must be a stable self-maintaining (steady state) system producing one cell per day without large fluctuations in the number of basal cells; 2) it must maintain a reserve division potential at all times to ensure an effective and rapid response to injury without necessarily relying on the stem cells; 3) the cells must migrate from the basal layer in order of age with the most mature cells located at the periphery. The immature central sub-unit (see Fig. 1) usually contains 3 cells, the outer two perhaps playing some role in restraining or controlling the behaviour of the central cell (Allen and Potten, '74; Potten and Allen, '75a); 4) a central sub-unit of 2 cells, whether induced by wounding or due to the decline in proliferative potential of the unit with time, is unstable and the trigger for stem cell division and thus the revitalization of the EPU. For schematic purposes a final rule was devised which stated that the space left by migration was filled by the division of a neighbouring cell. It seems likely that this approximates to what actually happens since cells could not dramatically change their positions beneath a column. It is possible that immediate neighbours might close in, thus displacing the space left by migration which is then filled by a neighbour dividing. It has been suggested that the stimulus for general basal cell division might be a general basal cell depletion (Potten and Allen, '75b; Etoh et al., '73, '75). The model presented here suggests that local depletion might play a role in the control of local divisions.

This scheme produces one cell a day with a stem cell (and thus EPU) cycle of about 8 days which agrees with the continuous labelling data for the central region which indicated inter-mitotic times of 7 days or more for these cells (Potten, '74). There will be two or more post-mitotic cells present awaiting migration at any time with the remaining proliferative cells dividing at a rate consistent with inter-mitotic times of about 3-4 days. Under conditions of continuous labelling all cells will become labelled and under PLM conditions the maximum number of cells that could be expected to reach a second peak would be 50%. The role played by the Langerhans cell in this scheme is uncertain at present. It may be entirely independent of the keratinocytes, play a role in controlling their activity or be more intimately involved

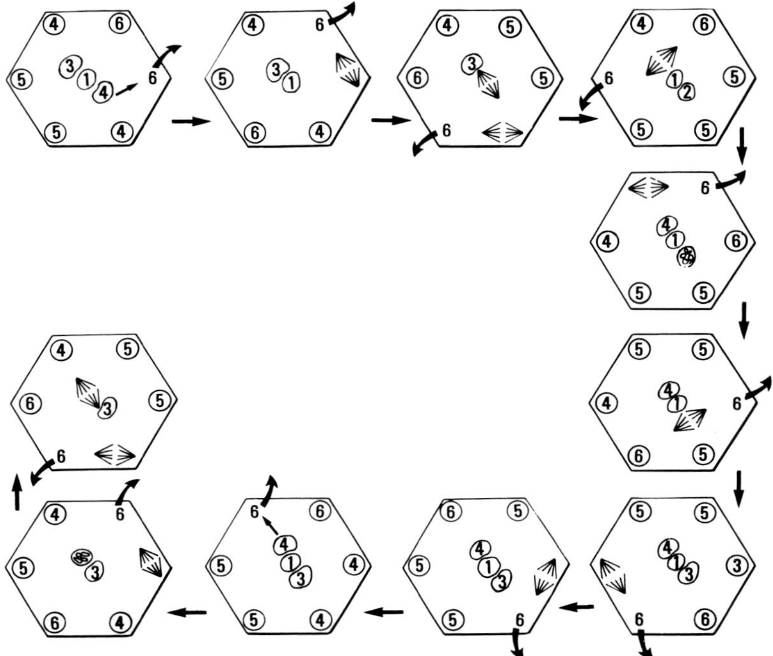

Fig. 5. Schematic representation of the operation of an EPU based on a series of progressively ageing cell divisions (Fig. 4). Cell numbering corresponds to that shown in Fig. 4. Only cells numbered 1-4 are capable of division, those numbered 5 and 6 are maturing or migrating (an arrow out of the hexagon). Each hexagon has a cell (arrow) leaving and this represents a time interval of one day. The process is cyclic and could begin at any stage. For convenience the first hexagon shows a cell migrating. This leaves a space which can be filled by either a lateral movement or by a neighbouring mitosis. The oldest central cell migrates and then undergoes its final division in the second hexagon. This migration leaves the central sub-group in a depleted stage and thus instigates a stem cell division (3rd hexagon). An 8 day stem cell cycle is thus represented by the sequence from hexagon 3 to hexagon 11. See text for further details. Note the positional relationship between cells 1 and 3 is reversed at the end of the stem cell cycle. Compare hexagons 3 and 11.

by itself acting as the stem cell (Potten and Allen, '75). In all but the last situation (which has been assumed for the basis of the diagrams) the schematic representation would need modification to incorporate an additional non-proliferative cell in the central sub-unit.

ACKNOWLEDGEMENTS

This work was supported by grants from the Cancer Research Campaign and the Medical Research Council.

LITERATURE CITED

Allen, T. D. and Potten, C. S. 1974. Fine structural identification and organisation of the epidermal proliferative unit. *J. Cell Sci. 15:* 291-319.

Allen, T. D. and Potten, C. S. 1975. Desmosomal form, fate and function in mammalian epidermis. *J. Ultrastruct. Res. 51:* 94-105.

Christophers, E. 1970. Eine neue methode zure Darstellung des Stratum Corneum. *Arch. Klin. exp. Derm., 237:* 717-721.

Christophers, E. 1971a. Cellular architecture of the stratum corneum. *J. Invest. Derm. 56:* 165-169.

Christophers, E. 1971b. The architecture of stratum corneum after wounding. *J. Invest. Derm. 57:* 241-246.

Christophers, E., H. H. Wolff and E. B. Lawrence. 1974. The formation of epidermal cell columns. *J. Invest. Derm. 62:* 555-559.

Etoh, H., Y. H. Taguchi and J. Tabachnick. 1973. Movement of epidermal basal cells to the spinous layer in the absence of division pressure. *J. Cell Biol. 59:* 95a.

Etoh, H., Y. H. Taguchi and J. Tabachnick. 1975. Movement of beta-irradiated epidermal basal cells to the spinous-granular layers in the absence of cell division. *J. Invest. Derm. 64:* 431-435.

Goerttler, K., M. Reuter and H. E. Stahmer. 1973. Morphologische Untersuchungen zur Proliferationskinetik der Mausehaut Standardisierte Methodik und Ergebnisse. *Z. Zellforsch. 142:* 131-146.

Hamilton, E. and C. S. Potten. 1972. Influence of hair plucking on the turnover time of the epidermal basal layer. *Cell Tissue Kinet. 5:* 505-517.

Hegazy, M.A.H. and J. F. Fowler. 1973. Cell population kinetics of plucked and unplucked mouse skin. I. Unirradiated skin. *Cell Tissue Kinet. 6:* 17-33.

Iversen, O. H., R. Bjerknes and F. Devik. 1968. Kinetics of cell renewal, cell migration and cell loss in the hairless mouse dorsal epidermis. *Cell Tissue Kinet. 1:* 351-367.

MacKenzie, I. C. 1969. Ordered structure of the stratum corneum of mammalian skin. *Nature (Lond.) 222:* 881-882.

MacKenzie, I. C. 1972. The ordered structure of mammalian epidermis. In: *Epidermal Wound Healing.* H. I. Maibach and D. T. Rovee, eds. Year Book Med. Publ., Chicago, pp. 5-25.

MacKenzie, I. C. 1975. Spatial distribution of mitosis in mouse epidermis. *Anat. Rec. 181*: 705-710.

Menton, D. N. and A. Z. Eisen. 1971. Structure and organization of mammalian stratum corneum. *J. Ultrastruct. Res. 35:* 247-264.

Potten, C. S. 1974. The epidermal proliferative unit: The possible role of the central basal cell. *Cell Tissue Kinet. 7:* 77-88.

Potten, C. S. 1975a. Epidermal cell production rates. *J. Invest. Derm.* In press.

Potten, C. S. 1975b. Epidermal transit times. *Brit. J. Derm.* In press.

Potten, C. S. and T. D. Allen. 1975a. A model implicating the Langerhans cell in keratinocyte proliferation control. *Differentiation.* In press.

Potten, C. S. and T. D. Allen. 1975b. The fine structure and cell kinetics of mouse epidermis after wounding. *J. Cell Sci. 17:* 413-447.

Potten, C. S. and T. D. Allen. 1975c. Control of epidermal proliferative units (EPUs): An hypothesis based on the arrangement of neighbouring differentiated cells. *Differentiation 3:* 161-165.

Potten, C. S. and J. H. Hendry. 1973. Clonogenic cells and stem cells in epidermis. *I. Rad. Biol. 24:* 537-540.

Potten, C. S., L. Kovacs and E. Hamilton. 1974. Continuous labelling studies on mouse skin and intestine. *Cell Tissue Kinet. 7:* 271-283.

The Influence of Epidermal Chalone on Cell Proliferation

EDNA B. LAURENCE AND ALAN L. THORNLEY

Mitosis Research Laboratory,
Birkbeck College, University of London,
London, WC1E 7HX, England

I. THE TWO DISTINCTIVE PROPERTIES OF CHALONES

The two distinctive properties of chalones are: 1) they are produced by the tissue on which they act, 2) they inhibit cell proliferation. It has been claimed that, in all vertebrate tissues so far examined for chalones, a tissue specific endogenous inhibitor of cell proliferation has been found (Bullough, '75). However, the process of cell division is a feature common to all multicellular organisms. Hence, it is plausible that chalones have evolved from a common ancestral molecule, in which case, the uniqueness of a particular chalone may lie mainly in the tissue specific reactive site.

The most telling criticism of all chalone research has been the inconsistent approach taken by biochemists and morphologists alike to the question of testing tissue specificity of these inhibitors. But, this is not the only test that has to be applied.

The first epidermal chalone to be isolated (Hondius Boldingh and Laurence, '68) acted at G_2 in the cell cycle, and it was assumed to reflect a much wider spectrum of action and also inhibit earlier in the cell cycle, thus regulating tissue homeostasis (Bullough, '62, '65, '75; Iversen, '65, '69; Laurence, '73). Recently, several workers, e.g. Bichel ('71, '73), Brugal and Pelmont ('74, '75), have completely dissociated inhibitors acting at G_1 and G_2. This cell cycle phase-specificity, which had been suggested by earlier work (Hall, '69; Elgjo *et al.*, '71, '72; Elgjo, '73), has become a powerful tool for testing the purity of chalone preparations.

II. SPECIFICITY OF CHALONE ACTION

A. *ONTOGENY DICTATES TISSUE AND CELL LINE SPECIFICITY*

Tissue specificity is the all important feature which separates chalones from other inhibitory substances. Too often this test is undertaken in a haphazard way, the only logic behind the choice of tissues being their ease of assay or accessibility (be it in the animal or the laboratory). However, the development of a vertebrate is a logical and ordered sequence of events which stipulates the choice of tissues to be studied.

In all vertebrates an embryo is formed from the fertilised egg. After gastrulation this consists of three germinal layers: ectoderm, mesoderm and endoderm. Each of these layers gives rise to its own prospective tissues; e.g. from the ectoderm develops the nervous system, the lens and cornea, the oral epithelium and the lining of the rectum and anus, etc.; the mesoderm gives rise to the blood and skeletal tissues, the muscles, etc.; and the gut etc. is formed from endoderm. These developmental processes are considered to reflect the evolutionary history of the vertebrates. Thus chalones synthesised by tissues derived from different germinal layers may have far fewer features in common (e.g. in their amino acid sequences, etc.) than those which develop from the same germinal layer. For example, the pilary unit in mammals is formed from the foetal or neonatal epidermis (Pinkus, '58), and the sebaceous gland chalone, while being similar to epidermal chalone may be very different from e.g. the granulocyte chalone. Therefore, using granulocytes as a target for epidermal chalone tissue specificity tests may inevitably give a negative result but this indicates nothing concerning the specificity of action within the ectodermally derived group of tissues. The futility of basing this test only on tissues derived from different germinal layers is obvious. Epidermal homogenates could contain inhibitors common to a variety of ectodermally derived organs! Thus tests must be made on closely related embryological tissues in order to discriminate between chalones, i.e. *cell line* specificity also must be demonstrated.

Recently it has been found to be practicable to use sebaceous glands for testing cell-line specificity of epidermal chalone action. These glands are a comparatively recent development and are absent from non-mammalian skin. Hence it is not surprising that crude extracts of Codfish skin have no inhibitory effect on mitosis in mouse sebaceous glands (Bullough and Laurence, '70). Codfish skin extracts inhibited mouse epidermal mitosis in the same way as mammalian extracts (Bullough *et al.*, '67). The latter result is particularly interesting as it leads to the assumption that chalone extracted from codfish skin has similar reactive sites as the chalone extracted from pig skin although these animals are separated by 350 million years of evolutionary history.

Thus it is postulated that closely related tissues have very similar chalones and should be used in crosschecking chalone tissue specificity. However, it must be emphasized that testing cell proliferation in tissue derived from different germinal layers serves the useful purpose of detecting common degradation products of the chalone; since, if in the purification process, the tissue specific portion of the molecule is lost a general inhibitory effect (as distinct from a toxic effect) may be detectable.

B. *ANOTHER TEST FOR HOMOGENEITY - CELL CYCLE PHASE-SPECIFICITY OF ACTION*

The process of cell proliferation is complicated and involves biochemical, physical and mechanical processes. At present modern technology allows us to study only two of these processes with "comparative" ease although newer methods, e.g. cell structuredness (Lord, '76) show other aspects which should be exploited. Chalones have been precisely defined as "an internal secretion produced by a tissue for the purpose of controlling by inhibition the mitotic activity of the cells of that same tissue" (Bullough, '67). In terms of this definition chalones could form a group of unrelated substances, each acting in its own peculiar manner and each with its own site of action in the cell cycle. In fact, at present only two types of chalones have been extracted from the epidermis so far, and they are distinguished by their action on the cell cycle (Hondius Boldingh and Laurence, '68; Laurence, '73a,b; Marks, '73, '76), but it is only recently that the individual cell cycle phase-specificity of their action has been rigorously demonstrated (Thornley and Laurence, '75, '76).

The assay of chalones has been very ably reviewed by Lord ('76) but the following points need stressing when the particular case of epidermis is considered. At present there are two main methods of examining cell proliferation in epidermis: 1) by measuring incorporation of radioactive precursors into DNA, and 2) by stathmokinetic methods using such substances as colcemid and vinblastine to arrest at the metaphase of mitosis (M-phase).

1) DNA-synthesis. Two methods have been used for studying DNA synthesis in the epidermis: a) liquid scintillation counting of incorporated precursors into DNA, and b) detection of labelled cells by autoradiography. Each method has its advantages and its limitations. Neither method by itself is adequate to determine the precise site of chalone action in the cell cycle. Liquid scintillation counting gives no indication of the histological state of the tissue. It cannot be used to determine a change in the number of proliferating cells as opposed to changes in the rate of DNA synthesis, but it is extremely useful for determining the "labelled pool" size. Autoradiography, however, while showing the number of DNA synthesising cells, gives very little indication of whether the radioactive precursors have been temporarily impeded from entering the cell (i.e. the "labelled pool" size). This may or may not be solved by grain counting and it must be emphasised that the autoradiographic interpretations are subjective. The advantage of this method is that histologically abnormal tissue can be eliminated.

The method of administration of radioactive precursors and chalone is very important. For example, there are indications that subcutaneous flank injection of ^{3}H-thymidine results in a slightly higher incorporation of the label into ear epidermal DNA than intraperitoneal injection (Spargo, unpublished). The rate of ^{3}H-thymidine uptake by epidermis as opposed to incorporation into DNA is an important consideration in cytokinetic calculations. For example, the "labelled pool" in ear epidermis reaches a maximum at 15-20 minutes whereas the incorporation peaks at approx. 70-90 minutes (Spargo, unpublished). Also, not enough attention is usually paid to sample size and randomisation of groups of animals.

2) Mitosis. The pitfalls of using stathmokinetic agents for measuring the mitotic rates have been ably summarised by Nome ('74) and Lord ('76). Extrapolation of results from one laboratory to another can give misleading information if circadian rhythms have not been taken into account (e.g. Burns and Scheving, '73, '75). This also applies to measurements of ^{3}H-thymidine incorporation.

C. EXPERIMENTS TO DEMONSTRATE THE TWO CHALONE PROPERTIES OF EPIDERMAL EXTRACTS

Methods

1) ^{3}H-thymidine Incorporation: A large series of experiments has been performed on male Swiss S mice, 3 months old. One or more injections of epidermal G_1 and G_2 chalones and the 71-80% ethanol precipitate from which they were extracted were used. The experiments were undertaken so as to take into account the fact that the diurnal rhythm in DNA synthesis may affect a chalone-induced response. Since it was found that the cells respond to these extracts in the same way over different parts of the circadian rhythm, composite tables have been presented here showing the recorded % depression in each case. The full technique and results are being published elsewhere (Thornley and Laurence, '75, '76; Thornley, Spargo and Laurence, '76).

Briefly, the technique used here was to inject extract and ^{3}H-thymidine at the appropriate time into mice which had been rigorously acclimatised as well as the boxes containing the mice being randomised. The mice were killed at the appropriate times, buried in ice and processed as quickly as possible. A central peripheral piece was taken from both ears for histological preparation before the ears were depilated with Immac (Ann French). A piece of ear 5 mm diameter, was punched from either side of the gap formed by the removal of the piece for histology. Together with the duodenum, these pieces were frozen at -10°C until required. The epidermis was stripped from the underlying dermis which contained the lower part of the pilary unit including the sebaceous glands, by immersing the punch in a mixture of EDTA and NaBr solutions (1.2 M NaBr, 10^{-3}M EDTA, pH 7.0). The acid soluble "labelled pool" was extracted and the radioactivity determined. After drying, the epidermal and dermal punches were weighed and then dissolved in Soluene (Packard & Co. Inc.). The chemiluminescence was reduced before adding the scintillation cocktail. The incorporation of ^{3}H-thymidine was determined by counting in the Packard Tricarb Liquid Scintillation Counter, 2425. The results were expressed as dpm/mg dry weight epidermis and dermis. The femora were taken from the mice and the "labelled pools" removed before dissolving in Soluene. The incorporation of ^{3}H-thymidine was determined and expressed as dpm/femur. The DNA was extracted from the duodenum (Marks, '73) and the incorporation of ^{3}H-thymidine was estimated by liquid scintillation counting and expressed as dpm/mg DNA.

The piece of ear that had been removed for histology was fixed in alcoholic Bouin, and prepared for wax sectioning at 3μ. After coating with Ilford G5 emulsion, the slides were exposed for three weeks at 5°C, developed and stained in Mayer's haemalum. The labelling index was determined by counting the number of basal cells with more than 5 silver grains overlying the nucleus in a 10 mm length of ear epidermis.

2) Mitosis: The experiments with the ethanol precipitate and epidermal G_2 chalone on sebaceous glands were taken from Bullough and Laurence ('70). The epidermal results were obtained during the normal sleep period by injecting the extracts and Colcemid (Ciba) subcutaneously at 11.30 h. and killing the animals at 15.30 h. The ears were fixed in alcoholic Bouin, prepared for wax sectioning at 7 μ and stained in Ehrlich's haematoxylin and eosin. The number of mitoses in metaphase was recorded in 10 mm lengths of epidermis.

Results

1) To demonstrate tissue-, cell-line and cell-cycle phase-specificity. The results in Table I show that none of the extracts depressed the ^{3}H-thymidine incorporation into femora or duodenum. In the epidermis, a significant depression was recorded approximately 5 and 6½ hours after the administration of the ethanol precipitate and the G_1 chalone. This demonstrates the tissue specificity of these two extracts. In contrast to the G_1 chalone, where no significant depression was found, the ethanol precipitate depressed the ^{3}H-thymidine incorporation in sebaceous glands. This demonstrates the cell-line specificity of the G_1 chalone which is lacking in the ethanol precipitate. The G_2 chalone had no effect on the ^{3}H-thymidine incorporation in any tissue over the 6½ hour experimental period. In addition to these results none of the extracts had any inhibitory effect on the uptake of ^{3}H-thymidine into the cells measured in the acid-soluble "labelled pool".

Table II shows that whereas the ethanol precipitate depressed the mitosis of epidermis and sebaceous glands, the G_2 chalone only inhibited the mitosis in the epidermis. The G_1 chalone had no effect on mitosis in either tissue.

Thus it may be concluded that the ethanol precipitate is tissue specific but it is neither cell-line specific nor cell-cycle phase specific in its action. Whereas the G_1 chalone may be said to be tissue specific, cell-line specific and cell-cycle phase specific, the purified G_2 chalone, with regard to its antimitotic effects, has not been so rigorously tested for tissue specificity; but it does show cell-line and cell-cycle phase specificity.

2) To demonstrate the point of action in the cell-cycle and confirmation of cell-line specificity. The results in Table III confirm, over many points, the cell-cycle phase-specificity of the two chalones and the expected lack of specificity of the ethanol precipitate from which they were extracted. By comparing the labelling index with the scin-

TABLE I

Results from a series of experiments to demonstrate tissue-, cell-line, and cell-cycle phase-specificity properties of epidermal G_1 and G_2 chalones and the ethanol ppt. from which they were derived, using ^3H-thymidine incorporation as a parameter.

	% depression ^{3}H-thymidine incorporation				
N > 10 animals	treatment h.	Epidermis	Sebaceous glands	Femora	Duodenum
Ethanol ppt.	6½	41†	39†	-	0
G_1 chalone	5	26†	12	0	11
G_2 chalone	6½	7	0	-	10

† $P < 0.01$

TABLE II

Results from experiments to demonstrate an example of cell line specificity of epidermal G_1 and G_2 chalone and the ethanol precipitate from which they were derived using the mitotic index as a parameter.

		% depression in 4h Colcemid mitotic index	
N = 10 animals	treatment h.	epidermis	sebaceous glands
Ethanol ppt.	4	61†	58†
G_1 chalone	4	0	-
G_2 chalone	4	47†	0

† $P < 0.01$

TABLE III

Results from an extensive series of experiments to demonstrate the cell cycle phase specificity of the epidermal G_1 and G_2 chalones, and the absence of this property in the 71-80% ethanol ppt. from which they were isolated, using three parameters.

EPIDERMIS	% depression										
No. = 10 animals	^{3}H-TdR incorporation					labelling index					mitotic index
No: hours	1½	2½	4	5	6½	1½	2½	4	5	6½	4
Ethanol ppt.	0	7	25†	-	41†	0	0	13†	-	33†	51†
G_1 chalone	-	-	-	26†	-	-	-	-	23†	-	0
G_2 chalone	9	0	0	0	0	0	0	7	0	7	46†
SEBACEOUS GLANDS											
Ethanol ppt.	0	17†	8	-	39†	-	-	-	-	-	58†
G_1 chalone	-	-	-	12	-	-	-	-	-	-	-
G_2 chalone	10	8	0	0	0	-	-	-	-	-	0

†$P < 0.01$

tillation results, it will be seen that, taking into account an S-phase length of around 18 hours (Laurence, '73; Laurence and Christophers, '76), the point of action in the cell cycle appears to be at the G_1/S-phase boundary.

D. CONCLUSIONS FROM EXPERIMENTAL DATA

It has been suggested that the G_1 chalone acts several hours before entry into S-phase, when the cells are preparing for DNA synthesis (Hall, '69; Hennings *et al.*, '69; Marks, '76). However, by carefully controlled sensitive experimentation with a full knowledge of the S-phase duration, it is possible to show a more precise point of action. With an S-phase duration of 18 hours, a 25% depression of ^{3}H-thymidine incorporation at 4 hours is explicable only on the assumption that a block occurs at the G_1/S-phase boundary. This indicates that the cells already prepared to divide are sensitive to G_1 chalone. There is a general assumption that many normal cells go through a preparatory period (enzyme synthesis etc.) before entry into S-phase (Baserga, '68, Frankfurt, '68). Hence this substance does not appear to act at the points suggested by Bullough's model ('75) and may not be involved directly in the "decision" of a cell to proceed along the maturation pathway. It also is apparent that the G_2 chalone acts as rapidly as the G_1 chalone and its sole detectable effect is to inhibit the flow of cells into mitosis. Until further epidermal chalones are found, the main conclusion from these experiments is that the two known epidermal chalones inhibit transit through the cell cycle.

These results with ethanol precipitate demonstrate that conclusions about specificity of action based on experiments with crude homogenates can, at best, be tentative. It is necessary to do tissue-, cell-line, and cell-cycle phase-specificity tests at each major step in the purification process. A formidable task and one which is not yet complete for the three extracts used here! A major consideration is a knowledge of the kinetic parameters of the target cell-line since this is the only way that non-specific and toxic effects can be monitored. It is obviously very helpful to have other chalones derived from the same and different germinal layers to verify the sensitivity of the assay procedure.

E. APPARENT EXCEPTIONS TO A GERMINAL LAYER THEORY OF TISSUE SPECIFICITY

Not all evidence points towards a germinal layer theory of tissue specificity. For example, Frankfurt ('71) and Nome ('74) found that crude extracts of epidermis depressed cell proliferation in forestomach epithelium. Bullough and Laurence found crude extracts of lung and oesophagus inhibited mitosis in epidermis (unpublished results) and crude extracts of epidermis also inhibited mitosis in oesophagus ('64). Laurence ('73) could not confirm these results on oesophagus using the more purified ethanol precipitate. The outcome of these conflicting results is dependent upon a thorough investigation using the appropriate highly purified extracts. There is now sufficient evidence to show that crude or partially purified extracts contain (toxic) degradation products and non-specific inhibitors which can easily mask or summate with chalone action (Thornley and Laurence, unpublished). Thus, unfortunately, in the light of later extensive work with purified chalones, all earlier results with crude extracts should be viewed with caution.

III. CHALONES AND STEM CELLS

The tables presented here suggest indirect evidence for a sebaceous gland chalone system which in turn implies that only once the function of a cell population is determined do chalones manifest themselves. This is an important concept when applied to stem cells. The term stem cell seems to be applied in two contexts: 1) pluripotential stem cell, i.e. a single cell present in the adult which forms separate distinct tissues or cell lines such as granulocytes, erythrocytes, macrophages; and 2) the committed stem cell which will continue to form only one cell line, e.g. cells of the erythrocyte system only.

1) The Pluripotential Stem Cell. Each of the tissues formed from a pluripotential stem cell may have its own chalone system (e.g. granulocyte chalone,erythrocyte chalones, etc.). If there is a feedback mechanism from all these tissues to the stem cell (Moore, '75) the substances could not be called chalones since they are not cell-line specific. However, this type of stem cell may have its own endogenous mitotic inhibitory system.

Although epidermis is a heterogeneous tissue in the adult, being composed of such cells as e.g. Langerhan cells, melanocytes etc., there is no evidence for a pluripotential stem cell at present. In the foetus (e.g. guinea pig) or the neonate (e.g. mouse), depending on species, the epidermis contains cells which will form the pilary unit consisting of hair, its follicle and sebaceous gland; i.e. at this stage in development the epidermis may contain the pluripotential stem cell of the pilary unit as well. Thus in interpreting results using epidermis of e.g. newborn rats and mice, it should be remembered that a heterogeneous group of cells is present. Nothing is known about the functional determination of chalone mechanisms at this stage but there is some evidence that neonatal mouse epidermal cells may be insensitive to the inhibitor of DNA synthesis (Marks, '74; Elgjo, personal communication) although they apparently can produce an inhibitor which acts on adult epidermis (Marrs and Voorhees, '71).

As already shown, there is some evidence for the presence of a sebaceous gland chalone in the adult. However, the hair unit is complex and reforms cyclically throughout the life of the adult animal. During anagen stage of this hair cycle, a stem cell divides many times to form a column of apparently undifferentiated dividing cells which ultimately are concentrated around the dermal papilla (Bullough and Laurence, '58). These cells differentiate to form the cell-lines of the hair, i.e. the outer root sheath, the Henry and Huxley layers, the inner root sheath, the hair proper etc. (Montagna and Scott, '58). Each layer apparently has a finite length of life (about three weeks in mouse) so that at catagen when all cell division has ceased, the hair suddenly retracts towards the surface through the degenerating outer sheath of cells. The cycle is repeated after a period in telogen where only a very few or even only one stem cell apparently remains. Thus it may be expected that many chalone systems are operating in the hair follicle. However, despite much research the mechanism underlying this cycle of hair growth still remains obscure. The meagre attempts to detect chalone systems active in hair bulbs were unsuccessful (Bullough and Laurence, unpublished).

Thus it seems that the pluripotential stem cell may be regarded as the persistence in the adult of an embryonic type of cell.

2) The Committed Stem Cell. It may be regarded as one which retains its power of division to form a clone of cells which follow the usual life cycle of division, maturation and death. The existence of such stem cells in the epidermis has been postulated by Potten ('74), but the evidence is not conclusive since their results are subject to several interpretations (e.g. Christophers and Laurence, '76). This type of cell may be regulated by its own chalone mechanism as it is a functionally committed cell. It is not known precisely which cells in the epidermis synthesise chalones or whether all progenitor cells are responsive to these substances.

IV. ACKNOWLEDGEMENTS

We wish to thank Dr. W. Hondius Boldingh, N. V. Organon for the gift of the ethanol precipitate and G_2 chalone; and Dr. F. Marks for the G_1 chalone. We are indebted to Mrs. S. Knibb, Mr. D. Spargo and Mrs. G. Brown for their excellent technical skill and help. The work was supported by grants from the Cancer Research Fund, Medical Research Council and Wellcome Trust.

LITERATURE CITED

Baserga, R. 1968. Biochemistry of the cell cycle. *Cell Tissue Kinet. 1:* 167-191.

Bertsch, S. and Marks, F. 1974. Lack of an effect of tumour-promoting phorbol esters and of epidermal G_1 chalone on DNA synthesis in the epidermis of newborn mice. *Cancer Res. 34:* 3283-3288.

Bichel, P. 1971. Autoregulation of ascites tumour growth by inhibition of the G_1 and the G_2 phase. *Europ. J. Cancer 7:* 349-355.

Bichel, P. 1973. Self-limitation of Ascites tumor growth: A possible chalone regulation. *Natl. Cancer Inst. Monogr. 38:* 197-203.

Brugal, G. and Pelmont, J. 1974. Presence, dans l'intestin du Triton adulte Pleurodeles waltlii Michah., de deux facteurs antimitotiques naturels (chalones) actifs sur la proliferation cellulaire de l'intestin embryonnaire. *C.R. Acad. Sci. Paris 278:* 2831-2834.

Brugal, G. and Pelmont, J. 1975. Existence of two chalone-like substances in intestinal extract from the adult newt, inhibiting embryonic intestinal cell proliferation. *Cell Tissue Kinet. 8:* 171-187.

Bullough, W. S. 1962. The control of mitotic activity in adult mammalian tissues. *Biol. Rev. 37:* 307-342.

Bullough, W. S. 1965. Mitotic and functional homeostasis. *Cancer Res. 25:* 1683-1727.

Bullough, W. S. 1975. Mitotic control in adult mammalian tissues. *Biol. Rev. 50:* 99-127.

Bullough, W. S. and Laurence, E. B. 1958. Mitotic activity of the follicle. In *The Biology of Hairgrowth.* Edit. Montagna W. and Ellis, R. A. Academic Press.

Bullough, W. S. and Laurence, E. B. 1964. Mitotic control by internal secretion: the role of the chalone-adrenalin complex. *Exp. Cell Res. 33:* 192-200.

Bullough, W. S. and Laurence, E. B. 1970. Chalone control of mitotic activity in sebaceous glands. *Cell and Tissue Kinet. 1:* 5-10.

Bullough, W. S., Laurence, E. B., Iversen, O. H. and Elgjo, K. 1967. The vertebrate epidermal chalone. *Nature 214:* 578-580.

Burns, E. R. and Scheving, L. E. 1973. Circadian influence on the wave form of the frequency of labeled mitoses in mouse corneal epithelium. *Cell and Tissue Kinetics 8:* 61-66.

Burns, E. R. and Scheving, L. E. 1973. Isoproterenol-induced phase shifts in circadian rhythm of mitosis in murine corneal epithelium. *J. Cell Biol. 56:* 605-608.

Christophers, E. and Laurence, E. B. 1976. Kinetics and structural development of the epidermis. In *Progress in Dermatology*. In press.

Elgjo, K. 1973. Epidermal chalone: Cell cycle specificity of two epidermal growth inhibitors. *Natl. Cancer Inst. Monogr. 38:* 71-76.

Elgjo, K., Laerum, O. D. and Edgehill, W. 1971. Growth regulation in mouse epidermis. I. G_2-inhibitor present in the basal cell layer. *Virchows Arch. Zellpath. 8:* 277-283.

Elgjo, K., Laerum, O. D. and Edgehill, W. 1972. Growth regulation in mouse epidermis. II. G_1 inhibitor present in the differentiating cells. *Virchows Arch. Zellpath. 4:* 45-53.

Frankfurt, O. S. 1968. Effect of hydrocortisone, adrenalin and actinomycin D on transition of cells to the DNA synthesis phase. *Exp. Cell Res. 52:* 220-232.

Frankfurt, O. S. 1971. Epidermal chalone. Effect on cell cycle and on development of hyperplasia. *Exp. Cell Res. 65:* 297-306.

Hall, R. G. 1969. DNA synthesis in organ cultures of the hamster cheek pouch. *Exptl. Cell Res. 58:* 429-431.

Hennings, H., Elgjo, K. and Iversen, O. H. 1971. Delayed inhibition of epidermal DNA synthesis after injection of aqueous skin extract (chalone). *Virchows Arch. Zellpath. 4:* 45-53.

Hondius Boldingh, W. and Laurence, E. B. 1968. Extraction, purification and preliminary characterisation of the epidermal chalone. *Europ. J. Biochem. 5:* 191-198.

Iversen, O. H. 1965. Cybernetic aspects of the cancer problem. *Prog. Cybernetics 2:* 76-110.

Iversen, O. H. 1969. Chalones of the skin. In *Ciba Foundation Symposium on Homeostatic Regulators*. Wolstenholme, G. E. and Knight, J. C. pp. 29-53.

Laurence, E. B. 1973a. Experimental approach to the epidermal chalone. *Natl. Cancer Inst. Monogr. 38:* 37-45.

Laurence, E. B. 1973b. The epidermal chalone and keratinising epithelium. *Natl. Cancer Inst. Monogr. 38:* 61-68.

Laurence, E. B. and Christophers, E. 1976. Selective action of cortisol on postmitotic epidermal cells *in vivo*. Submitted for publication.

Lord, B. I. 1976. The assay of cell proliferation inhibitors. In *Chalones*. Edit. Houck, J. C. Elsevier, Excerta Medica. North Holland.

Marks, F. 1973. A tissue-specific factor inhibiting DNA synthesis in mouse epidermis. *Natl. Cancer Inst. Monogr. 38:* 79-90.

Marks, F. 1976. The epidermal chalones. In *Chalones*. Edit. Houck, J. Elsevier, Excerta Media. North-Holland Press.

Marrs, J. M. and Voorhees, J. J. 1971. Preliminary characterisation of an epidermal chalone-like inhibitor. *J. invest. Dermatol. 56:* 353-358.

Montagna, W. and Scott, E. J. van. 1958. The Anatomy of the Hair Follicle. In *The Biology of Hair Growth*. Edit. Montagna W. and Ellis, R. A. Academic Press.

Moore, M. 1975. Humoral control and cellular interactions in granulopoiesis. *Xth Paterson Symposium on Cellular Interactions*. Manchester.

Nome, O. 1974. Tissue specificity of epidermal chalones. Thesis submitted to the University of Oslo.

Pinkus, H. 1958. Embryology of hair. In *The Biology of Hair Growth*. Montagna, W. and Ellis, R. A. Academic Press.

Potten, C. S. 1974. The epidermal proliferative unit: the possible role of the central basal cell. *Cell Tissue Kinet. 7:* 77-88.

Thornley, A. L. and Laurence, E. B. 1975. Chalone regulation of the epidermal cell cycle. *Experientia 31:* 1024-6.

Thornley, A. L. and Laurence, E. B. 1976. Further evidence for two epidermal chalone systems in the epidermis. Submitted for publication.

Thornley, A. L., Spargo, D. and Laurence, E. B. 1976. Mouse ear epidermis as a chalone assay system. Submitted for publication.

Regulation of Epidermal Stem Cells

JACK FOWLER AND JULIANA DENEKAMP

Cancer Research Campaign Gray Laboratory
Mount Vernon Hospital, Northwood, Middx.

The epidermis of all mammals is maintained at an appropriate thickness on different parts of the body, both under normal conditions and after injury. This steady state must be the result of a delicately balanced control system. After most forms of mechanical or chemical wounding or after surgical incision, compensatory proliferation usually begins within a day and rises to a peak within two days, as the damaged cells are replaced. After irradiation with ionising radiation or after thermal burns, however, compensatory proliferation is not usually seen until about 7 or 8 days.

For some years we have been interested in the effects of radiation on the epidermis in relation to fractionated radiotherapy. The effect of radiation on most mammalian cells is limited to proliferating cells, except at very high doses. This form of damage therefore differs from most other injuries where both differentiating and proliferating cells are affected. Hence, radiation may be a useful tool for studying control mechanisms, in particular for determining whether it is the loss of proliferating or of differentiated cells that acts as the stimulus for more rapid compensatory proliferation. We shall summarise our data on proliferation changes in mouse skin after single doses and after fractionated irradiation. We also have data on plucked skin, both before and after irradiation, which shed further light on the regulatory mechanisms in skin. These results will be related to the post-irradiation data of other workers and will also be compared and contrasted with the response after other forms of injury, including surgical incision, toxic chemicals, burns, cellotape stripping, and hair plucking.

Because the effect of radiation is expressed gradually, starting from the basal layer and moving up through the epidermis, it is a more useful method of analysing control mechanisms than the historical methods of surgical wounding and other types of trauma, which damage several layers simultaneously.

METHOD

For most proliferative studies *in vivo* histological techniques are used involving the scoring of mitotic indices, MI, with or without vinca alkaloid administration, or the scoring of labelled cells, LI, or per cent of labelled mitoses, PLM, after administration of tritiated thymidine. All these techniques are difficult to use on a population which is not in a steady state and the results can be further complicated by the known diurnal rhythms of both mitotic and labelling indices (Brown, '70). Radiation damage is usually expressed at mitosis, and radiation itself induces a mitotic delay which is both dose dependent and cell cycle dependent and which may be followed by an overshoot in MI as a synchronised cohort of cells is released into mitosis. In addition, radiation damaged cells may look viable for some time before they actually die and are included in any estimate of the labelling or mitotic indice of surviving clonogenic cells. For this reason there have been difficulties in interpreting the autoradiographic data on irradiated populations. However, because of the large volume of data on gross skin reactions (e.g. Fowler *et al.*, '65; Field *et al.*, '75; Moulder *et al.*, '75) and the correlation which has been shown to exist between skin reactions and epidermal cell survival (Denekamp *et al.*, '69; Emery *et al.*, '70) it is now possible to study proliferation in skin indirectly by means of functional tests of the level of cell depletion after a single dose or during a course of fractionated irradiation. This approach yields estimates of the additional dose necessary to compensate for repopulation during the interval being studied. Since the dose necessary to kill 50% of a cell population is known for epidermal cells (Withers, '67; Emery *et al.*, '70), the dose increments can be converted to cell doubling times (Denekamp, '73). Such indirect functional measurements together with our more conventional autoradiographic studies, are presented below.

RESULTS

Autoradiographic studies using the per cent labelled mitoses technique (together with other techniques) have shown that epidermal cells in unstimulated skin on the back of mice have a long inter-mitotic cycle time, T_C, of about 100 hours (Figure 1). There appears to be a rather narrow distribution of T_C judged from the precision of the second wave, indicating that the G_1 phase is well defined and does not represent a random stimulation of cells out of an ill-defined G_0 phase.

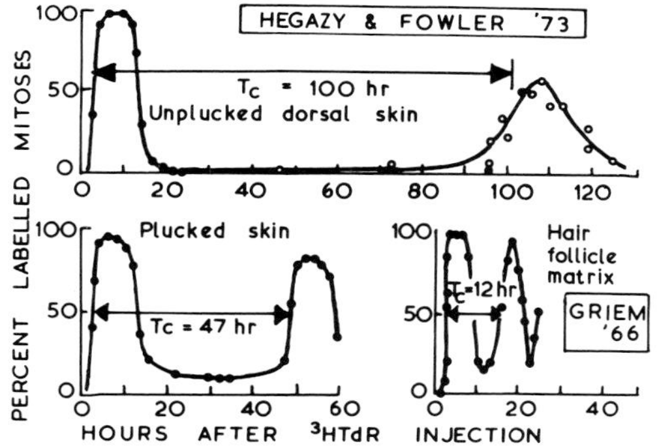

Figure 1. Mouse skin.

Plucking of the epidermis 24 hours before the injection of tritiated thymidine results in a shorter T_c of about 47 hours in the interfollicular epidermis (Hegazy and Fowler, '73a). Furthermore, in stimulated hair follicles, T_c can be as short as 12 hours (Figure 1). Thus it appears that the cell cycle time can be shortened in response to a stimulus, to a degree that varies in different situations.

An alternative method of compensatory faster proliferation would be to increase the proportion of cells in the growth fraction. Repeated injections of tritiated thymidine have been shown to label at least 95% of the basal cells of the mouse dorsum in the plucked and unplucked skin experiments of Hegazy and Fowler in a time equal to the cell cycle time minus the duration of DNA synthesis. This is good evidence for a growth fraction of close to unity in both unstimulated and plucked skin (Figure 2). Similarly, in continuous labelling

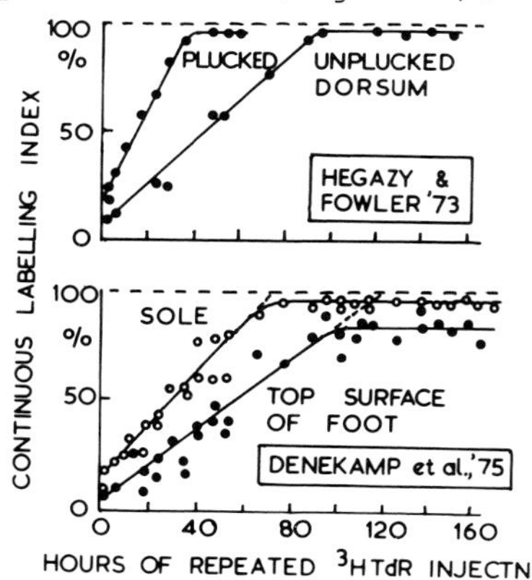

Figure 2.

experiments on the feet of mice more than 85% of the basal cells were labelled on the upper and lower surfaces, although the time of reaching maximum labelling in these experiments indicated that the normal cycle time in these two locations was somewhat different, being faster on the sole where there is more "wear and tear" (Denekamp *et al.*, '75). These high growth fractions in unstimulated skin do not allow for much more rapid proliferation simply by stimulation of more cells into the cell cycle. The control is mainly through a variation in the length of the well defined G_1 phase. However, in regenerating epidermis, the proliferating compartment may expand by cell division occurring in the suprabasal as well as the basal layer.

PROLIFERATION IN IRRADIATED SKIN

When graded test doses of X-rays are given to a tissue previously depleted of cells (e.g. by a single dose of X-rays) the size of the test dose required to bring the sub-threshold damage up to a constant level of skin reaction can be used as a measure of the level of depletion. If these test doses are applied at various times after the pre-treatment, repopulation of the depleted skin can be estimated, as more dose is then necessary to produce the same level of skin reaction.

After irradiation with single doses of 1,000 rads no additional dose was found necessary to compensate for proliferation in unplucked mouse skin for about one week, using equal skin reactions as the functional test (Denekamp *et al.*, '69). After that, approximately 50 rads per day were necessary for the next week. Since the D_{50} (i.e. dose required to sterilise 50% of the cells on the exponential part of an X-ray survival curve) is about 100 rads, this corresponds to a cell doubling time of about 2 days (Figure 3a). 1,000 rads will sterilise about 90% of the basal cells, and this results in a halving of the cell cycle time, but only after about 7 days have elapsed. A similar conclusion about the delay in compensatory proliferation is drawn from the increase in mitotic and labelling indices in the basal layer (Hegazy and Fowler, '73b; Figure 3a). The LI rises above normal at about 7-9 days after 1,000 rads, soon after the peak of degenerate pyknotic cells occurs in the basal layer. This delayed appearance of degenerate cells and of raised proliferative indices is due to the radiation-induced mitotic delay and is even longer for higher doses (Figure 3b and Table I).

Gross damage is scarcely visible after a dose of 1,000 or 1,500 rads; no dry or moist desquamation occurs at any

TABLE 1

METHOD & SYSTEM	INCREASE BEGINS		PEAK		REF.
	MI	LI	MI	LI	
CELLOTAPE STRIPPING					
Dorsum haired mice					
Slight; 1 or 2 strip.	<18h	12h	18h	24h	Pot (75)
Human forearm	36h	-	72h	-	Pin (52)
30 strippings	24-48h	-	48-72h	-	Wil (57)
Dorsum hairless mice					
5 strippings	10h	5-6h	24h	24h	Hen (70)
Bat's wing. Centre:	6h	-	48h	-	Ive (74)
20 strippings. edge:	10h	-	60h	-	
HAIR PLUCKING					
Dorsal skin of mouse	18h	10h	-	20h*	Pot (71)
Dorsal skin of mouse	-	<18h	-	-	Heg (73a)
HAIR CLIPPING					
Guinea pig flank	<24h	<24h	24h	24h	Tag (74)
SKIN MASSAGE					
Ear of mouse	-	-	1-2d	-	Bul (59)
Dorsum mouse	-	-	1-2d	-	Bul (59)
CARCINOGENS					
1% MCA (once)	12h	10h	14d	14d	Ive (62)
0.5% Croton oil (once)	2-4h	2-4h	2d	2d	Ive (62)
CAUTERY					
$AgNO_3$ on rat tongue	-	<18h	-	24h	Blo (63)
Slight destr.					
SOLID CO_2 BURN					
Guinea pig skin	1-2d	-	4d	-	Fir (50)
1 cm^2, 3 mins.					
BURNS					
Pig skin, 2 cm^2					
60°C for 60 sec	8d	-	9-15d	-	Win (74)
80°C for 15 sec					

Footnote: * Followed by waves at 12-18 hr intervals, later 20-36 hr.

TABLE 1 (continued)

METHOD & SYSTEM	INCREASE BEGINS		PEAK		REF.
	MI	LI	MI	LI	
TEMPERATURE Room 35° not 22° Mouse ear (G_2)	5h	-	-	-	Gel (75)
WOUNDING Mouse ear, cut (vitro)	<5h	-	2d	-	Gel (59)
Guinea pig ear, cut (vivo/vitro)	8h	4h	-	24h*	Hel (63)
Rat dorsum, cut	18h*	-	30h	-	Ber (65)
Rat ear, cut	24h*	-	42h	-	Ber (65)
Rat abdom, cut	-	1h	-	24h	Blo (63)
Pig skin 25x25mm 0.2 mm thick	18h (migration)		2-7d	-	Win (64)
Human skin, cuts	30h*	-	40-60h	-	Eps (64)
Mouse ear, cuts	36h	-	48h	-	Bul (57)
G-pig, needle injecn.	-	12h	-	24h	Tag (75)
X OR β RAYS Hamster cheek pouch:					
1000 rads (6pm)	-	2d	-	4-10d	Bro (70)
1000 rads (10am)	-	9d	-	12-16d	Bro (70)
Flank of rat:					
500 rads	<5d	-	7d	-	Van (66)
1000 rads	8d	-	10d	-	Van (66)
Dorsum of mouse:					
Unplucked: 500 rads	-	6d	8d	-	Heg (73b)
(Tc=100h) 1000 rads	-	7d	10d	8d	Heg (73b)
1500 rads	9d	18d	12d	20d	Heg (73b)
2000 rads	9d	>28d	10d	>28d	Heg (73b)
At margin: 2000 rads	-	5d	-	18d	Heg (73b)

Footnote: * Followed by waves at 8 to 24 hr intervals

TABLE 1 (continued)

METHOD & SYSTEM		INCREASE BEGINS		PEAK		REF.
		MI	LI	MI	LI	
Plucked:	500 rads	-	-	-	8d	Heg (73b)
(Tc=47h)	1000 rads	3d	8d	4d	8d	Heg (73b)
	1500 rads	4d	18d	12d	20d	Heg (73b)
	2000 rads	6d	-	12d	28d	Heg (73b)
Margin:	2000 rads	-	<7d	-	20d	(Tc=12h)
Guinea pig flank:						
25x25 mm:	1000 rads	<20d	6d	20d	12-20d	Eto (75)
	2200 rads	-	13d	-	15d	Eto (75)
	3000 rads	18d	15d	22d	18d	Eto (75)
At margin:	3000 rads	8d	10d	20d	17-20d	(Tc=17h)

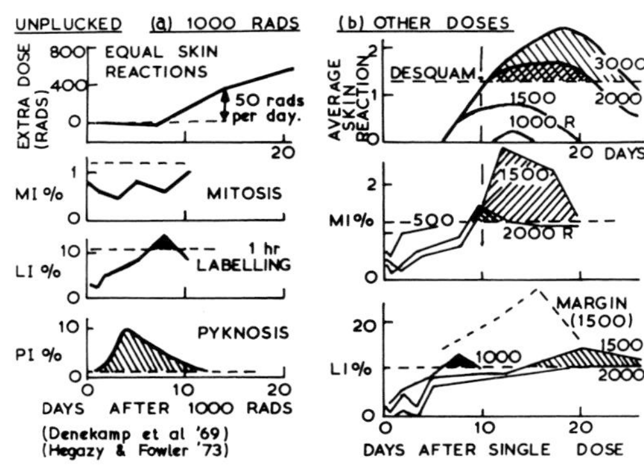

Figure 3.

time. Therefore, depletion of the keratin layer cannot be a factor in causing the enhanced proliferation which occurs. Doses of 1,700 rads or more are necessary to produce visible desquamation, which then commences at 9-12 days in previously normal mouse skin and reaches a peak at 18-20 days (Figure 3b). The compensatory proliferation after such large doses, although delayed by 7 or 9 days after irradiation, still begins 2 or 3 days before loss of function of the keratinised layer, confirming that loss of keratin cannot provide the feedback stimulus.

In plucked skin, where the cycle time is 47 hours before irradiation, instead of 100 hours as in unplucked skin, the proliferative responses occur more quickly (Figure 4). More rapid proliferation occurs within 2 days, as judged from the

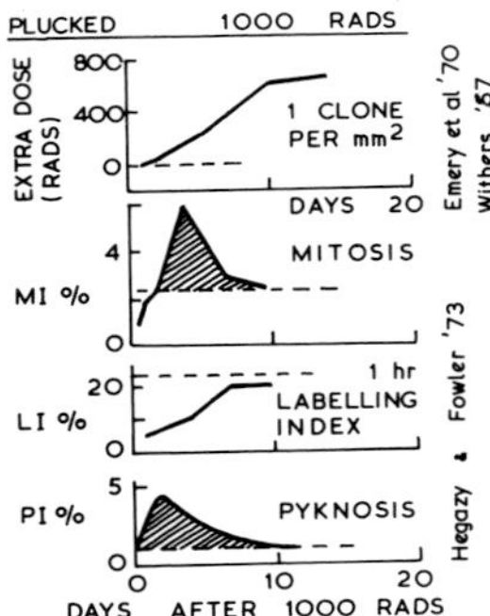

Figure 4.

dose increments needed in epidermal clone experiments in plucked skin. This is somewhat faster than expected from the changes in MI and LI but correlates with the earlier appearance of pyknotic cells which peak at 2 days (Figure 4). This emphasises the problem of interpreting kinetic and autoradiographic data such as MI and LI in grossly disturbed tissues. Functional tests, such as the determination of iso-effect doses, are therefore necessary and are more informative.

After larger single doses of 3,000 rads of beta rays to guinea pig epidermis, Tabachnick and his co-workers observed little change in the rate of movement of basal cells into the more superficial layers, in spite of the total cessation of cell division for at least 4 days (Etoh *et al.*, '75). Hence, a cell-age related or programmed migration appears to precede cell division, although cells accumulating at a G_2 block will have an average size that is larger than that of cells uniformly distributed throughout the cell cycle. Squeezing-out cannot be excluded but it is not the result of cell division (Leblond *et al.*, '64).

Figure 5 shows that the level of cell depletion that developed before compensatory proliferation in the basal layer is dose dependent and occurred later for higher doses. After 1,000 and 2,200 rads the compensatory proliferation restored the basal layer to 100% about 5 days after reaching the maximum depletion (arrow in Fig. 5). After 3,000 rads this restoration resulted from rapid proliferation and migration from the unirradiated margin. Therefore, the basal layer must be depleted by an appreciable proportion, possibly some 20-30%, before compensatory proliferation becomes rapid. In the interval between reaching this critical level and the attainment of rapid proliferation, the basal cell population falls still further, at a rate just less than normal progres-

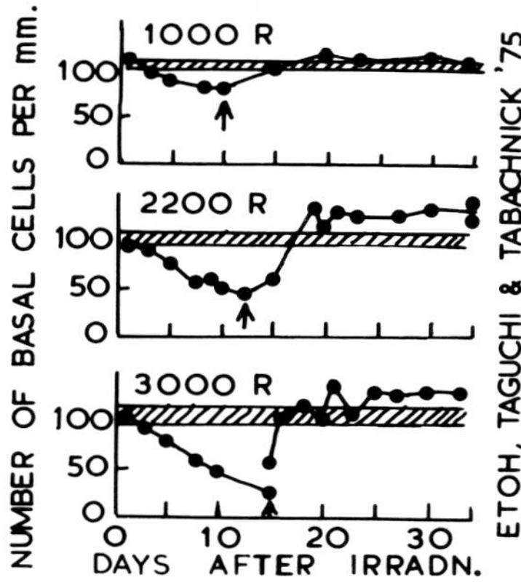

Figure 5.

sion.

Cells in the margin of an irradiated area respond several days earlier than the irradiated cells (Devik '62; Hegazy and Fowler, '73b; Etoh *et al.*, '75); Fig. 3b. This shows that signals to proliferate are provided at about 5-7 days from the irradiated area. Further studies of the time course of such "edge proliferation" would be useful in determining the earliest time at which such stimulation occurs in the absence of mitotic delay.

FRACTIONATED IRRADIATION

Although the earliest time of stimulation of rapid proliferation after large single doses is of interest, another approach is to ask 'how long can enhanced proliferation be delayed?'. If it did not begin until a long time after loss of cells from the basal layer, then basal cell loss would be unlikely to be the stimulus. Experiments using repeated small doses of radiation can provide this information. An indication of the effect of proliferation during fractionated treatments can be obtained by giving several fractions, e.g. five, either at daily or weekly intervals. More dose is then necessary to produce a particular level of depletion in the longer treatment and an average dose per day to counteract repopulation can be measured. Such experiments have yielded 30 rads per day averaged over 3-4 weeks of "daily" irradiations, indicating an average cell doubling time of about 3 days.

It is not possible from these equal-dose experiments to tell whether the more rapid proliferation occurred at the same rate throughout the period of study. Other skin reaction experiments were therefore designed to determine the time profile of proliferation after one, two or three weeks of daily 300 rad irradiations (Denekamp, '73). The extent of repopulation in the two weeks after each fractionated treatment was estimated, using large test doses to bring the sub-threshold skin reaction up to a constant level. The conclusions from this study are summarised in Figure 6. No more rapid proliferation occurred after 4 fractions, but an additional 60 rads per day were required after 9 fractions in 10 days (corresponding to 1½-2 days doubling time), and after 14 fractions 100-150 rads per day were necessary to compensate for the very rapid proliferation (doubling time of 16-24 hours). The conclusions were confirmed using continuous labelling by repeated injections of thymidine, as shown in Figure 7.

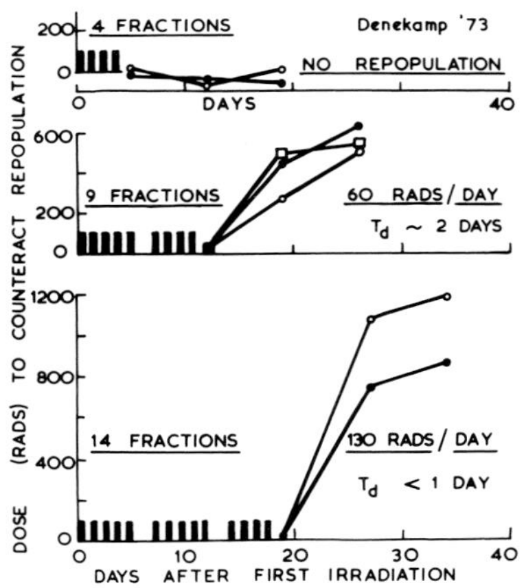

Figure 6.

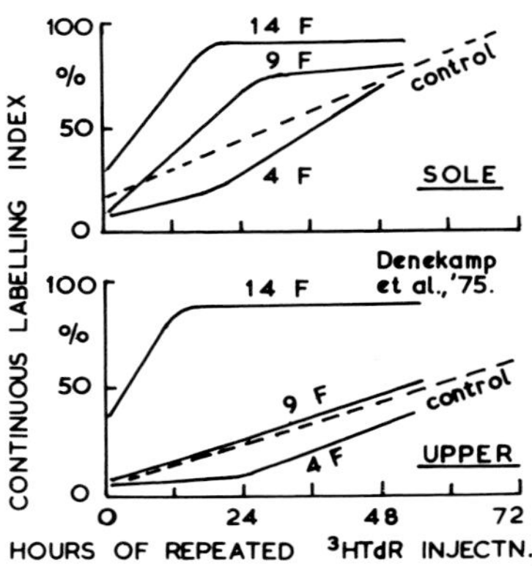

Figure 7.

Proliferation had become faster by 10 days (i.e. 9F) on the sole but not on the upper surface of the foot, which has a longer cell cycle time and hence experiences longer mitotic delay. Figure 7 shows that the LI at 1 hour is a poor indication of the proliferation rate, except after 14 fractions. It appears that the proliferative response to daily doses of 300 rads starts only slightly later than after single X-ray doses (compare Figures 3a and 6). Faster proliferation than normal was not found until about 10 days after starting the daily doses; it had become very fast at 18-25 days after starting, reaching a cycle time of less than 1 day. In these multiple dose experiments the irradiations were not continued into the period of testing, so that the enhanced proliferation diminished after a week, when it was calculated that the basal cell population would have been replenished

(Denekamp, '73).

These results on the timing of enhanced proliferation agree well with other multifraction data on rat skin, when the irradiations were continued longer (Van den Brenk, '66; Moulder *et al.*, '75). Figure 8 shows that in all these instances the compensatory proliferation became significant about 9-14 days after starting the course of irradiation. Sufficiently rapid proliferation can thus be induced so that the reactions heal during the course of irradiation (Field, personal comm.; Moulder, personal comm.), when the daily repopulation exceeds the daily killing caused by each dose.

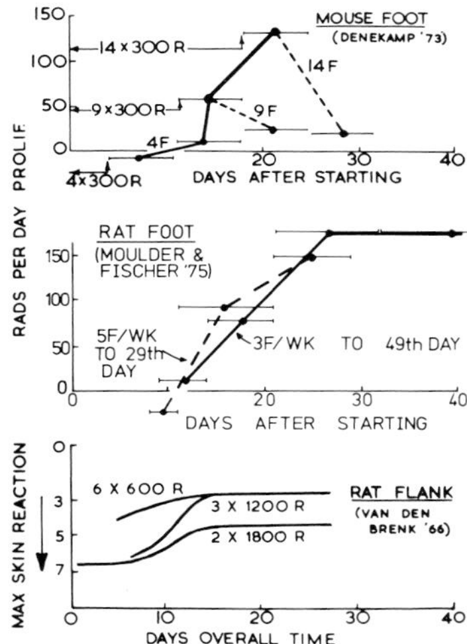

Figure 8.

These data from functional tests, particularly supported by the continuous labelling data, are less prone to misinterpretation than simple mitotic and labelling indices because of the effect on MI of radiation induced synchrony, and on both indices of the continuing presence of doomed radiation-injured cells. However, the conclusions from both types of study are the same:

1. After large single doses of ionising radiation the compensatory proliferation in previously normal rodent skin does not become faster than normal until 7-9 days later.
2. After multiple small dose fractions, this compensatory proliferation begins 9-14 days after starting the irradiation, i.e. at a similar time after appreciable radiation dose has accumulated.
3. The rate of compensatory proliferation is graded according to the severity of the damage, i.e. to the level of cellular depletion.

4. Compensatory proliferation precedes desquamation after irradiation and occurs after doses too low to produce any gross desquamation at all.
5. Since the initial radiation damage is restricted to the proliferating basal layer it seems likely that it is loss of basal cells, not of keratin, that triggers the compensatory response.

DISCUSSION

The results from irradiated skin shed some light on the mechanism of homeostatic control. It seems likely that in this case the stimulus for repopulation is depopulation of the *basal* layer. Although cells continue to leave the basal layer in the absence of cell proliferation immediately after irradiation, this occurs at slightly less than the "normal" rate of several days halving time (Etoh *et al.*, '75). Sufficient cells, therefore, remain to maintain an apparently functional basal layer until they are replenished by the delayed repopulation (see Figure 5). This differs markedly from some other forms of trauma, e.g. cellotape stripping to remove the superficial layers, plucking of hair, thermal or chemical burns, and surgical incisions. With most of these injuries cell damage or removal is immediate, is not restricted to a single cell layer, and the compensatory response is much more rapid than after irradiation. The timing of the changes in MI and LI for various forms of injury are summarised in Table 1. After all forms of mechanical or chemical trauma so far recorded the LI and MI rise within a day and peak in about 2 days. Only radiation and thermal burns demonstrate the delayed responses.

CELLOTAPE STRIPPING

Cellotape stripping has often been used specifically to remove the superficial layers of epidermis, but the pressure and tension involved in applying and removing the cellotape causes gross mechanical damage to the underlying layers too (Mishima and Pinkus, '68). The number of strippings has been varied in order to vary the number of keratin layers removed from human arms and a graded response was found. After 4-14, 22 or 32 cellotape strippings the 1-hour LI values at 48 hrs were 1, 2 and 4% respectively (Pinkus, '52). Potten and Allen ('75) have shown that there is a rapid transit of cells from the basal into the more superficial layers shortly after light cellotape stripping. Thus, the rapid proliferative response after stripping cannot be interpreted as simply due to loss of keratinised layers since the basal layer is rapidly depleted to replace the differentiating layers. Increased labelling indices have been observed in mouse skin within 5-6 hours after stripping by Hennings and Elgjo ('70) and at 12 hours by Potten and Allen ('75), with both groups observing a peak at 24 hours. The mitotic index also increased rapidly and peaks have been observed at 18-72 hours (Table 1). Cellotape stripping of one surface of a bat's wing resulted in an increased mitotic index observed on the opposite epidermis within 6 hours with a peak at 48 hours (Iversen *et al.*, '74). Since all layers, including the basal cells, were removed by the 20 strippings it does not indicate whether the stimulus - possibly lack of a chalone - was of basal or differentiating cell origin.

OTHER MECHANICAL INJURY

Table 1 shows that raised labelling and/or mitotic indices have also been observed within the first 24 hours after hair plucking (Potten, '71; Hamilton and Potten, '72), after hair clipping (Taguchi and Tabachnick, '74), and after skin massage (Bullough and Laurence, '59). The LI was also found to be increased locally at 12 and 24 hours after subcutaneous injection of a small volume of saline or of tritiated thymidine, and even after insertion of a needle with no fluid injected (Taguchi *et al.*, '75). Surgical incisions have been studied in great detail and again a rapid response is observed (Table 1). The labelling index can rise within 1-4 hours at about 1 mm away from the incision, with the wave of stimulated proliferation then progressing toward the wound (Hell and Cruickshank, '63; Block *et al.*, '63). Mitotic activity also increases rapidly, usually being raised within 24 hours and often reaching a peak at about 2 days (Bullough, '57; Gelfant, '59). There is some evidence of cyclic fluctuations at 8-24 hour intervals in rat and human skin (Bertalanffy *et al.*, '65; Epstein and Sullivan, '64).

If the wound is more extensive than a simple incision, migration occurs across the wound surface at up to 0.5 mm per day in moist conditions, starting within 8-18 hours (Winter, '64). Significant migration can also occur from the hair follicles in a shallow wound, as indeed after doses of radiation less than about 2,500 rads (Van den Brenk, '66; Yamaguchi and Tabachnick, '72).

CHEMICAL "BURNS"

Carcinogens and other chemicals which induce hyperplasia cause raised LI and MI within 1-3 days after painting them the epidermis. These chemicals cause rapid degeneration of cells of all layers. Croton oil is remarkable in that it induces an increased MI and LI within 2-4 hours (Iversen and Evensen, '62). The proliferative activity remains high until the damaged cells have been replaced, but in the case of carcinogenic chemicals the rapid proliferative period is prolonged because of the persistence of the chemical. Even chemical burns which cause damage to only a few superficial layers, e.g. silver nitrate on the tongue of rats, result in more rapid proliferation in the basal layer within 18-24 hrs (Block *et al.,* '63). This is the only experimental result which does not fit the concept of basal cell depletion as the stimulus to faster proliferation.

THERMAL BURNS

Cold burns, e.g. caused by solid carbon dioxide, can elicit a rapid response, with the mitotic index being raised within 1-2 days, as after wounding (Firket, '50). However, the notable feature about the literature on heat burns and scalds is that compensatory proliferation from the edge of the burn is not reported to occur before about 8 days. This applies to burns involving the dermis to a greater or lesser extent and contrasts sharply with surgical injury even when the surgical excision involved all the dermis (Winter, '64, '71, '72, '74). This slow response (Table 1) may be peculiar to thermal burns because of the coagulation of the dermis that is produced, or because the transition from damaged epidermis occurs gradually over several mm, but the lack of a more rapid feedback is not clearly understood. The continued presence of the apparently intact epidermis, although killed by heat, may be the reason for the delayed response. In this respect the heat burn resembles skin "burned" by X-rays, and it may be this aspect that gives them a similarly slow response time.

CONCLUSIONS

Radiation studies have shown that in a situation where death of the proliferating cells of the basal layer is delayed several days because of radiation induced mitotic delay, and where little or no intermitotic death occurs, the stimulus for increased proliferation is also delayed. Basal cells continue to move into the upper layers within the period of mitotic delay, suggesting that this transit is not dependent on pushing out by other dividing cells, but is a programmed event independent of cell division. The compensatory proliferation, although delayed, nevertheless occurs before gross loss of function, i.e. before desquamation. After surgical incision, however, migration from the cut edge occurs within about one day, followed soon after by a burst of proliferative activity which starts at some distance from the wound edge. A similar rapid proliferative burst follows plucking, massage, cellotape stripping and painting with strong chemicals. The only other form of injury which elicits a delayed proliferative response is a heat burn involving the dermis. The delay is similar to that after irradiation, i.e. about 8 days.

The evidence from these other different forms of trauma does not allow one to distinguish which cells provide the controlling influence, because all layers of the epidermis are damaged simultaneously. However, none of the experimental results (except that for silver nitrate) is inconsistent with the conclusions from the radiation experiments that it is a loss of basal cells that stimulates more rapid proliferation.

ACKNOWLEDGEMENTS

We have pleasure in thanking Drs. J. Tabachnick, O. H. Iversen, J. E. Moulder, S. B. Field and C. S. Potten for helpful discussions and for allowing us to quote some of their unpublished results. We thank Mrs. Enid Trevaskis for her great help in the preparation of the typescript.

LITERATURE CITED

Bertalanffy, F. D., V. Pusey and M. O. Abbott. 1965. Mitotic rates of rat epidermis. *Arch. Derm. 92:* 91-102.

Block, P., I. Seiter and W. Oehlert. 1963. Autoradiographic studies of the initial cellular response to injury. *Exp. Cell Res. 30:* 311-321.

Brown, J. M. 1970. The effect of acute X-irradiation on the cell proliferation kinetics of induced carcinomas and their normal counterpart. *Rad. Res. 43:* 627-653.

Bullough, W. S. and E. B. Laurence. 1957. Mitotic activity in healing wounds: a technique for the study of small epidermal wounds. *Brit. J. Exp. Path. 38:* 273-283.

Bullough, W. S. and E. B. Laurence. 1959. The control of epidermal mitotic activity in the mouse. *Proc. Roy. Soc. B 151:* 517-536.

Denekamp, J. 1973. Changes in the rate of repopulation during multifraction irradiation of mouse skin. *Brit. J. Radiol. 46:* 381-387.

Denekamp, J., M. M. Ball and J. F. Fowler. 1969. Recovery and repopulation in mouse skin as a function of time after X-irradiation. *Rad. Res. 37:* 361-370.

Denekamp, J., F. A. Stewart and B. G. Douglas. 1975. Changes in the proliferation rate of mouse epidermis after irradiation: continuous labelling studies. *Cell Tissue Kinet.* (in press).

Devik, F. 1962. Studies on the duration of DNA-synthesis and mitosis in irradiated and regenerating epidermis cells in mice, by means of tritium-labelled thymidine. *Int. J. Rad. Biol. 5:* 59-66.

Emery, E. W., J. Denekamp and S. B. Field. 1970. Survival of mouse skin epithelial cells following single and divided doses of X-rays. *Rad. Res. 41:* 450-466.

Epstein, W. L. and D. J. Sullivan. 1964. Epidermal mitotic activity in wounded human skin. In: *Advances in Biology of Skin. V. Wound Healing.* W. Montagna and R. E. Billingham, eds. Pergamon, Oxford. pp. 68-75.

Etoh, H., Y. H. Taguchi and J. Tabachnick. 1975. Movement of beta-irradiated epidermal basal cells to the spinous-granular layers in the absence of cell division. *J. Invest. Derm. 64:* 431-435.

Field, S. B., C. Morris, J. Denekamp and J. F. Fowler. 1975. The response of mouse skin to fractionated X-rays. *Europ. J. Cancer 11:* 291-299.

Firket, H. 1950. Etude histologique de la peau de cobaye lesee par la neige carbonique. *Comptes Rendus Soc. Biol. 144:* 1715-1718.

Gelfant, S. 1959. A study of mitosis in mouse ear epidermis *in vitro.* 1 - Cutting of the ear as a mitotic stimulant. *Exp. Cell Res. 16:* 527-537.

Gelfant, S. 1975. Temperature-induced cell proliferation in mouse ear epidermis *in vivo.* *Exp. Cell Res. 90:* 458-461.

Greim, M. L. 1966. Use of multiple biopsies for the study of the cell cycle of the mouse hair follicle. *Nature 210:* 213-214.

Hamilton, E. and C. S. Potten. 1972. Influence of hair plucking on the turnover time of the epidermal basal layer. *Cell Tissue Kinet. 5:* 505-517.

Hegazy, M.A.H. and J. F. Fowler. 1973a. Cell population kinetics of plucked and unplucked mouse skin. I. Unirradiated skin. *Cell Tissue Kinet. 6:* 17-33.

Hell, E. A. and C.N.D. Cruickshank. 1963. The effect of injury upon the uptake of ^{3}H-thymidine by guinea pig (ear) epidermis. *Exp. Cell Res. 31:* 128-139.

Hennings, H. and K. Elgjo. 1970. Epidermal regeneration after cellophane tape stripping of hairless mouse skin. *Cell Tissue Kinet. 3:* 243-252.

Iversen, O. H. and A. Evenson. 1962. Experimental skin carcinogenesis in mice. *Norwegian Monographs on Medical Science.* Norwegian Universities Press, Oslo. 184 pp.

Iversen, O. H., R. Bjerknes and F. Devik. 1968. Kinetics of cell renewal, cell migration and cell loss in the hairless mouse dorsal epidermis. *Cell Tissue Kinet. 1:* 351-367.

Iversen, O. H., K. S. Bhangoo and K. Hansen. 1974. Control of epidermal renewal in the bat web. *Virchows Arch. B. Cell Path. 16:* 157-179.

Leblond, C. P., R. C. Greulich and J.P.M. Pereira. 1964. Relationship of cell formation and cell migration in the renewal of stratified squamous epithelia. In: *Advances in Biology of Skin. V. Wound Healing.* W. Montagna and R. E. Billingham, eds. Pergamon, Oxford. pp. 39-67.

Mishima, Y. and H. Pinkus. 1968. Electron microscopy of keratin-layer stripped human epidermis. *J. Invest. Derm. 50:* 89-102.

Moulder, J. E., J. J. Fischer and A. Casey. 1975. Dose-time relationships for skin reactions and structural damage in rat feet exposed to 250-kVp X-rays. *Radiology 115:* 465-470.

Pinkus, H. 1952. Examination of the epidermis by the strip method. II. Biometric data on regeneration of the human epidermis. *J. Invest. Derm. 19:* 431-447.

Potten, C. S. 1971. Tritiated thymidine incorporation into hair follicle matrix and epidermal basal cells after stimulation. *J. Invest. Derm. 56:* 311-317.

Potten, C. S. and T. D. Allen. 1975. Fine structure and cell kinetics of mouse epidermis after wounding. *J. Cell Sci. 17:* 413-447.

Taguchi, Y. H. and J. Tabachnick. 1974. The effect of clipping guinea pig hair and chronic radio-dermatitis on diurnal (circadian) rhythms in epidermal labelling and mitotic indices. *Arch. Derm. Forsch. 249:* 167-177.

Taguchi, Y. H., J. Tabachnick and K. Manaka. 1975. Effect of needle puncture and intradermal fluid injection on epidermal cell kinetics of albino guinea-pig skin. *Arch. Derm. Forsch.* (in press).

Williams, M. G. and R. Hunter. 1957. Studies of epidermal regeneration by means of the strip method. *J. Invest.*

Derm. 29: 407-413.

Winter, G. D. 1964. Movement of epidermal cells over the wound surface. In: *Advances in Biology of Skin. V. Wound Healing.* W. Montagna and R. E. Billingham, eds. Pergamon, Oxford. pp. 113-127.

Winter, G. D. 1971. The poor healing of burns - a histological study of the repair of burns compared with surgical wounds in the skin of the pig. In: *Research in Burns.* P. Matter, T. L. Barclay and Z. Konickova, eds. Huber, Bern. pp. 614-620.

Winter, G. D. 1972. Epidermal regeneration studied in the domestic pig. In: *Epidermal Wound Healing.* H. I. Maibach and D. T. Rovee, eds. Year Book Med. Pub. Inc., Chicago. pp. 71-112.

Winter, G. D. 1974. The dose-survival relationship for irradiation of epithelial cells of mouse skin. *Brit. J. Radiol. 40:* 187-194.

Van den Brenk, H.A.S. 1966. Relation of dose-fractionation of X-rays and its spacing in time to skin damage. Experimental studies. *Amer. J. Roentgenol. 97:* 1023-1031.

Yamaguchi, T. and J. Tabachnick. 1972. Cell kinetics of epidermal repopulation and persistent hyperplasia in locally beta-irradiated guinea pig skin. *Rad. Res. 50:* 158-180.

Discussion, Session II - Stem Cells in the Epidermis

Dr. Potten's presentation

Dr. Lamerton asked about the nature of the stimulus for changing the rate of production of mature cells in the epidermis. Dr. Potten suggested that migration precedes cell division, and that migration of cells off the basal layer leaves a local space that needs to be filled by local divisions. The central position in the basal layer under the columns of differentiating and cornified cells is probably protected by its immediate neighbors and shielded from cell movement, thus preserving the cell in the central position from maturation. The superstructure of more mature cells probably also imposes some sort of functional limitations on cells in the basal layer. However, in rapidly dividing tissue, this sort of organization is not seen, nor can functional groups of cells in the basal layer be identified.

Dr. Oakberg asked about the cell cycle of the central cell and pointed out that in the testis it appears that about 10% of type A spermatogonia serve a stem cell role and also have long cycle times. Dr. Potten replied that the information presently available seems to suggest that the outer ring of cells have cycle times of 4 or 5 days, while the ones toward the center have longer cycle times of perhaps 7 days or more. In reply to a question by Dr. Leblond, Dr. Potten discussed attempts to identify the presumptive stem cells of the epidermis by electron microscopy. For mouse epidermis, each group of 10 basal cells lying beneath a column has at a central position a cell that is morphologically distinguishable from the rest. However, it is not possible to say definitely whether the presumptive stem cell is indeed the central one or another of the group of 3 or 4 cells toward the center.

In reply to another question from Dr. Leblond, Dr. Potten said that the approaches developed for epidermis have also been applied to other stratified epithelia of the mucosa, esophagus and forestomach, but it has not yet been possible to make them work effectively except for the epidermis.

Dr. Laurence's presentation

Drs. Potten and Blackett raised the question of possible effects of epidermal chalones on differentiation as well as proliferation. Dr. Laurence suggested that chalones may promote differentiation and only secondarily inhibit cell proliferation, although proof of this is lacking at present. She also suggested that each class of differentiated cells should have its own chalone system, such as epidermal versus sebaceous gland chalones, and that it would be interesting to be able to isolate and compare the two chalone systems.

Dr. Patt expressed concern about what appears to be a marginal effect of chalones, and asked about dose-response relationships for the chalone effects, and the maximum effects that could be achieved. Dr. Laurence replied that there has not been enough purified material to permit proper dose-response studies to be done. Work with impure materials has been complicated by a toxic reaction which showed up in the intestine, even though the maximum inhibition in epidermis was no more than 50%.

Dr. Fowler's presentation

Dr. Baker asked whether intercellular communication had been investigated in the epidermis; Dr. Fowler agreed that such studies would be worthwhile but difficult to carry out in this system. In reply to another question, he suggested that the evidence that the stimulus for proliferation is a release from chalone inhibition is quite good, but, because there is a lack of information about biochemical mechanisms by which the depletion of cells operates, one cannot exclude the possibility of stimulation by a "wound hormone". In reply to a question by Dr. Lala, Dr. Fowler could see no difficulty in reconciling his own results with Dr. Potten's model of cell proliferation in the epidermis.

Dr. Bruce asked about the time required for cells in the epidermis to go through the differentiation process, and whether this time was independent of radiation dose. Dr. Fowler replied that, apart from radiation induced delay, about 12 days are required. This time is independent of radiation dose, and the results fit very well with a depletion of basal cells moving up through several layers. The rate is a little bit quicker on the sole of the mouse foot than on the upper part of the foot.

There followed a discussion with Drs. Bruce, Withers and Laurence about the number of clonogenic cells in the hair follicles compared to interfollicular areas. Although no quantitative data are available, perhaps overall there may be more clonogenic cells in the hair follicles than in the interfollicular areas. Dr. Withers suggested that the interfollicular cells and the hair follicle cells are interchangeable and can be regarded as one population of clonogenic cells, although hair growth occurs after regeneration only if repopulation of the base of the follicle occurs quickly enough.

Dr. Potten pointed out that ultraviolet radiation does not cause cell depletion, yet results in a good proliferative reaction after a short time interval; this presents a problem for the hypothesis that depletion precedes mitotis. Dr. Fowler agreed that the high compensatory proliferation following slight burns such as sunburn merited investigation.

In reply to a question from Dr. Miller about the identification of cells belonging to a skin clone, Dr. Fowler pointed out that, although the proliferative units discussed by Dr. Potten are very tidy, this is not the case for the skin clones seen in the Withers type of experiment. In the latter case no much structure is apparent until the clones spread out and epithelium is regenerated. Drs. Lamerton, Leblond, Denekamp and Fowler then discussed the possible effects of population pressure on the rate of removal of cells into the maturing pathway. Dr. Fowler felt that population pressure could still continue for 3 or 4 days after radiation damage because irradiated cells can still undergo a few abortive divisions during this limited interval. He felt that population pressure should not be excluded as a factor affecting the rate of removal of cells into the maturing pathway after irradiation.

General discussion

With regard to the question raised in the discussion of Dr. Laurence's paper whether or not chalones or chalone-like substances can act on undifferentiated and differentiated cells, Dr. Clermont described some unpublished work done with Dr. Mauger. In young (35 days old) growing rats, both type A and type B spermatogonia are proliferating fairly actively. Drs. Clermont and Mauger found that testicular extracts had a differential effect on these two cell populations; undifferentiated type A spermatogonia were affected, but not the differentiating intermediate and type B cells. In adult animals neither population was affected. Dr. Withers asked whether the effect could be indirect; for example, the testicular extracts might suppress pituitary activity. In reply, Dr. Clermont pointed out that, since adult type A spermatogonia were not affected, they serve as useful controls in this experiment. If a hormone is involved, it must affect the type A population of growing animals but not adults.

Drs. Denekamp, Fowler and Till then asked Dr. Laurence about some of the properties of the epidermal chalones. The G_1 chalone is pronase resistant and heat resistant, and its apparent size has decreased as purification has proceeded. The G_2 chalone is heat-sensitive and trypsin-sensitive, and appears to be much larger than the G_1 chalone, although its apparent size may also decrease as it is purified. The G_2 chalone does not appear to be bound to RNA although this may be the case for most chalones. The chalones gradually lose activity over 3 weeks *in vitro,* but the half-life *in vivo* is not known. In reply to a question from Dr. Denekamp about whether the chalone-containing extracts were derived from differentiated epidermal cells or from whole skin preparations, Dr. Laurence replied that they were from whole skin preparations. She also indicated a difficulty in determining whether or not the G_2 chalone is produced by differentiated cells. Because the G_2 chalone is trypsin-sensitive, it could be difficult to obtain from populations of differentiated cells prepared using trypsin.

With respect to the heat stability of chalones, Dr. Brugal pointed out that the opposite situation is found for intestinal chalones in the newt, where the G_1 chalone is heat labile and the G_2 chalone is heat stable. The precise conditions of measurement, and in particular the pH and the duration of exposure to heat, may be of critical importance. Dr. Laurence felt that measurements of heat stability on impure substances may not be very meaningful. Dr. Brugal also emphasized the usefulness of cytophotometric methods for the study of cell cycle-dependent effects.

Dr. Bruce raised the question of the presence of a cell class with a "managerial" role in various tissues such as intestine, skin or bone marrow. As an example, he cited the Sertoli cells of the testis. Dr. Potten did not think there was firm evidence for any cell playing a corresponding role in the intestine or the skin, although the rather mysterious Langerhans cells in the skin might be of interest. Dr. Trentin suggested that in the hemopoietic system there is a microenvironmental influence that is an apparent function of stromal cells of various types. He described the work of Friedenstein on cultures of cells from marrow or spleen. The cultured cells resemble fibroblasts but when transplanted reconstruct a stroma which, for cells of marrow origin, produces bone which harbors hemopoiesis. Cells of spleen origin reconstruct a stroma which resembles spleen and which supports lymphopoiesis. Dr. Trentin also summarized work in his own laboratory designed to obtain evidence for an erythroid hemopoietic inductive microenvironment which induces pluripotent stem cells to become erythropoietin-sensitive. Dr. Cairnie discussed the role of Paneth cells in the intestinal crypts. Crypts are not found without at least one Paneth cell down in the bottom of the crypt. In neonatal mice, no Paneth cells are seen until about the time crypts appear; then, the Paneth cells appear and are found at the bottom of the crypts. In crypt fission after irradiation, Paneth cells are found at the bottom of each limb of the crypt before it completes the fission process. For these reasons Dr. Cairnie felt that Paneth cells might be a possible candidate for managerial cells in the crypts.

Session III

Stem Cells in Haemopoietic and Lymphoid Tissue I

Regulation of Hemopoietic Stem Cells

J. E. TILL

Ontario Cancer Institute
Toronto, Canada

Although the concept of pluripotent stem cells as the source of a variety of cellular subpopulations of the hemopoietic system has been accepted for many years, no reliable basis is yet available for the recognition of such cells by the detection of stem cell-specific markers. It has been necessary to use other criteria as the basis for recognition of hemopoietic stem cells. The major criterion has been a developmental one; a cell able to renew itself and to give rise to large numbers of progeny cells of more than one differentiated type is considered to be a pluripotent stem cell. The spleen colony assay (Till and McCulloch, '61) utilizes this criterion; it detects a class of colony forming units (CFU-S) with the properties of pluripotent stem cells (Siminovitch et al., '63). The colony assays for granulopoietic progenitor cells (CFU-C) (Bradley and Metcalf, '66; Pluznik and Sachs, '65) and for erythropoietic progenitor cells (BFU-E)(Axelrad et al., '74; Iscove and Sieber, '75) also utilize a developmental criterion. Application of this approach has yielded a generally-accepted model of cell lineage relationships in the hemopoietic system (Chervenick et al., '75). This model involves three main levels of differentiation, the pluripotent stem cells, the committed progenitor cells, and their more mature progeny. Each of these major levels can be subdivided into recognizable subclasses, and it is important to bear in mind that the subclasses represent an abstraction useful mainly for purposes of classification; cellular differentiation is likely to be a process of continuous change rather than a series of discrete steps, except perhaps at the level of the genome. Thus, each cellular subclass is likely to be heterogeneous in composition.

Each of these three levels of differentiation is subject to regulation. Of particular interest are those regulatory mechanisms acting at the level of the pluripotent stem cells, since such mechanisms necessarily influence the entire "differon", composed not only of the stem cells themselves, but also all their more differentiated progeny. The regulation of pluripotent stem cells appears to involve short range cell-cell interactions (McCulloch et al., '73; Gidali and Lajtha, '72) between the stem cells and a class of "managerial" cells not as yet identified. The molecular basis for these interactions needs to be clarified; presumably, molecular messages are exchanged between managerial cells and stem cells, and stem cells bear "markers" which permit these messengers to recognize stem cells and interact specifically with them.

The purpose of this paper is to consider some aspects of stem cell regulation, with emphasis on the possible presence of "markers" characteristic of pluripotent stem cells.

REGULATION OF STEM CELL PROLIFERATION

Pluripotent hemopoietic stem cells (CFU-S) in the marrow or spleen of adult mice are normally in a relatively quiescent state; only a minority of cells appear to be proliferating, as measured by the proportion of cells in the DNA synthesis phase of the cell cycle (Becker et al., '65). Initial evidence concerning the presence of markers on stem cells has come from the work of Byron ('75), on mechanisms involved in the enhancement of the rate of proliferation of CFU-S. On the basis of studies carried out *in vitro*, he has identified three mechanisms involved in the initiation of DNA synthesis in quiescent CFU-S. These are β_1 adrenergic receptor stimulation, cholinergic receptor stimulation, and steroid stimulation. The latter two forms of stimulation have also been studied *in vivo*. The *in vitro* assay that was used has limitations because the effects observed are often small in magnitude and careful control of experimental conditions is essential, but the results that have been obtained are sufficiently novel to provoke a re-examination of current concepts concerned with the regulation of hemopoiesis. As mentioned above, these concepts are based on the view that cells, or cellular domains, in the vicinity of CFU-S are responsible for their regulation (McCulloch et al. '73; Curry and Trentin, '67). Although the nature of the cells with regulatory responsibilities is unknown, it has not usually been assumed that they would produce stimulators of adrenergic or cholinergic receptor sites on CFU-S. The apparent absence of β-adrenergic receptors on CFU-C (Moore and Williams, '74) would be compatible with a specific regulatory role of these markers on CFU-S. Whether or not these markers play a functional role on CFU-S, their presence on this class of cells is compatible with the view that they may be useful for distinguishing between CFU-S and their more differentiated progeny.

An immunological approach to the comparison of markers on CFU-S and CFU-C has been reported by van den Engh and Golub ('74). They used an anti-brain antiserum to detect antigenic differences between CFU-S and CFU-C. A major source of concern about these experiments is the possibility that the apparent antigenic differences are a reflection of differences in the assay procedures used rather than a reflection of actual differences in the binding of antibody by the two cell types. Controls were done to show that inefficient lysis *in vitro* was not responsible for the lack of reduction of CFU-C seen after *in vitro* treatment with antiserum plus complement. However, it is still possible that antibody-coated CFU-S may be sequestered in vivo in splenic sites unsuitable for colony formation, whereas antibody-coated CFU-C might still be capable of colony formation *in vitro*. With this reservation, their work provides additional evidence for a marker present on CFU-S but absent on more differentiated CFU-C. This interpretation of the evidence is compatible with the view that CFU-S may carry markers not present on the more differentiated cells derived from them.

REGULATION OF STEM CELL DIFFERENTIATION

CFU-S were shown to be pluripotent with respect to their capacity for differentiation on the basis of the heterogeneous composition of colonies formed from single cytogenetically-marked CFU-S (Wu et al., '67). At present, this approach is the most direct one available for studying the capacity for differentiation of hemopoietic stem cells. It is obviously not entirely satisfactory, since the properties of the parental stem cell must be deduced from those of its (often distant) descendants. For this reason, disagreement persists over the nature of the cell-cell interactions involved in the regulation of the earliest steps in the differentiation of CFU-S. Models have been put forward including a stochastic model (Till et al., '64; Korn et al., '73) based on chance interactions between CFU-S and unspecified regulatory mechanisms that govern the probability of differentiation along a particular pathway, and a "HIM" model (Curry and Trentin, '67; Trentin, '75) based on unique domains or "hemopoietic inductive microenvironments" within hemopoietic tissues, which are the source of inductive interactions between CFU-S and unspecified regulatory mechanisms. A third model is based on the induction of differentiation of CFU-S by direct interaction with known regulatory factors such as erythropoietin (see, for example, Krantz and Jacobson, '70). Only the third of these models involves the assumption that any known markers such as receptors for erythropoietin are present on CFU-S. Because of the present lack of direct methods for constructing, step by step, the family history of a single pluripotent stem cell as a function of time, it seems unlikely that the differences between these models will be resolved by direct experimental analysis of cell lineage relationships. Instead, indirect evidence must be used to provide support for one or other of the models. Recent studies of erythropoietin-responsiveness in culture (Axelrad et al., '74 ; Iscove and Sieber, '75; Gregory, '75) provide an example of such evidence. The erythropoietin dose-response curves obtained for CFU-E, Day 3 BFU-E and Day 8 BFU-E indicate that these classes of cells require different concentrations of erythropoietin for their formation (Gregory, '75). The larger the colony formed (and, presumably, the more primitive the progenitor) the higher the concentration of erythropoietin required for stimulation of colony formation. On the basis of these studies, it has been suggested that very high concentrations of erythropoietin, much higher than those used in the past, may be required to distinguish between an indetectable response of CFU-S to erythropoietin, and an inability to respond (Gregory, '75). If it should prove to be possible to stimulate CFU-S with very high concentrations of erythropoietin, then it would be necessary to obtain evidence that such stimulation involved a specific interaction with erythropoietin receptors and not some non-specific effect of erythropoietin on, for example, the cell membrane. However, this approach does provide a means to test for the presence of known differentiation markers on CFU-S. In this case they are markers (receptors for erythropoietin) of the type predicted by the direct interaction model of stem cell regulation.

OTHER ASPECTS OF STEM CELL REGULATION

Some aspects of stem cell regulation have already been considered above. Four other phenomena associated with stem cell regulation will be discussed briefly in relation to the presence of markers on pluripotent stem cells. These are the decline in the growth potential of serially transplanted stem cells (Siminovitch et al., '64), the mobility of the stem cell pool (Goodman and Hodgson, '62; Micklem, '66), the defective nature of the stem cell function in mice with genetically-determined anemias (McCulloch et al., '64; McCulloch et al., '65) and the repression of growth of parental stem cells in F_1 hybrid hosts (McCulloch and Till, '63). These phenomena are discussed not because they provide strong evidence for markers on pluripotent stem cells, but because they must be taken into account in any consideration of mechanisms of stem cell regulation.

a) Decline in Growth Potential of CFU-S: Transplantation of hemopoietic cells in vivo results in an enhanced rate of proliferation of CFU-S together with a strong differentiation stimulus; for reasons not yet clarified, a protracted exposure of CFU-S to such stimulation as a result of serial transplantation results in a diminution or loss of detectable CFU-S (Siminovitch et al., '64; Lajtha and Schofield, '71; Metcalf and Moore, '71; Pozzi et al., '73). This phenomenon could be analogous to clonal senescence (Hayflick, '65), although no evidence which clearly relates the phenomenon to aging *in vivo* has yet been obtained (Harrison, '72). A possible explanation for this phenomenon is based on heterogeneity within the population of CFU-S. Perhaps stem cells which have been called upon to proliferate rapidly for an extended period in the presence of a strong differentiation stimulus have suffered some diminution of stem cell potential and thus occupy a more differentiated extreme of the pluripotent stem cell class. Differentiation stimuli could have a direct or an indirect effect on stem cell function. An indirect mechanism would involve depletion of the pool of committed progenitors as a result of strong differentiation stimuli, followed by a feedback effect on stem cells via some as yet unspecified mechanism. A direct effect of differentiation stimuli requires the presence of markers in pluripotent stem cells, such as receptors of the type predicted by the direct interaction model.

b) Mobility of the Stem Cell Pool: It is well established that CFU-S circulate in the peripheral blood (Micklem, '66; Micklem et al., '75), although the cells that circulate may not be a random sample of those present in marrow. Evidence has been obtained that blood CFU-S may have a reduced capacity for proliferation and self-renewal, and it has been suggested that the CFU-S normally found in blood have undergone clonal senescence and, for this reason, have been expelled from the marrow (Micklem et al., '75). If this view is correct, then the presence of CFU-S in the blood could reflect a secondary consequence of the same phenomenon seen following serial transplantation, and would have a similar basis.

c) Genetically-determined Defects: Mice with genetically-determined anemia resulting from mutation at the W locus exhibit a defective function of CFU-S (McCulloch et al., '64); those with anemia resulting from mutation at the *Sl* locus have a reduced capacity to support the proliferation of CFU-S (McCulloch et al., '65; Sutherland et al., '70), presumably because of a defect in one or more classes of cells with regulatory functions. A possible explanation for these defects is based on the assumption that CFU-S carry a recognition marker that is utilized in regulatory interactions. It may then be proposed that the W locus affects the ability of CFU-S to express this recognition marker, while the *Sl* locus affects the ability of the regulatory mechanism to detect the recognition marker on CFU-S.

d) Repressed Growth of Parental CFU-S in F_1 Hybrid Hosts: Parental hemopoietic cells transplanted into irradiated F_1 hybrid hosts of suitable genotype show a defective proliferation of CFU-S (McCulloch and Till, '63; McCulloch et al., '73). Extensive studies by Cudkowicz ('75) have provided evidence that the genetic control of the phenomenon involves two classes of genes. One class appears to specify cell surface components phenotypically expressed only on hemopoietic cells; these genes map near or within the end of the major histocompatibility complex. The other set of genes appears to have a regulatory function; these genes determine whether or not the hybrid hosts will recognize and repress the growth of the transplanted cells. The latter genes appear to segregate independently of the major histocompatibility region.

It has been suggested that the repression phenomenon reflects an involvement of gene products of the major histocompatibility region in the normal physiological regulation of hemopoietic stem cells (Till and McCulloch,'71; McCulloch et al., '73). This hypothesis, like others concerned with the biological functions of the various gene products of the major histocompatibility complex (see, for example, Doherty and Zinkernagel, '75; Lennox, '75), remains speculative. It does, however, raise the possibility that certain products of the major histocompatibility region might be used for the recognition of hemopoietic stem cells.

Although none of these phenomena provide direct evidence for markers on pluripotent stem cells, they serve to emphasize the need for markers permitting the recognition of stem cells in order for selective regulation to occur at the stem cell level.

MULTIPLY-MARKED STEM CELLS

Because of the very limited evidence for stem cell-specific markers, it has generally been assumed that stem cells are "null" cells in the sense used by immunologists to define "null" lymphocytes (see, for example, Davis, '75). This concept of null stem cells has tended to dominate thinking in the field, and alternatives are seldom discussed. One alternative view is that <u>a characteristic of stem cells may be the presence of a multiplicity rather than a paucity of markers</u>. None of the markers need necessarily be unique, nor need they be expressed at other than low levels. Some could be "differentiation markers" characteristic of later stages of hemopoietic differentiation. From the latter viewpoint, the differentiation of stem cells could involve a progressive restriction in the expression of markers, with a specialization in the expression of the particular markers characteristic of a single pathway of differentiation. Application of this concept to the current model of hemopoietic cell lineage relationships is illustrated diagrammatically in Figure 1. This concept of multiply-marked stem cells allows somewhat different interpretations to be placed on the results of experiments designed to investigate stem cell regulation. One example concerns the three different models of stem cell regulation outlined above. It is possible, on the basis of the concept of multiply-marked stem cells, to reconcile these three models.

Suppose that stem cells possess receptors of the types appropriate for biologically-significant interaction with a variety of regulatory factors, possibly including those factors whose primary function appears to be the regulation of later stages of differentiation; examples of regulators of the latter class are erythropoietin, and stimulators of granulopoiesis detected by their effects on CFU-C. Suppose also that a basal rate of differentiation of the stem cells into committed erythropoietic, granulopoietic and other progenitors, together with self-renewal of stem cells, can occur prior to any significant interaction with later-stage regulatory factors, and that the capacity of cells to

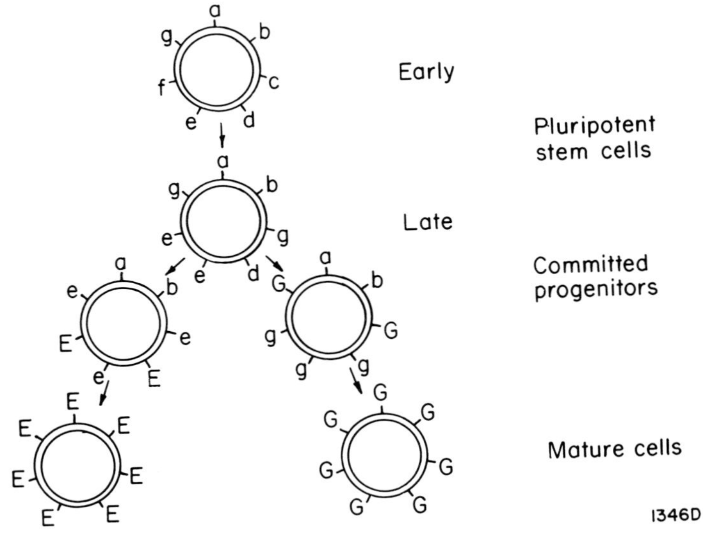

Fig. 1. A diagrammatic illustration of the concept of multiply-marked stem cells, as applied to the three-level model of hemopoietic cell lineage relationships. Lower case letters represent markers present at low levels, or of limited functional capability; upper case letters represent the converse. For simplicity, only cell surface markers are shown, and in equal numbers on all cell types; however, cytoplasmic or nuclear markers, and increases or decreases in numbers of markers, are possible. Only two of the possible pathways of differentiation from pluripotent stem cells are shown.

interact with later-stage factors increases as differentiation proceeds, but becomes restricted to one pathway of differentiation. This means that differentiation of pluripotent stem cells into progenitors committed to erythropoiesis involves a restriction of receptors for granulopoietic stimulators, and an enhanced sensitivity or density of erythropoietin receptors. One could then visualize three different situations:

(i) A very high local concentration of one factor, such

as a stimulator of erythropoiesis, together with a low local concentration of the other factors. This situation could lead to a direct induction of stem cells into the erythropoietic pathway, as expected from the direct interaction model (Krantz and Jacobson, '70).

(ii) A moderate local concentration of one factor, such as an erythropoietic stimulator, together with a low local concentration of the other factors. In this situation, a direct stimulation of stem cell differentiation might be unlikely, because of an insufficient local concentration of erythropoietic stimulator. Instead, the stimulators would act primarily at the level of committed progenitors; once such cells more responsive to stimulation were present, the relatively greater local concentration of erythropoietic stimulator would result in a local focus of erythropoiesis, with little evidence of granulopoiesis or other forms of differentiation, as expected from the HIM model (Trentin, '75).

(iii) Similar local concentrations of different factors, such as an erythropoietic stimulator and a granulopoietic stimulator. This situation would be somewhat analogous to case (ii) in that stimulation might occur primarily at the level of committed erythropoietic and granulopoietic progenitors. However, the equivalent level of stimulation of erythropoiesis and granulopoiesis would result in heterogeneous clones of descendants, containing both erythropoietic and granulopoietic elements, whose composition could fluctuate as a result of random perturbations in the multiplicity of steps involved in the formation of each clone. This is the result expected from the stochastic model (Till et al., '64; Korn et al., '73).

Thus, the concept of multi-marked stem cells is compatible with each of the three major current explanations for the induction of stem cell differentiation.

CONCLUDING REMARKS

Markers on hemopoietic stem cells could be:

(a) Specific for such cells,

(b) Characteristic of relatively undifferentiated cells in general,

(c) Characteristic of one or another pathway of hemopoietic differentiation,

(d) Characteristic of regulatory mechanisms shared with other cell types,

(e) Nonspecific, but present in varying amounts in different cell types,

(f) Completely nonspecific.

Any of these classes of markers except type (f) could be useful for the detection and characterization of stem cells. The example discussed in the previous section was based on the assumption that pluripotent stem cells may carry differentiation markers of type (c). The work of Byron('75) and van den Engh and Golub ('74) indicates the presence of other types of markers, perhaps of type (d). Whatever the nature of the markers, it should be possible to detect their presence on stem cells even at low levels. Thus, the concept of multiply-marked stem cells should be testable by experiment, and it also makes the prediction that enrichment for pluripotent stem cells in the presence of their more differentiated descendants might be possible by means of sequential selection for more than one differentiation marker, even when each of these markers is characteristic of one of the differentiated descendants, and none are unique for pluripotent stem cells. This approach to the isolation of pluripotent stem cells, if successful, should yield populations of significantly greater purity than those that can be obtained by present methods based on physical properties of cells such as size, density and surface charge (Shortman, '72).

In summary, the concept of multiply-marked stem cells is compatible with current information on the regulation of pluripotent hemopoietic stem cells, and is testable by experiment in that it predicts a means for the identification and selection of such cells.

LITERATURE CITED

Axelrad, A. A., D. L. McLeod, M. M. Shreeve and D. S. Heath. 1974. Properties of cells that produce erythrocytic colonies *in vitro*. In: *Hemopoiesis in Culture*. W. A. Robinson, ed. U.S. Government Printing Office, Washington, pp. 226-237.

Becker, A. J., E. A. McCulloch, L. Siminovitch and J. E. Till 1965. The effect of differing demands for blood cell production on DNA synthesis by hemopoietic colony-forming cells of mice. *Blood 26:* 296-308.

Bradley, T. R. and D. Metcalf. 1966. The growth of mouse bone marrow cells *in vitro*. *Aust. J. Exp. Biol. Med. Sci. 44:* 287-299.

Byron, J. W. 1975. Manipulation of the cell cycle of the hemopoietic stem cell. *Exp. Hemat. 3:* 44-53.

Chervenick, P. A., P. Ernst, L. G. Lajtha, D. Metcalf, M. A. S. Moore, J. E. Till and J. J. Trentin. 1975. Hematopoiesis (working paper). In: *Advances in the Biosciences* Vol. 16. R. Burkhardt, C. L. Conley, K. Lennert, S. S. Adler, T. Pincus and J. E. Till, eds. Pergamon Press, Oxford, pp. 255-271.

Cudkowicz, G. 1975. Genetic control of resistance to allogeneic and xenogeneic bone-marrow grafts in mice. *Transplant. Proc. 7:* 155-159.

Curry, J. L. and J. J. Trentin. 1967. Hemopoietic spleen colony studies. I. Growth and differentiation. *Devel. Biol. 15:* 395-413.

Davis, S. 1975. Hypothesis: differentiation of the human lymphoid system based on cell surface markers. *Blood 45:* 871-880.

Doherty, P. C. and R. M. Zinkernagel. 1975. A biological role for the major histocompatibility antigens. *Lancet 1:* 1406-1409.

Gidali, J. and L. G. Lajtha. 1972. Regulation of haemopoietic stem cell turnover in partially irradiated mice. *Cell Tissue Kinet. 5:* 147-157.

Goodman, J. W. and G. S. Hodgson. 1962. Evidence for stem cells in the peripheral blood of mice. *Blood 19:* 702-714.

Gregory, C. J. 1975. Erythropoietin sensitivity as a differentiation marker in the hemopoietic system: studies of three erythropoietic colony responses in culture. In preparation.

Harrison, D. E. 1972. Normal function of transplanted mouse erythrocyte precursors for 21 months beyond donor life spans. *Nature New Biol. 237:* 220-222.

Hayflick, L. 1965. The limited *in vitro* lifetime of human diploid cell strains. *Exp. Cell Res. 37:* 614-636.

Iscove, N. N. and F. Sieber. 1975. Erythroid progenitors in mouse bone marrow detected by macroscopic colony formation in culture. *Exp. Hemat. 3:* 32-43.

Korn, A. P., R. M. Henkelman, F. P. Ottensmeyer and J. E. Till. 1973. Investigations of a stochastic model of haemopoiesis. *Exp. Hemat. 1:* 362-375.

Krantz, S. B. and L. O. Jacobson. 1970. Models for the control of hematopoietic stem cell differentiation. In: *Erythropoietin and the Regulation of Erythropoiesis.* University of Chicago Press, Chicago, pp. 144-148.

Lajtha, L. G. and R. Schofield. 1971. Regulation of stem cell renewal and differentiation: possible significance in aging. In: *Advances in Gerontological Research,* vol. 3. B. L. Strehler, ed. Academic Press, New York, pp. 131-146.

Lennox, E. 1975. Viruses and histocompatibility antigens: an unexpected interaction. *Nature 256:* 7-8.

McCulloch, E. A., C. J. Gregory and J. E. Till. 1973. Cellular communication early in haemopoietic differentiation. In: *Haemopoietic Stem Cells,* Ciba Foundation Symposium, vol. 13. ASP, Amsterdam, pp. 183-199.

McCulloch, E. A., L. Siminovitch and J. E. Till. 1964. Spleen-colony formation in anemic mice of genotype WW^v. *Science 144:* 844-846.

McCulloch, E. A., L. Siminovitch, J. E. Till, E. S. Russell and S. E. Bernstein. 1965. The cellular basis of the genetically determined hemopoietic defect in anemic mice of genotype Sl/Sl^d. *Blood 26:* 399-410.

McCulloch, E. A. and J. E. Till. 1963. Repression of colony -forming ability of C57BL hematopoietic cells transplanted into non-isologous hosts. *J. Cell. Comp. Physiol. 61:* 301-308.

Metcalf, D. and M.A.S. Moore. 1971. Haemopoietic Cells. North-Holland, Amsterdam.

Micklem, H. S. 1966. Effect of phytohemagglutinin-M (PHA) on the spleen-colony-forming capacity of mouse lymph node and blood cells. *Transplantation 4:* 732-741.

Micklem, H. S., N. Anderson and E. Ross. 1975. Limited potential of circulating haemopoietic stem cells. *Nature 256:* 41-43.

Moore, M.A.S. and N. Williams. 1974. Functional, morphologic, and kinetic analysis of the granulocyte-macrophage progenitor cell. In: *Hemopoiesis in Culture.* W. A. Robinson, ed. U.S. Government Printing Office, Washington, pp. 17-29.

Pluznik, D. H. and L. Sachs. 1965. The cloning of normal "mast" cells in tissue culture. *J. Cell. Comp. Physiol. 66:* 319-324.

Pozzi, L. V., U. Andreozzi and G. Silini. 1973. Serial transplantation of bone-marrow cells in irradiated isogeneic mice. *Current Topics in Radiation Research Quarterly 8:* 259-302.

Shortman, K. 1972. Physical procedures for the separation of animal cells. *Ann. Rev. Biophys. Bioeng. 1:* 93-130.

Siminovitch, L., E. A. McCulloch and J. E. Till. 1963. The distribution of colony-forming cells among spleen colonies. *J. Cell. Comp. Physiol. 62:* 327-336.

Siminovitch, L., J. E. Till and E. A. McCulloch. 1964. Decline in colony-forming ability of marrow cells subjected to serial transplantation into irradiated mice. *J. Cell. Comp. Physiol. 64:* 23-31.

Sutherland, D.J.A., J. E. Till and E. A. McCulloch. 1970. A kinetic study of the genetic control of hemopoietic progenitor cells assayed in culture and *in vivo*. *J. Cell. Physiol. 75:* 267-274.

Till, J. E. and E. A. McCulloch. 1961. A direct measurement of the radiation sensitivity of normal mouse bone marrow cells. *Radiat. Res. 14:* 213-222.

Till, J. E. and E. A. McCulloch. 1971. Initial stages of cellular differentiation in the blood-forming system of the mouse. In: *Developmental Aspects of the Cell Cycle*. I. L. Cameron, G. M. Padilla and A. M. Zimmerman, eds. Academic Press, New York, pp. 297-313.

Till, J. E., E. A. McCulloch and L. Siminovitch. 1964. A stochastic model of stem cell proliferation, based on the growth of spleen colony-forming cells. *Proc. Nat. Acad. Sci. U.S.A. 51:* 29-36.

Trentin, J. J. 1975. Hemopoietic inductive microenvironments (HIM): implications for myelofibrosis. In: *Advances in Biosciences,* vol. 16. R. Burkhardt, C. L. Conley, K. Lennert, S. S. Adler, T. Pincus and J. E. Till, eds. Pergamon Press, Oxford, pp. 77-86.

van den Engh, G. J. and E. S. Golub. 1974. Antigenic differences between hemopoietic stem cells and myeloid progenitors. *J. Exp. Med. 139:* 1621-1627.

Wu, A. M., J. E. Till, L. Siminovitch and E. A. McCulloch. 1967. A cytological study of the capacity for differentiation of normal hemopoietic colony-forming cells. *J. Cell Physiol. 69:* 177-184.

Cell Cycle Characteristics of Haemopoietic Stem Cells

N. M. BLACKETT

Division of Biophysics, Institute of Cancer Research, Sutton, Surrey, England

The existence of 'resting' haemopoietic stem cells has been inferred largely from the many reports that in normal mice there is only a small reduction in stem cell survival when tritiated thymidine is incorporated into cells synthesising DNA in amounts sufficient to kill these cells by the ensuing self-irradiation ('thymidine suicide'). The large capacity for stem cell regeneration as shown by the rapid increase in cell numbers in developing spleen colonies is a further indication that in normal animals the rate of production of stem cells is low. In several strains of mice the 'thymidine suicide' technique has been reported to give a cell kill for normal animals as low as 5% although other strains give values up to about 20%. If one uses the 5% value and one assumes that the duration of DNA synthesis is unlikely to be less than 5 h one obtains a minimum average intermitotic time of 100 h. Since it would seem unlikely that cells can have a cell cycle where $t_{G1} + t_{G2} = 95$ h, it seems reasonable to assume that the cell cycle time is much shorter, say about 15 h, so that 85% of the cells can be considered as 'resting' cells, or 'out of cycle' cells, generally referred to as cells in a G_o phase.

One may consider the G_o phase to be outside the cell cycle so that cells after mitosis enter G_1 (Fig. 1). This assumption, however, causes certain problems since it implies that only the small proportion of stem cells which is cycling can differentiate. If the G_o cells could differentiate they would be depleted by cell differentiation without replacement. Furthermore, changes in cell production would have to occur as a result of 'recruitment' of G_o cells, since the estimated cell cycle time is already short, and recruitment would have to be followed by replacement of G_o cells from the cycling cells. This would result in G_o cells being in effect part of the cell cycle.

Alternatively, G_o cells can be assumed to be part of the cell cycle under all conditions and situated between mitosis and the start of G_1, although other positions might be assumed (Fig. 1). The G_o phase must differ in some way from the

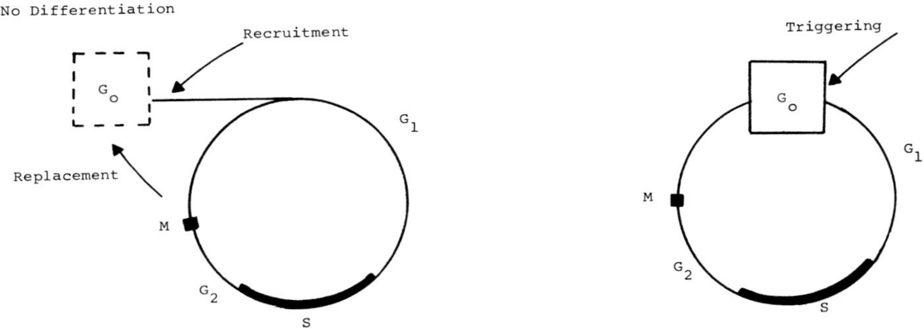

Fig. 1. Diagrammatic representation of possible relationship between 'resting' G_o cells and cells preparing for DNA synthesis and mitosis.

G_1 phase and the usual assumption is that cells can only leave G_o if they receive a specific stimulus with the probability of cells responding to this stimulus being independent of the time that the cells have been in G_o. This implies the existence of 'resting' cells and contrasts with G_1 cells for which there is a continuous process which allows cells to progress steadily towards DNA synthesis. This model has the advantage of allowing all cells to be available for differentiation, and changes in cell production can be met by altering the probability of cells leaving G_o in a given time. Furthermore, it avoids the need for replacement (the control of post mitotic cells into G_o).

If the every-day maintenance of haemopoiesis were provided by changes in the kinetics of maturing haemopoietic cells so that stem cells were only involved in 'panic' situations, it might be reasonable to assume that G_o cells could be outside the cell cycle. However, there is evidence that this is not the case since the proliferation rate of haemopoietic stem cells can be altered by stimuli such as neutropenia and anaemia. Consequently, it would seem likely that stem cells respond continuously to changes in the production of mature cells and that G_0 cells should be considered as part of the cell cycle. The difference between a G_0 phase and a G_1 phase is, however, quite small because in terms of cell kinetics they differ only in the shape of the distributions for the time spent in each phase and Burns and Tannock ('70) demonstrated the difficulty of distinguishing experimentally between these two phases. Biologically, however, the existence of a G_0 phase could be important since it implies a control by cells being 'triggered' into G_1 while without a G_0 phase cell proliferation would be controlled by the rate at which they pass through G_1. Moreover one might expect the action of cytotoxic agents to differ for cells in G_0 compared to cells in G_1.

Although many thymidine suicide studies have been made on stem cells in mice under various physiological conditions these have been primarily aimed at detecting changes in the proliferation rate of stem cells. Very little work has been reported with the specific purpose of investigating the cell cycle characteristics. What is required is an estimate of the rate of progression of cells around the cell cycle. This can in principle be achieved by determining the rate at which cells enter DNA synthesis after cells already in this phase have been killed. Multiple administrations of tritiated thymidine or a phase specific cytotoxic agent provide a means for obtaining such information.

Vassort, Winterholer, Frindel and Tubiana ('73) have carried out experiments of this sort. They injected mice with either hydroxyurea or tritiated thymidine, then at various times afterwards measured the thymidine suicide by incubating the cells *in vitro* with tritiated thymidine, and then assayed for the number of spleen colony forming cells. Two strains of mice were used, C3H mice which normally have a high thymidine suicide (~20%), and C57BL mice with a low suicide (~5%). The results obtained for C57BL mice which show a low initial kill with tritiated thymidine are shown in Fig. 2. No further kill was obtained until about 6 h but then there was a rapid rise to 40% at 12 h which was interpreted in terms of a rapid recruitment of cells into the cell cycle. The response following injection of 400 mg/Kg hydroxyurea was essentially the same as for 1 mCi (~ 0.05 μg) tritiated thymidine. For C3H mice the initial kill was 20% (Fig. 3), decreasing at 1 h but back to normal at 4 h, and then showing peaks at 12 h and

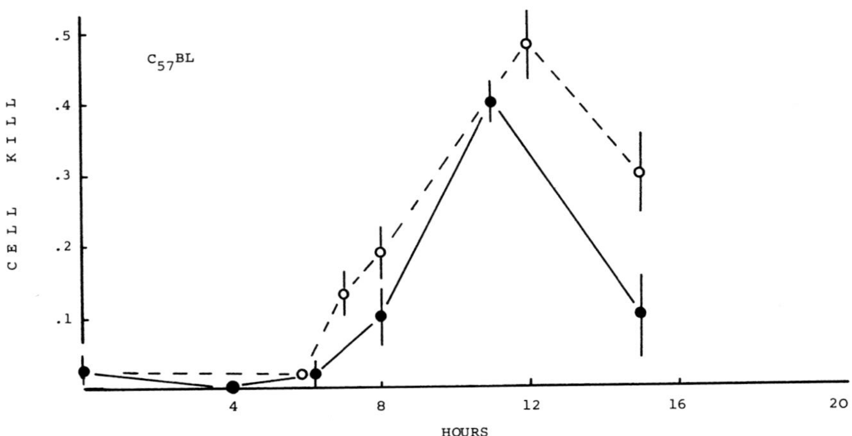

Fig. 2. Tritiated thymidine suicide of CFU-S measured by incubation in vitro at different times after injection of either 1 mCi tritiated thymidine (●) or, 400 mg/Kg hydroxyurea (0) in C57BL mice (Vassort et al., '73).

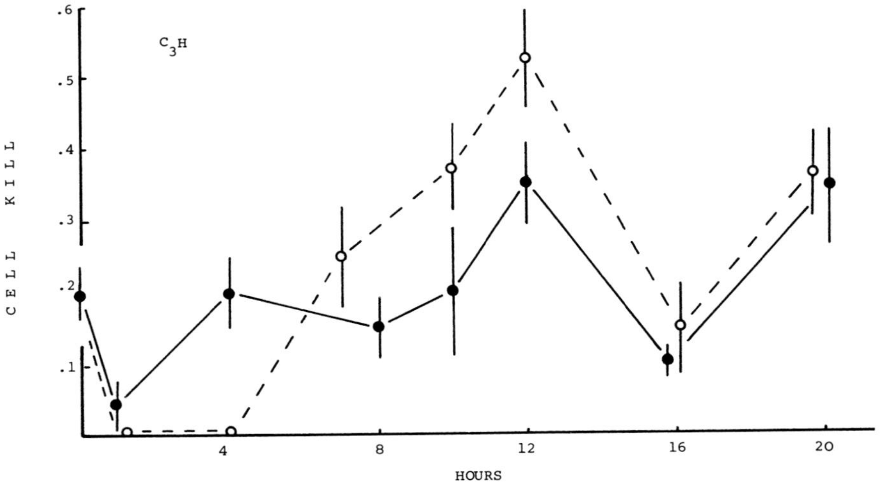

Fig. 3. Tritiated thymidine suicide of CFU-S measured by incubation in vitro at different times after injection of either 1 mCi tritiated thymidine (●) or, 400 mg/Kg hydroxyurea (0) in C3H mice (Vassort et al., '73).

20 h. This was interpreted as indicating a cell cycle time of 8 h and a duration of DNA synthesis of 4 h. Thus for an overall labelling index of 20% one would estimate that about 60% of the stem cells were in G_O. The results for hydroxyurea suggest that there is a hold up in the progression of cells around the cell cycle since the increase in thymidine suicide occurred later and reached a higher peak value at 12 hours. The greater than normal suicide at 12 h was also interpreted as evidence for recruitment.

We have carried out similar studies, but using two doses of the phase-specific agent cytosine arabinoside (Fig. 4) (Blackett, Millard and Belcher, '74). The initial dose gave a 26% kill which is consistent with our thymidine suicide studies. The second dose had no further effect until 12 h and

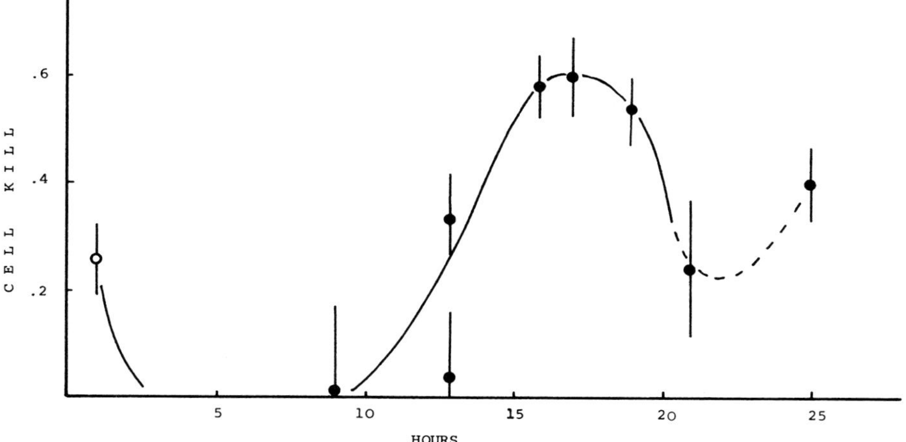

Fig. 4. Effect of a second dose of cytosine arabinoside 200 mg/Kg (●) on CFU-S at different times after a first dose of the same amount (O) for C57BL mice.

showed a peak at 16 h which was greater and also later than the peak observed by Vassort for C3H mice. No evidence for more than one peak was observed. Presenting the results in this way does not give the whole picture, however, because it ignores the changes in stem cell level after the initial dose. Figure 5 shows our results as the actual survival measured and includes results for CFU-C as well as CFU-S, there being no apparent difference between the results for the two assays. During the time that the second injection had no effect there was a decrease in the number of stem cells which can be explained by continued differentiation in the absence of cell

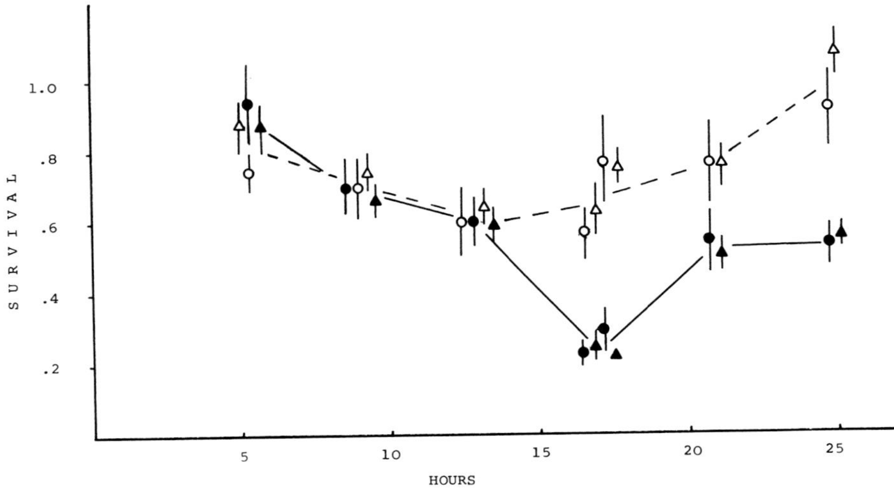

Fig. 5. The survival of CFU-S and CFU-C at different times after a single dose of cytosine arabinoside (200 mg/Kg) (0 CFU-S; Δ CFU-C) and after a second similar dose administered 2 h before assaying for colony forming cells (● CFU-S; ▲ CFU-C).

division due to the arrest of cells in the cell cycle. The beginning of CFU-S recovery after the single dose at about the time that a second injection was found to have an effect supports this suggestion as it indicates a resumption of cell proliferation. The height of the peak at 16 h can be explained to a large extent by the release of cells held up in the cell cycle. Twelve hours after injection of cytosine arabinoside about 60% of the original population survived, 20% killed by the cytotoxic agent and the remainder lost by differentiation. Assuming a mean intermitotic time of 40 h there will be about 30% of the original number of cells held up near the G_1/S boundary and when these are released a kill of about 50% (30%/60%) will be expected. If one postulates that the cells held up in the cell cycle continue to differentiate then the proportion of cells killed will be somewhat smaller. So long as one avoids assuming an extremely long mean intermitotic time it is possible to explain the results without invoking massive recruitment of cells into the cell cycle.

A similar argument can explain the lower but earlier peak value obtained by Vassort *et al.* for C3H mice. The results for C57BL mice obtained by Vassort cannot be explained in this way since it is unlikely that either the small amount of thymidine injected or the radiation to cells that do not incorporate tritiated thymidine is sufficient to arrest the cycling cells. Furthermore, the mean intermitotic time would be too long for significant accumulation of cells over a period of 10 h. Consequently the similarity between the effects of tritiated thymidine and hydroxyurea suggests that there has been no hold up of the cells by the hydroxyurea. Since the stem cell kill is not sufficient to induce a marked increase in stem cell proliferation, Vassort *et al.* have suggested that the increased proliferation is the result of a feedback control from the maturing cells which have a much higher labelling index and therefore are likely to be appreciably depleted. However this presupposes that such a feedback control exists, which is very uncertain, since there appears to be no experimental evidence for increased stem cell proliferation following depletion of maturing cells where there has been no depletion of the stem cells themselves. Furthermore the maturing cells are killed by continuous irradiation from the incorporated tritium and it would seem unlikely that cell death, feedback control and entry of cells into S could all occur within a period of 12 hours.

In conclusion it would seem that the evidence for the existence of haemopoietic stem cells in a 'resting' or G_0 phase rests largely on the low thymidine suicide measurements obtained for some strains of mice. For other strains the stem cell proliferation level is higher. It should be recognised, however, that accurate measurements of low values of thymidine suicide, < 10% (survival > 90%) require large numbers of recipient mice for the spleen colony assay due to the statistical variation of colony numbers (Blackett, '74). Consequently it is difficult to establish that doses on the plateau of the dose response curve are being used. Furthermore, for the low thymidine suicide in C57BL mice studied by Vassort which appear to have a large G_0 stem cell population it is difficult to explain the apparent massive recruitment from only 10% of the cells in the cell cycle to 90% within only a few hours of administering a tracer dose of thymidine. It is difficult to find alternative explanations for this observation but the possibility of a marked diurnal variation in suicide should perhaps be considered.

The C57BL data of Vassort *et al.* do not give an estimate of the mean intermitotic time, but their C3H data and the cytosine arabinoside results for our C57BL mice, both of which have a high thymidine suicide, are consistent with a mean intermitotic time of about 40 h and a duration of DNA synthesis of about 8 h.

LITERATURE CITED

Blackett, N. M. 1974. Statistical accuracy to be expected from cell colony assays; with special reference to the spleen colony assay. *Cell and Tissue Kinetics 7:* 407-412.

Blackett, N. M., Millard, R. E. and Belcher, H. M. 1974. Thymidine suicide '*in vivo*' and '*in vitro*' of spleen colony forming and agar colony forming cells of mouse bone marrow. *Cell and Tissue Kinetics 7:* 309-318.

Burns, F. J. and Tannock, I. F. 1970. On the existence of a G_0 phase in the cell cycle. *Cell and Tissue Kinetics 3:* 321-334.

Vassort, F., Winterholer, M., Frindel, E., Tubiana, M. 1973. Kinetic parameters of bone marrow stem cells using '*in vivo*' suicide by tritiated thymidine and hydroxyurea. *Blood 41:* 789-796.

Stem Cell Reserve and its Control

B. I. LORD

Paterson Laboratories
Christie Hospital and Holt Radium Institute
Manchester, England

It is inconceivable that nature could have foreseen the day when research workers would subject animals to such stresses as, for example, lethal irradiation and then expect the injection of as little as one tenth of one per cent of the animal's normal quota of haemopoietic stem cells to enable that animal to recover. That such a recovery can take place against all evolutionary requirements which, though sometimes severe in themselves, have been minimal by comparison, is ample evidence that the haemopoietic stem cells carry an enormous reserve capacity for both self replication and differentiation.

It is rather more obvious that those cells should be well controlled and under normal conditions all the haemopoietic cell populations are maintained at the same level with respect to each other. Since each cell population is dependent on its precursor cell and the output of mature cells is normally dependent upon the animal's requirements it is therefore clear that some control process must operate at each cell level including that of the stem cell.

It is difficult to confine oneself to a specific pluripotent stem cell since a definitive morphological recognition of them still remains somewhat elusive. Functional tests permit the evaluation of cells fulfilling the role of stem cells but, as will be shown below, the homogeneity of the cell population normally considered to fill that role is now being called into some doubt. Consequently, it is more realistic to consider the stem cell complex as a whole including the committed, though still morphologically unrecognizable precursor cells.

ORGANIZATION OF THE STEM CELL COMPLEX

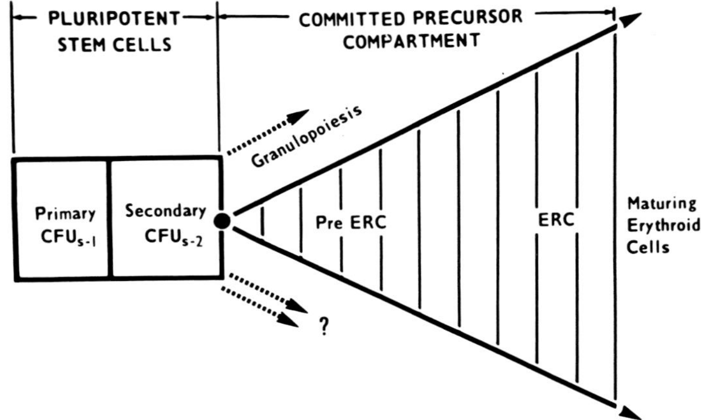

Fig. 1. Cell populations comprising the stem cell complex.

Figure 1 represents this complex (as related to erythropoiesis) showing the pluripotent stem cell compartment divided into two parts, (i) primary stem cells; cells which are fully capable of maintaining their own population and (ii) secondary stem cells; cells which are still pluripotent in that they may give rise to more than one differentiated cell type but cannot maintain their own population. They are however, still recognizable by their ability to produce spleen colonies in irradiated mice (Till and McCulloch, '61). The two classes of spleen colony forming units (CFU) have been designated as CFU_{S-1}', CFU_{S-2} by Schofield and Lajtha ('73) but probably represent a simplification of an age structure within the pluripotent stem cell compartment. As such, it is probable that the committed precursor cells are derived from the CFU_{S-2} but the possibility of random differentiation taking place from both the CFU_{S-1} and CFU_{S-2} must not be overlooked.

In the case of erythropoiesis, the committed precursor compartment (EPC) also is divisible into the pre-erythropoietin responsive cells (preERC) and erythropoietin responsive cells (ERC) from which the recognizable erythroid cells are formed (Reissmann and Samorapoompichit, '70; Lajtha et al., '71). It is known that a committed precursor compartment exists also for granulopoiesis and comparable systems may well apply to other lines of haemopoietic cell development though relatively little is known about them.

PROLIFERATIVE CAPACITY OF THE COMMITTED PRECURSOR CELLS

From data obtained using the techniques available for the assay of spleen colony forming cells (CFC_S): the CFU_S assay (Till and McCulloch, '61), the proportion of CFU_S in DNA synthesis (Becker et al., '65; Lord et al., '74), the fraction of injected CFC_S which seed the spleen and appear there as CFU_S (Siminovitch et al., '63; Lord, '71); a knowledge of the distribution of active bone marrow throughout the skeleton (Schofield and Cole, '63) and an estimate of the nature of the CFU_S cell cycle, Schofield ('74) estimated that the number of CFC_S being produced for differentiation into the committed precursor compartment of a mouse is approximately 2×10^5 per day. Similarly working from the mature cells, he calculated that these cells must give rise to about 10^9 mature cells and this represents an amplification through the committed and recognizable stages of haemopoiesis of about 12 cell cycles. It is known that there are about six divisions in the recognizable erythroid compartment (Tarbutt and Blackett, '68) so that at least six divisions must also arise in the committed compartment.

Under a variety of experimental conditions, it is known that the ERC population is maintained and is proliferating at a fast rate even though the demand for erythropoiesis has been removed. This occurs for example in mice which have been hypertransfused with red blood cells and in post hypoxic conditions. In both cases, the circulating red cell mass is increased, thus eliminating the erythropoietin stimulus required to enable the ERC to move into the recognizable erythroid compartments. The erythroid precursor cell compartment (EPC) is, therefore, not limited to matching exactly the animal's erythroid requirements and from the rapidity with which it can meet demand under conditions such as acute haemorrhage, it is clear that the demand must be met from a reserve of ERC which have been overproduced to meet just such a need. Under normal conditions, therefore, one must expect one or two more divisions to occur in the EPC compartment thus allowing a large ready reserve of cells.

The limit of this reserve, however, is probably not seen even under these conditions. During conditions of extreme demand as for example arises during spleen colony formation, Lajtha has calculated that in order to produce sufficient cells to make a colony, there must be about ten divisions in the EPC compartment. Experimentally, Reissmann and Samorapoompichit ('70) have shown that these extra cell divisions can be induced in the preERC compartment by stimulation with suitably high doses of erythropoietin.

Further evidence of the reserve capacity of the EPC compartment arises out of continuous irradiation studies. Low dose rate chronic irradiation (70 rads/day) leads to over 95% reduction in stem cells in the mouse after about four days and thereafter remains very low (Wu Chu-Tse and Lajtha, '75a). At the same time, after an initial drop in erythropoiesis, haemoglobin levels in the circulating blood (Lamerton et al., '60) iron incorporation and ERC numbers (Porteous and Lajtha, '66; Blackett, '67) return to and remain normal. To maintain this amount of blood cell production and a bone marrow cell population of 20 to 30% of normal (Lord, '64) in spite of stem cell numbers remaining at about 2 to 3% of normal, Wu Chu-Tse and Lajtha ('75) calculated that the amplification required would have to add at least five cell cycles to those already occurring during transit; three or four of these arising in the EPC compartment since some extra divisions can also be seen in the recognizable erythroid compartments (Lord, '65a, Blackett, '68).

The EPC compartment, therefore, exhibits two types of reserve capacity (i) production of ERC in excess of the normal demand requirements under normal conditions and (ii) a considerable capacity to expand the production of ERC under conditions of increased demand.

PROLIFERATIVE CAPACITY OF THE PLURIPOTENT STEM CELLS

Stem cells (CFU_S) clearly have the capacity to produce different committed cell lines as evidenced by the production of erythroid, granulocytic and megakaryocytic colonies in mouse spleens but it is not yet possible to make the same kind of calculations for CFU_S as for the EPC, for example. Consequently one can only draw generalized conclusions concerning the amount of proliferation they may be capable of. Certainly they do have the capacity to increase proliferation because their normal low rate as shown by tritiated thymidine suicide studies (< 10% in DNA synthesis) (Becker et al., '65; Lord et al., '74) can be increased under a number of experimental conditions. Most perturbations calling for an increased rate of cell production lead directly or indirectly to an increase in proliferation rate of the CFU_S. Simple regrowth following irradiation or transplantation of stem cells into irradiated animals manifests increased CFU_S proliferation as does that following treatment with a variety of cytotoxic drugs (Blackett, previous paper). During protracted regimes of continuous irradiation, they have the capacity to maintain an adequate supply of cells to the committed precursor compartments although as pointed out above, their own population is reduced to something less than 5% of normal. In the face of the high differentiation pressures caused by these procedures, the proliferation rate of the stem cells is necessarily very much increased and figures of up to 40% of the cells in DNA synthesis are not uncommon. In addition there is some evidence from studies on the distribution of CFU_S in the marrow spaces that some CFU_S might have a thymidine suicide value of about 75% and it is possible that this may represent the limit of their proliferative capacity (Lord et al., '75).

SELF MAINTENANCE CAPACITY OF THE PLURIPOTENT STEM CELLS

Something of the reserve capacity of CFU_S must also be apparent from assays of their behaviour in young and old mice. It has been shown that the CFU_S from old mice have the same colony forming capacity and growth doubling time in the spleens of irradiated mice as those from young mice (Lajtha and Schofield, '71). Harrison ('72) took the same problem a stage further and transplanted +/+ marrow into genetically anaemic W/W^v mice. This grew and cured the anaemia. At the end of its lifespan, the "cured" marrow from these mice was capable of "curing" the anaemia in further W/W^v recipients. Clearly, therefore, the CFU_S population has the capacity to maintain itself for considerably longer than the normal lifespan of the animal.

The question now arises of whether these very substantial reserves are unlimited. Experimentally, Siminovitch et al. ('64) successively transferred repopulated spleens to further irradiated mice and found a progressive decline in the ability of the CFU_S to repopulate an irradiated mouse. Lajtha and Schofield ('71) and Pozzi et al. ('73) refined and amply confirmed these observations. Similarly, repeated damage to stem cells impaired their ability to recover. Hendry and Lajtha ('72) and Hendry et al. ('74) showed that radiation doses of 450 rads repeated every 24 days led to a progressive decline in the animals' capacity for haemopoietic repopulation. These studies suggest that each time the stem cell population is damaged or stressed, it loses a little of its self maintenance capacity; it becomes an "older" population. Some years ago, Worton et al. ('69) physically separated two classes of CFU_S; those with a high and those with a low self reproducing capacity. More recently, Schofield ('74) has determined the self replicating ability of individual colony forming cells and has shown that about 70% of the CFU_S

(CFU_{S-1}) have a very good self maintenance capacity whilst the remaining 30% (CFU_{S-2}) are relatively poor. In fact, a complete distribution exists ranging from those with a very high to those with a very low self maintenance capacity. This probably represents an age distribution of the CFU_S and only some of them (the younger ones?) appear to have this capacity which may well be lost some time before becoming committed to a specific line of development.

THE SPLEEN AS A RESERVE ORGAN

In most animals, the spleen contributes little to the production of granulocytes and erythrocytes though under severe conditions it does become a haemopoietic organ. For example, in conditions of acute haemorrhage, the increase in splenic erythropoiesis contributes over twice as much to the recovery as does the bone marrow (Lord, '67). Similarly, in the later stages of exposure to continuous irradiation when the bone marrow shows no further capacity to expand its production of mature cells, the spleen then begins to increase its output (Lord, '65b). Fruhman ('66a, '66b) also demonstrated this shunt in erythropoiesis from bone marrow to spleen following stimulation by endotoxin or zymosan.

CONTROL OF THE PRECURSOR CELL POPULATIONS

Control over the size and performance of each of the defined cell populations can clearly take place at each of the compartment boundaries; i.e. the points at which the pluripotent stem cell becomes committed to a specific line of development and a committed cell becomes a morphologically recognizable cell. In addition, a control may act on specific cells within these compartments and may be either a stimulator such as erythropoietin or an inhibitor of the "chalone" type. Little is known about controls over the pluripotent stem cells except what may be gleaned by inference from a number of indirect experiments. With the committed precursors, the situation is rather better, though even here it is by no means fully evaluated.

Most experiments designed for studying control processes have involved disturbing the equilibrium between the sequential cell compartments and observing the nature of the response. The ability of the populations to recover their normal equilibrium depends upon the nature of the perturbation which will dictate whether the precursor cell populations may be allowed to recover whilst only a minimal strain is placed on them for differentiation purposes or whether the rapid production of differentiated functional progeny is an urgent requirement. A major part of the control system therefore must operate over the balance between self maintenance of the precursor cell populations and their differentiation into the production of functional cells. A complete return to normal of the stem cell populations is clearly desirable but a rapid production of mature cells is often essential to ensure survival and as pointed out by Schofield ('74), "The fact that the stem cell population will eventually recover its productive capacity is of no comfort to a patient who had already died from infection."

Whether a CFU_S reproduces itself or differentiates, however, must depend not only on this differentiation pressure but also on the long term requirements of the animal. While it may be of no comfort to know that stem cells would have eventually recovered if the demand for differentiated cells could have been met, neither is there much advantage to producing nothing but differentiated cells. If the differentiation rate becomes greater than 50% then the stem cell population can only run down. Vogel et al. ('68) and Lajtha et al. ('71) both demonstrated that the maximum differentiation rate is about 40%, leaving about 60% self renewal.

CONTROL OF THE COMMITTED PRECURSOR CELL POPULATIONS

It has long been recognized that erythropoietin (EPO) is a stimulator of the committed erythroid precursor cells. The production of erythropoietin is related to oxygen tension which under most conditions is related to the circulating red cell mass. It has generally been accepted that erythropoietin induces the differentiation step, from the committed cell, initiating the haemoglobin synthesis which permits the earliest recognition of erythroid cells (Gurney et al., '62). Recent evidence has shown that this compartment can be subdivided into preERC and ERC and as discussed above, cell production in these compartments can be considerably amplified by increasing the number of cell divisions. Reissmann and Samorapoompichit ('70) maintained the pluripotential CFU_S at a negligible level for extended periods of time and studied the response of the EPC to large doses of EPO. They were thus able to demonstrate a considerable increase in the measurable number of ERC and came to the conclusion that EPO was directly stimulating proliferation and increasing the amplification of the preERC compartment, thus accounting for the extra divisions calculated by Lajtha et al. ('71). Does, however, EPO actually initiate haemoglobin synthesis in ERC? There is no direct evidence that it does and on the evidence that EPO is a preERC proliferation stimulator together with the evidence that it also promotes proliferation in the recognizable erythroid cells, it may be more reasonable to suggest that haemoglobin synthesis is programmed at the point of commitment. ERC then simply need the extra kick that a small dose of EPO can supply to stimulate their proliferation as recognizable erythroid cells. An inhibitor of erythroid cell proliferation has been suggested (Kivilaakso and Rytömaa, '71; Lord et al., '74) but my own studies have demonstrated it has no effect on the ERC compartment. In any event, there seems no place for an inhibitor of EPC proliferation unless EPO acts by effectively neutralizing that inhibitor.

Committed granulocytic and other cell line precursors are less completely resolved. The granulocyte precursor cells, known as CFU_C for their ability to produce granulocytic colonies in agar culture (Pluznik and Sachs, '65; Bradley and Metcalf, '66), require a colony stimulating factor (CSF) to grow but whether this factor can be considered as a physiological control process for granulopoiesis is unknown. Udupa and Reissmann ('75) have demonstrated that increased granulocytopoiesis is not always preceded by increased CSF levels. Furthermore, Milenkovic and Lord (unpublished observations) have noticed that serum from animals with the enhanced granulopoiesis as produced by Udupa and Reissmann is capable of reducing the degree of enhancement when injected into further stimulated animals. There is thus some evidence that the committed granulocyte precursors may be under the control of both stimulating and/or inhibiting factors.

CONTROL OF THE PLURIPOTENT CELL POPULATIONS

CFU_S normally have a very low average proliferation rate, studies with tritiated thymidine indicating that less than 10% are in DNA synthesis at any one time. Following depopulation by irradiation or drug treatment, recovery of the CFU_S population will occur at a rapid rate particularly if the initial depopulation has been large. During recovery, the net CFU_S proliferation rate is increased. When recovery has reached about 20-50% control, however, further recovery takes place only slowly (e.g. Till, '63; Valeriote et al., '68; Chan and Schooley, '70; Guzman and Lajtha, '70; Hendry and Lajtha, '72). The recovery of the EPC populations however has taken place much more rapidly and by the time the second and slower CFU_S recovery phase has started, the ERC populations are at normal levels. Similarly, under conditions of continuous low dose rate irradiation, normal levels of erythropoiesis are maintained in spite of the very low levels of stem cells surviving the continuous irradiation. A dose rate of 70 rads/day does not produce further CFU_S depopulation. At this stage, differentiation from the CFU_S to committed precursor cells exactly balances their proliferation. On stopping the continuous irradiation, the ERC return to normal within 5 days but, although the CFU_S population is still less than 50% of normal at 24 days, its net proliferation rate is low (Wu Chu-Tse and Lajtha, '75b). Thus, differentiation pressure on the CFU_S, effected by long range acting agents such as erythropoietin, appears to be of primary importance in governing the proliferative performance of CFU_S.

The work of Croizat et al. ('70) and Gidali and Lajtha ('72) using partial body irradiation and studying CFU_S depopulation, recovery and proliferation in both the shielded and irradiated parts of the body affords some clues as to the nature of the controlling factors. A study of the CFU_S population in the shielded limb of an otherwise totally irradiated (600 rads) mouse (Gidali and Lajtha, '72) showed that CFU_S numbers initially fell to about 20% of normal, presumably due to their mobilization into the totally irradiated bones but by 120 hours their number was again normal. In parallel with these changes, while CFU_S numbers were low their thymidine killing index went as high as 50%. As CFU_S numbers returned to normal, so did their thymidine index. In the irradiated limb, the CFU_S number remained very low (< 1%) and the kill remained high throughout the period of observation. Although haemopoietic tissue in 90% of the animal remained virtually non existent, that in the shielded limb was apparently perfectly normal both in number and proliferative status. Thus, while long range acting control factors undoubtedly play a part in controlling the CFU_S, it is clear from these experiments that the ultimate control must lie in local factors.

The nature of these local controls is not known. There is some evidence to suggest that cell-cell interaction processes may be involved. For example, Lord and Schofield ('73) demonstrated that live thymus cells may enhance the colony forming ability of sublethally damaged haemopoietic tissue. Frindel and Croizat ('73) produced evidence of a similar nature.

Environmental factors have been invoked as local controls. It has been suggested that the spleen microenvironment may well determine whether a CFU_S will express itself as a colony and into which differentiated cell line it will develop (Wolf and Trentin, '68; Wolf, '74). Recently it has been shown that CFU_S are not randomly distributed within the bone marrow spaces but that there is a higher concentration in the region of the bone surface (Lord and Hendry, '72). Furthermore those CFU_S close to the bone demonstrate a higher proliferation rate than those some distance away (Lord et al., '75). These data fit in well with a wealth of information involving the bone in the marrow regeneration processes so well reviewed recently by Patt and Maloney ('75).

In general, however, it appears that higher proliferation is induced in CFU_S when their population is significantly decreased, (perhaps below some critical level) and, in particular, when the committed precursor population is also decreased. It would appear, therefore, that the trigger might operate by some cell concentration feedback factor. In a pilot experiment carried out in our laboratories, Dr. Mori and I have found that a saline extract from normal bone marrow of molecular weight 5 x 10^4 to 10^5 daltons may be used to protect regenerating, rapidly cycling CFU_S from the killing effects of a large dose of tritiated thymidine when incubated with the cells for 5 hours (Table 1, unpublished observations).

TABLE 1

Percent tritiated thymidine kill of regenerating CFU_S after treatment with bone marrow extract for 5 hours. (Mori and Lord, unpublished observations, '75).

Bone Marrow Extract (Molecular Weight Range)	% CFU_S Killed by ^{3}HTdR
500 - 10000	41
10000 - 30000	29
30000 - 50000	11
50000 - 100000	4
CONTROL	33

This, then, is an effect comparable with the cell cycle modulatory effects produced by mature blood cell extracts on the recognizable precursors of granulocytes and erythrocytes and may well represent some aspect of CFU_S proliferation control.

SUMMARY

The morphologically unrecognizable haemopoietic precursor cells carry an extremely large and normally untapped reserve capacity to counter and recover from severe depletions of their population. This they do by increasing their rate of proliferation and in the case of the erythroid committed precursors, at least, by increasing the number of divisions and, therefore, amplification, during their transit to the recognizable stages. Under extreme conditions the spleen may become a reserve haemopoietic organ.

Variations in the balance between the cell compartments can affect the differentiation pressure on the earlier cells and long range acting factors, such as erythropoietin, can directly, or indirectly by inducing local controls to come into operation, affect the proliferation rate of the more primitive cells. The nature of the local controls is unknown but there is some evidence to indicate that microenvironmental factors, both geographical and local cell concentration, may be operative. Whether these factors are stimulatory or inhibitory is still an open question but there is now some slight evidence that an inhibitory factor may be operative over the proliferation rate of the CFU_S.

ACKNOWLEDGMENTS

The work was supported by grants from the Cancer Research Campaign and the Medical Research Council.

LITERATURE CITED

Becker, A. J., E. A. McCulloch, L. Siminovitch and J. E. Till. 1965. The effects of differing demands for blood cell production on DNA synthesis by haemopoietic colony forming cells of mice. *Blood 26:* 296-308.

Blackett, N. M. 1967. Erythropoiesis in the rat under continuous irradiation at 45 rads/day. *Brit. J. Haemat. 13:* 915-923.

Blackett, N. M. 1968. Changes in proliferation rate and maturation time of erythroid precursors in response to anaemia and ionizing radiation. In: *Effects of Radiation on Cellular Proliferation and Differentiation.* IAEA Symposium, Vienna, pp. 235-246.

Bradley, T. R. and D. Metcalf. 1966. The growth of mouse bone marrow cells *in vitro*. *Aust. J. exp. Biol. med.*

Sci. 44: 287-300.

Chen, M. C. and J. Schooley. 1970. Recovery of proliferative capacity of agar colony-forming cells and spleen colony-forming cells following ionizing radiation or vinblastine. *J. Cell. Physiol. 75:* 89-96.

Croizat, H., E. Frindel and M. Tubiana. 1970. Proliferative activity of the stem cells in the bone marrow of mice after single and multiple irradiations (total or partial-body exposure). *Int. J. Radiat. Biol. 18:* 347-358.

Frindel, E. and H. Croizat. 1973. The possible role of the thymus on CFU proliferation and differentiation. *Biomedicine Express 19:* 392-396.

Fruhman, G. J. 1966a. Shunting of erythropoiesis in mice following the injection of zymosan. *Life Sciences 5:* 1549-1556.

Fruhman, G. J. 1966b. Bacterial endotoxin: effects on erythropoiesis. *Blood 27:* 363-370.

Gidali, J. and L. G. Lajtha. 1972. Regulation of haemopoietic stem cell turnover in partially irradiated mice. *Cell Tissue Kinet. 5:* 147-157.

Gurney, C. W., L. G. Lajtha and R. Oliver. 1962. A method for investigating stem cell kinetics. *Brit. J. Haemat. 8:* 461-466.

Guzman, E. and L. G. Lajtha. 1970. Some comparisons of the kinetic properties of femoral and splenic haemopoietic stem cells. *Cell Tissue Kinet. 3:* 91-98.

Harrison, D. E. 1972. Normal function of transplanted mouse erythrocyte precursors for 21 months beyond donor life-span. *Nature New Biology 237:* 220-222.

Hendry, J. H. and L. G. Lajtha. 1972. The response of hemopoietic colony-forming units to repeated doses of x-rays. *Radiat. Res. 52:* 309-315.

Hendry, J. H., N. G. Testa and L. G. Lajtha. 1974. Effect of repeated doses of X-rays or 14 MeV neutrons on mouse bone marrow. *Radiat. Res. 59:* 645-652.

Kivilaakso, E. and T. Rytomaa. 1971. Erythrocyte chalone, a tissue specific inhibitor of cell proliferation in the erythron. *Cell Tissue Kinet. 4:* 1-9.

Lamerton, L. F., A. H. Pontifex, N. M. Blackett and K. Adams. 1960. Effects of protracted irradiation on the blood forming organs of the rat. Part I. Continuous exposure. *Brit. J. Radiol. 33:* 287-301.

Lajtha, L. G., C. W. Gilbert and E. Guzman. 1971. Kinetics of haemopoietic colony growth. *Brit. J. Haemat. 20:* 343-354.

Lajtha, L. G. and R. Schofield. 1971. Regulation of stem cell renewal and differentiation: Possible significance in ageing. In: *Advances in Gerontological Research.* B. L. Strehler, ed. Academic Press, New York and London, vol. 3, pp. 131-146.

Lord, B. I. 1964. The effects of continuous irradiation on cell proliferation in rat bone marrow. *Brit. J. Haemat. 10:* 496-507.

Lord, B. I. 1965a. Cellular proliferation in normal and continuously irradiated rat bone marrow studied by repeated labelling with tritiated thymidine. *Brit. J. Haemat. 11:* 130-143.

Lord, B. I. 1965b. Haemopoietic changes in the rat during growth and during continuous γ-irradiation of the adult animal. *Brit. J. Haemat. 11:* 525-536.

Lord, B. I. 1967. Erythropoietic cell proliferation during recovery from acute haemorrhage. *Brit. J. Haemat. 13:* 160-167.

Lord, B. I. 1971. The relationship between spleen colony production and spleen cellularity. *Cell Tissue Kinet. 4:* 211-216.

Lord, B. I., L. Cercek, B. Cercek, G. P. Sah, T. M. Dexter and L. G. Lajtha. 1974. Inhibitors of haemopoietic cell proliferation?: Specificity of action within the haemopoietic system. *Brit. J. Cancer 29:* 168-175.

Lord, B. I. and J. H. Hendry. 1972. The distribution of haemopoietic colony-forming units in the mouse femur and its modification by X-rays. *Brit. J. Radiol. 45:* 110-115.

Lord, B. I., L. G. Lajtha and J. Gidali. 1974. Measurement of the kinetic status of bone marrow precursor cells: Three cautionary tales. *Cell Tissue Kinet. 7:* 507-515.

Lord, B. I. and R. Schofield. 1973. The influence of thymus cells in haemopoiesis: Stimulation of haemopoietic stem cells in a syngeneic, *in vivo,* situation. *Blood 42:* 395-404.

Lord, B. I., N. G. Testa and J. H. Hendry. 1975. The relative spatial distributions of CFU_S and CFU_C in the normal mouse femur. *Blood 46:* 65-72.

Mori, J. and B. I. Lord. 1975. Unpublished observations.

Patt, H. M. and M. A. Maloney. 1975. Bone marrow regeneration after local injury: A review. *Exp. Hemat. 3:* 135-148.

Pluznik, D. H. and L. Sachs. 1965. The cloning of normal "mast" cells in tissue culture. *J. cell. comp. Physiol. 66:* 319-324.

Porteous, D. D. and L. G. Lajtha. 1966. On stem cell recovery after irradiation. *Brit. J. Haemat. 12:* 177-188.

Pozzi, L. V., U. Andreozzi and G. Silini. 1973. Serial transplantation of bone-marrow cells in irradiated isogeneic mice. In: *Current Topics in Radiation Research.* E. Ebert and A. Howard, eds. Elsevier North-Holland, Amsterdam, vol. 8, pp. 259-302.

Reissmann, K. R. and S. Samorapoompichet. 1970. Effect of erythropoietin on proliferation of erythroid stem cells in the absence of transplantable colony forming units. *Blood 36:* 287-296.

Siminovitch, L., E. A. McCulloch and J. E. Till. 1963. The distribution of colony-forming cells among spleen colonies. *J. cell. comp. Physiol. 62:* 327-336.

Siminovitch, L., J. E. Till and E. A. McCulloch. 1964. Decline in the colony forming ability of marrow cells subjected to serial transplantation into irradiated mice. *J. cell. comp. Physiol. 64:* 23-32.

Schofield, R. 1975. Haemopoietic cell kinetics. In: *Proceedings XI International Cancer Congress,* Florence, 1974. Elsevier/Excerpta Medica/North-Holland, Amsterdam, vol. 1, Cell Biology and Tumour Immunology, pp. 18-33.

Schofield, R. and L. J. Cole. 1968. An erythrocyte defect in splenectomized X-irradiated mice restored with spleen colony cells. *Brit. J. Haemat. 14:* 131-140.

Schofield, R. and L. G. Lajtha. 1973. Effect of isopropyl methane sulphonate (IMS) on haemopoietic colony-forming cells. *Brit. J. Haemat. 25:* 195-202.

Tarbutt, R. G. and N. M. Blackett. 1968. Cell population kinetics of the recognizable erythroid cells in the rat. *Cell Tissue Kinet. 1:* 65-80.

Till, J. E. 1963. Quantitative aspects of radiation lethality at the cellular level. *Am. J. Roent. Rad. Therapy 90:* 917-921.

Till, J. E. and E. A. McCulloch. 1961. A direct measurement of the radiation sensitivity of normal mouse bone marrow cells. *Radiat. Res. 14:* 213-222.

Udupa, K. B. and K. R. Reissmann. 1975. Stimulation of granulopoiesis by androgens without concomitant increase in the serum level of colony stimulating factor. *Exp. Hemat. 3:* 26-43.

Valeriote, F. A., D. C. Collins and W. R. Bruce. 1968. Hematological recovery in the mouse following single doses of gamma radiation and cyclophosphamide. *Radiat. Res. 33:* 501-510.

Vogel, H., H. Niewisch and G. Matioli. 1968. The self renewal probability of hemopoietic stem cells. *J. Cell Physiol. 72:* 221-228.

Wolf, N. S. 1974. Dissecting the hematopoietic microenvironment. I. Stem cell lodgement and commitment, and the proliferation and differentiation of erythropoietic descendants in the Sl/Sl^d mouse. *Cell Tissue Kinet. 7:* 89-98.

Wolf, N. S. and J. J. Trentin. 1968. Hemopoietic colony studies. V. Effect of hemopoietic organ stroma on differentiation of pluripotent stem cells. *J. exp. Med. 127:* 205-214.

Worton, R. G., E. A. McCulloch and J. E. Till. 1969. Physical separation of hemopoietic stem cells from cells forming colonies in culture. *J. Cell Physiol. 74:* 171-182.

Wu Chu-Tse, and L. G. Lajtha. 1975a. Haemopoietic stem-cell kinetics during continuous irradiation. *Int. J. Radiat. Biol. 27:* 41-50.

Wu Chu-Tse, and L. G. Lajtha. 1975b. Recovery of haemopoietic stem cell and precursor cell populations. In: *Paterson Laboratories' Annual Report, 1974-1975.* Christie Hospital and Holt Radium Insitute, Manchester, p. 147.

The Granulocytic and Monocytic Stem Cell

M.A.S. MOORE, J. KURLAND AND H.E. BROXMEYER

Sloan-Kettering Institute for Cancer Research, New York, New York 10021, U.S.A.

The development of a semi-solid agar culture system which supports the clonal proliferation of murine and human granulocytes and macrophages (Bradley and Metcalf, '66; Pike and Robinson, '70) resulted in extensive characterization of a class of morphologically undifferentiated stem cells committed to granulocytic and monocytic differentiation. The colony forming cells (CFU-c) have been distinguished from their immediate precursors, the multipotential stem cells (CFU-s) by physical separation, antigenic differences and differential kinetics following perturbation of hemopoiesis (Moore, '74).

In the majority of species, physical separation procedures have revealed an heterogeneous size and density distribution of CFU-c; however, in Rhesus monkey bone marrow, CFU-c were distributed in a relatively narrow density region of lighter density than the majority of nucleated cells. By the use of two stage density separation Moore et al ('72) obtained populations containing over 50% of the total CFU-c present in unfractionated marrow where one cell in three was capable of colony or cluster formation. Morphological and autoradiographic studies of such enriched cell fractions revealed that the majority of CFU-c were mononuclear cells of diameter 9-11μ possessing irregular leptochromatic nuclei and with approximately 40% of the population in DNA synthesis. The CFU-c were morphologically distinct from marrow small lymphocytes and myeloblasts and resembled transitional "lymphocytes". Proof of the clonal origin of *in vitro* colonies, particularly those of mixed granulocyte-macrophage type was obtained by single cell micromanipulation using marrow fractions highly enriched

---

This work was supported in part by grants CA-17353, NCI-08748 and the Gar Reichman Foundation.

for CFU-c by density separation (Moore et al., '72).

Essential to the cloning technique is the presence of regulatory macromolecules which have been given the operational title "colony stimulating factor" (CSF) (Bradley and Metcalf, '66). A sigmoid dose-response relationship exists between CSF concentration and the number of colonies developing in marrow culture and this provides a sensitive bioassay system capable of detecting as little as 10^{-11} - 10^{-12}M CSF/ml (Stanley and Metcalf, '72). The action of CSF *in vitro* is complex and it cannot be considered simply as an inducer of progenitor cell differentiation. The factor is required not only to initiate colony formation but is necessary for every cell division and differentiation step in the granulocyte-macrophage pathway *in vitro* (Metcalf and Moore, '71). Recently, colony formation by activated peritoneal macrophages has been described (Lin and Stewart, '74). These macrophage colonies develop after a 7-10 day lag period in culture and are derived from cells with the adherence and phagocytic characteristics of macrophages. It appears that this proliferation of mature macrophages is dependent upon a factor which is similar, if not identical to CSF (Moore and Kurland, '75).

Colony stimulating factors are heterogeneous with respect to molecular weight and charge but appear to share a common specificity for stimulation of granulopoiesis and monocytopoiesis. The most extensive characterization of the chemical nature of CSF has been performed on the material present in human urine where 100,000 - 200,000 fold purification has been obtained with respect to protein in unfractionated urine (Stanley and Metcalf, '72). CSF from this source was characterized as a neuraminic acid containing glycoprotein of 45,000 molecular weight.

Exogenous CSF is not an absolute requirement for *in vitro* colony formation and spontaneous colony formation characterizes marrow cultures of all species. Detailed analysis of this phenomena has shown that in every species studied, a sigmoid relationship exists between cell concentration and colony incidence in unstimulated cultures which contrasts with the linear relationship seen in cultures stimulated by optimum concentrations of exogenous CSF (Moore and Williams, '72). Each species has a particular threshold cell concentration below which colony formation exhibits an absolute requirement for exogenous CSF. This concentration can be as low as 2×10^4 marrow cells/ml in monkey or guinea pig marrow culture and as high as 5×10^5 - 1×10^6 in mouse cultures. Physical separation of marrow, spleen and blood using density, sedimentation or adherence criteria has shown that spontaneous colony formation is due to the endogenous elaboration of CSF by colony stimulating cells (CSC) present in hemopoietic tissue. CSF has been extracted from a broad spectrum of tissues both normal and neoplastic, however the potential physiological role of CSF in the regulation of hemopoiesis is probably dependent on the interaction between colony stimulating cells and CFU-c. The former have been identified as phagocytic mononuclear cells, both monocytes and fixed tissue macrophages (Moore and Williams, '72; Cline et al., '74).

CELLULAR INTERACTIONS IN THE REGULATION OF GRANULOPOIESIS

The role of monocytes and macrophages in the elaboration of CSF suggests that granulopoiesis and monocyte production are controlled by a positive feedback mechanism limited by the differentiation of CFU-c into granulocytes when CSF levels are increased and balanced by preferential monocyte differentiation when CSF levels are depressed (Moore et al.,'74). In this context the monocyte-macrophage population would function as a sensor, detecting bacterial antigens, responding to local tissue destruction or inflammation and by increasing CSF production would promote increased granulopoiesis and local macrophage proliferation.

Regulation of granulopoiesis can be mediated at the level of the proliferative status of the CFU-c. Unlike the erythropoietin sensitive stem cell compartment, the fraction of CFU-c in DNA synthesis can be markedly influenced by demand for granulocyte production. This is most clearly seen in mid-gestation fetal liver of species with a long gestation period where minimal granulopoiesis is evident (Moore and Williams, '73). CFU-c in marrow and blood of untreated patients with acute and chronic myeloid leukemia are likewise in a non-cycling state (Moore, '74). In Table 1 it can be seen that brief 3 hour *in vitro* exposure of fetal or leukemic CFU-c to CSF results in a marked activation of the cells into DNA synthesis as revealed by increased sensitivity to killing by high doses of H^3TdR.

The actual concentration of CSF required to activate CFU-c into cycle *in vitro* (10-50 units/ml) was generally considerably lower than the concentration of CSF present in the serum of the tissue donor (50-700 units per ml). Consequently, the small fraction of CFU-c in DNA synthesis in untreated CML, AML and fetal liver cannot be attributed to the lack of CSF *per se* suggesting that inhibitory factors may play an important role *in vivo* counteracting the proliferative stimulus of CSF.

TABLE 1

Sources of Cells	No. of Observations	% CFU-c in DNA Synthesis Before Culture	3hr Culture - CSF	3hr Culture + CSF
Normal human marrow	3	40±3	42±3	43±2
Untreated chronic myeloid leukemia	13	4±3	3±3	71±4
Untreated acute myeloid leukemia	20	6±4	6±5	30±3
Mid-gestation fetal liver	6	4±3	2±2	42±2

In order to investigate the influence of marrow cell concentration on CFU-c proliferation, a new technique was developed in order to sustain human marrow hemopoiesis at cell concentrations comparable to those pertaining in normal or leukemic marrow *in situ*. Marrow cells were incubated at concentrations of 40 - 200 x 10^6 cells per ml in autologous plasma in dialysis bags implanted intraperitoneally in mice for 1-3 days. Absolute recovery of CFU-c was 50-100% in the first 24 hours (Fig. 1) but the cycle status of the colony and cluster forming cells was influenced by marrow cell concentration with non-cycling state which persisted for up to 3 days.

Further analysis of this cell concentration dependent phenomenon indicated that the concentration of mature granulocytes was primarily responsible for inhibition of CFU-c proliferation. Mature granulocytes have been implicated in mediating a mitotic inhibitory negative feedback (Rytoma and Kiviniemi, '68) possibly acting at the level of the CFU-c; however, recent observations by Broxmeyer *et al*. ('75a,b) suggest an alternative mechanism. Removal of mature PMN from human marrow by density and adherence procedures produced a significant enhancement of spontaneous colony formation whereas readdition of intact PMN, or PMN extracts or conditioned media suppressed colony formation. Further analysis of this inhibition revealed that mature PMN contain and release a thermo-

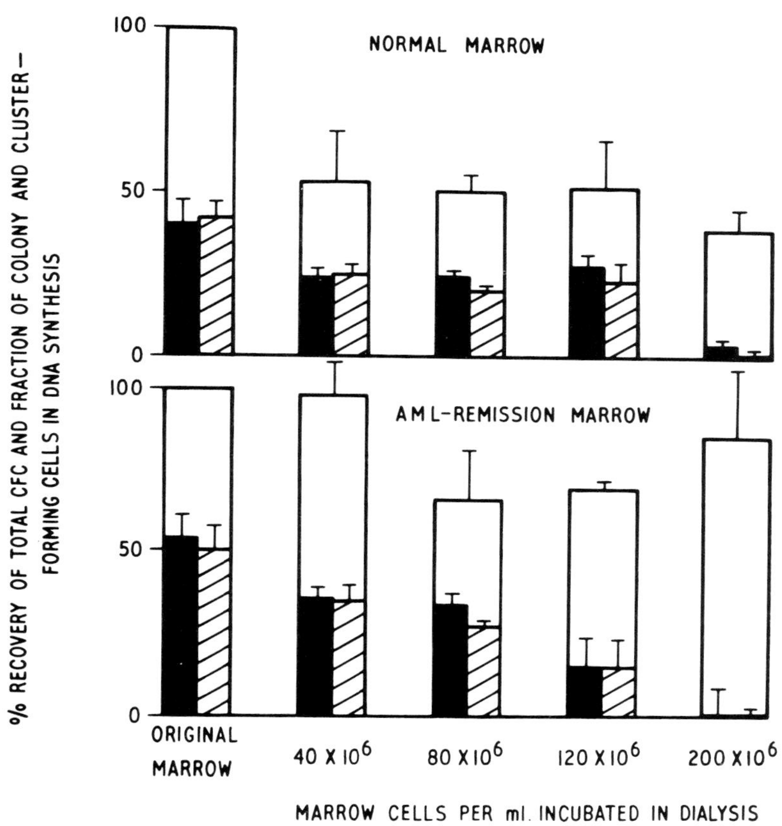

Fig. 1. Recovery and proliferative status of CFU-c following 18 hrs incubation of varying numbers of normal or AML-remission marrow cells in dialysis bags implanted i.p. in mice. The absolute recovery of CFU-c after incubation (open columns) is expressed as a percent of the original CFU-c incidence. The fraction of recovered colony forming (black columns) or cluster forming (hatched columns) cells killed by in vitro ^{3}*HTdr is shown.*

labile factor which suppresses the synthesis or release of CSF by colony stimulatory cells (monocytes and macrophages). This factor was not cytotoxic and had no inhibitory influence on CFU-c stimulated by an exogenous source of CSF. It is possible that this same mechanism accounts for the non-cycling status of CFU-c in dialysis bag cultures incubated at high concentrations since we have observed that endogenous CSF production in this system was inversely related to granulocyte cell concentration.

A further regulatory mechanism which can influence CFU-c proliferation is the synthesis and release of prostaglandins of the E series by hemopoietic cells. PGE_1 and PGE_2 (but not $F_{2\alpha}$) produced a very profound suppression of CFU-c proliferation and of B lymphocyte proliferation *in vitro* (Fig. 2). 50% inhibition of CFU-c proliferation was observed with concentrations as low as 10^{-7} - 10^{-8}M PGE_1 which is well within the physiological concentration in serum and locally within

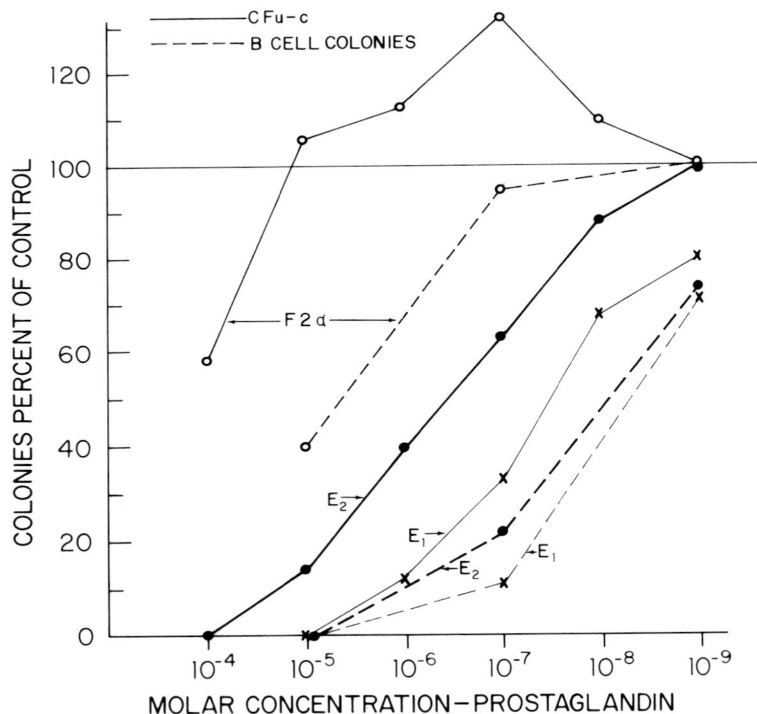

Fig. 2. The influence of prostaglandins of the E_1, E_2 and $F_{2\alpha}$ series on in vitro colony formation by mouse CFU-c stimulated by CSF and on mouse splenic B lymphocyte colony formation in agar culture in the presence of 2-Mercaptoethanol.

the bone marrow. The PGE inhibition seen at low concentrations can be overcome by increasing the concentration of CSF and it is conceivable that a homeostatic balance is established whereby prostaglandins modulate the responsiveness of CFU-c to CSF stimulation (Kurland and Moore, '75).

In vitro studies on the regulation of proliferation and differentiation of the granulocyte-monocyte stem cell reveal a complex interaction between the CFU-c, colony stimulatory cells (monocytes and macrophages) and colony inhibitory cells (mature PMN). This interaction involves the production and release of humoral regulators which mediate both positive and negative feedback operating either at the level of the CFU-c or the colony stimulating cell. No one factor, e.g. CSF, can satisfactorily account for regulation of granulopoiesis, hence attempts to observe simple correlations between CSF levels and neutrophil production have generally been unsuccessful. If such regulatory mechanisms are operative *in vivo* their analysis will require a more comprehensive understanding of the characteristics of the various regulatory factors, both biochemical and biological, in both steady-state hemopoiesis and following perturbation of granulopoiesis.

LITERATURE CITED

Bradley, T. R. and D. Metcalf. 1966. The growth of mouse bone marrow cells *in vitro*. *Aust. J. Exp. Biol. Med. Sci. 44:* 287.

Broxmeyer, H. E., F. L. Baker and P. R. Galbraith. 1975a. *In vitro* regulation of granulopoiesis in human leukemia: Application of an assay for colony-inhibiting cells. *Blood*. In press.

Broxmeyer, H. E., M.A.S. Moore and P. Ralph. 1975b. Soluble granulocytic colony inhibiting activity (CIA) derived from PMN. Manuscript in preparation.

Cline, M. J., B. Rothman and D. W. Golde. 1974. Effect of endotoxin on the production of colony-stimulating factor by human monocytes and macrophages. *J. Cell Physiol. 84:* 193.

Kurland, J. and M.A.S. Moore. 1975. Prostaglandins and colony stimulating factor as bioregulators of hematopoiesis. I. Specificity of action at the level of the committed granulocyte-macrophage stem cell. Manuscript in preparation.

Lin, H. S. and C. C. Stewart. 1973. Peritoneal exudate cells. I. Growth requirement of cells capable of forming colonies in soft agar. *J. Cell. Physiol. 83:* 369.

Metcalf, D. and M.A.S. Moore. 1971. *Haemopoietic Cells*. North Holland Publishing Co., Amsterdam.

Moore, M.A.S. 1974. *In vitro* studies in the myeloid leukaemias. In: *Advances in Acute Leukaemia*. Cleton, F. J., D. Crowther and J.S. Malpas, ed. ASP Biological and Medical Press, Amsterdam. p. 161.

Moore, M.A.S. and J. Kurland. 1975. Regulation of granulopoiesis. *Proceedings 2nd International Conference on Cell Differentiation*. Viza, D. and N. Muller-Berat, N. Ed. In press.

Moore, M.A.S. and N. Williams. 1972. Physical separation of colony-stimulating cells from *in vitro* colony forming cells in monkey hemopoietic tissue. *J. Cell Physiol. 80:* 195.

Moore, M.A.S., N. Williams and D. Metcalf. 1972. Purification and characterization of the *in vitro* colony forming cell in monkey haemopoietic tissue. *J. Cell Physiol. 79:* 283.

Pike, B. L. and W. A. Robinson. 1970. Human bone marrow colony growth *in vitro*. *J. Cell Physiol. 76:* 77.

Rytoma, T. and Kiviniemi, K. 1968. Control of granulocyte production. I. Chalone and antichalone, two specific humoral regulators. *Cell Tissue Kinet. 1:* 329.

Stanley, E. R. and D. Metcalf. 1972. Purification and properties of human urine colony stimulating factor (CSF). In: *Cell Differentiation, Proc. 1st Int. Conf.* R. Harris and D. Viz, ed. Munksgaard, Copenhagen. p. 149.

Discussion, Session III
Stem Cells in Haemopoietic and Lymphoid Tissue I

Dr. Till's presentation

Dr. Patt began the discussion by stating how improbable he considers the "null stem cell". Dr. Till responded to the effect that while the concept might seem improbable when stated explicitly, it was being used implicity in definitions such as the "null lymphocyte" and was inhibiting research of the early stem cell by defining these cells as undistinguishable. Dr. Everett and Dr. Osmond argued that the null lymphocyte was null really only from the standpoint of a certain well-defined surface antigen or receptor and this in no way excluded a whole variety of other possible markers on early stem cells. Dr. Osmond raised the possibility that the turnover of surface markers on the membrane might lead to problems in the detection of these primitive cells.

Dr. Leblond supported Dr. Till's point of view that stem cells are not null cells. He described the apparent stem cells of the pylorus and colon, for instance, as cells that contain differentiated vacuoles and these cells apparently function all through life with a degree of differentiation. He admitted that he had at first found the idea of a stem cell with differentiation properties uncomfortable.

Dr. Lord's presentation

Dr. Trentin asked whether, when looking at the response of shielded marrow to radiation, the important life-saving thing wasn't the showering out of stem cells from the shielded areas. Dr. Lord agreed but thought what was surprising was that this shielded marrow didn't knuckle down and contribute its share to the recovery process directly. This meant that the control of the proliferating CFU in that shielded area was entirely local.

Dr. Denekamp asked how long an animal takes to repopulate their stem cells following radiation. The levels were still very low at 10 days following 300 to 400 rads.

Dr. Till asked what the specificity controls for the inhibition studies were. Dr. Lord stated that the experiments were pilot at this time and there were no specificity controls, though some preliminary experiments suggest that the effect is on the whole bone marrow and not just one cell type.

Dr. Moore's discussion

Dr. Bruce asked whether there had been systematic studies of CSA or prostaglandins levels in patients with leukemia. Dr. Moore stated CSF or CSA levels had been measured in patients under a wide range of conditions, but the results had been rather a long frustrating experience, because although one can in certain studies see correlations, in most studies one cannot. In particular, there might be marked granulopoiesis in the absence of significant elevations of CSA. It was possible that the assay systems measure a complexity of inhibitors. The CSF may be a permissive substance whose production may be modulated by some negative control mechanism. Assays for prostaglandins raise other problems, for what one is interested in is their levels in a local marrow environment since they are very labile molecules.

Dr. Patt asked whether CSF was specific for the CFU-C type of cell and what evidence there was that CSF played a role *in vivo*. Dr. Moore said that CSF had no effect that he could observe on lymphoid proliferation, PHA-stimulated cultures, erythroid colonies, fibroblast colonies, or on the growth of a wide range of cell strains and cell lines. But the other question was more crucial - evidence of a physiological role. At present we do not have a sufficiently good *in vivo* assay to answer this question. Perhaps a time will come when an assay paralleling the polycythemic mouse for erythropoietin studies will be possible. A further problem is raised by the source of CSF. There is something very different between the human and mouse in terms of CSF production. In the mouse it is seen in many tissues, perhaps on different cell surfaces, or perhaps on ubiquitous fixed tissue macrophages. In the human it is difficult to get active CSF from cells other than hematopoietic, although embryonic kidney also produces it.

Dr. Blackett's presentation

In response to a query for information, Dr. Blackett stated that the response of C57 and C3H strains of mice to tritiated thymidine was significantly different. There was a 5% suicide with the former, 10% with the latter.

Dr. Lala cited data pertaining to the possibility of feedback to stem cells from maturing cells. These were transplantation experiments in which differentiated cells were believed to extinguish the growth of stem cells.

Dr. Denekamp raised a question concerning the G_0 and a long G_1 with ill-defined transit time. Couldn't the two possibilities be distinguished by a second wave in a percent labelled mitosis curve? Dr. Blackett thought the distinction between a G_0 and long G_1 was difficult at this time. It was difficult to establish a dose response curve for tritiated thymidine to plateau levels, and at these levels there might be artefacts from the high level of radioactivity itself.

Dr. Broxmeyer's presentation

Dr. Broxmeyer spoke briefly of his recent studies with colony inhibiting activity. This is a cell free, serum free, factor derived from mature granulocytes. It appears to have no effect on colony-forming cells or on colony stimulating activity, but it does appear to have a specificity for inhibiting the colony stimulating cell. It was suggested that the interaction of the colony stimulating and the colony inhibiting cell was complex - inhibition could be overcome with stimulation and vice versa.

Session IV

Stem Cells in Hemopoietic and Lymphoid Tissue II

Potentials of Bone Marrow Lymphocytes

D. G. OSMOND

Department of Anatomy, McGill University, Montreal, Quebec, Canada

Mammalian bone marrow contains a complex mixture of lymphoid cells, including lymphocyte progenitors, differentiating lymphocytes and recirculating immigrants. This presentation will review briefly some of our studies aimed to distinguish between these various cell populations in the bone marrow of inbred mice. The work uses a combination of cell markers, notably DNA-labeling together with several surface membrane markers, to examine the normal production and developmental potentials of bone marrow lymphoid cells under physiological steady state conditions *in vivo*. The conclusion is reached that a large-scale commitment of progenitor cells in normal bone marrow is concerned with the genesis of immunoglobulin (Ig)-bearing lymphocytes, i.e. "B" lymphocytes, potential precursors of antibody-forming cells (Osmond, '75). Some factors which may modify this process *in vivo* will also be considered.

PROLIFERATION AND RENEWAL OF BONE MARROW LYMPHOID CELLS

Smear preparations of mouse bone marrow reveal many lymphocyte-like cells, readily distinguishable from other bone marrow cell types (Miller and Osmond, '74, '75). Such lymphoid cells all possess minimal quantities of cytoplasm, lacking specific organelles or other morphological evidence of differentiation. Despite their apparent uniformity these cells have proved to be functionally heterogeneous.

Radioautographic smears of bone marrow taken from mice one hour after intraperitoneal administration of ^{3}H-thymidine allow a distinction to be drawn between dividing and non-dividing bone marrow lymphoid cells (Osmond and Nossal, '74b). The size distribution profile of all the bone marrow lymphoid cells, from measurements of mean nuclear diameter in smears prepared under standard conditions, ranges from 5μ to 14μ (mode 7μ). DNA-synthesis, as shown by direct incorporation of ^{3}H-thymidine, occurs in lymphoid cells larger than 8.0μ nuclear diameter; such cells have diffuse nuclear chromatin and are termed "large lymphoid" cells. They correspond with similar cells in guinea pig and rat bone marrow, in which they have also been called "transitional" cells (Hudson, Osmond and Roylance, '63; Osmond and Everett, '64; Yoffey, Hudson and Osmond, '65; Rosse, '70; Yoffey and Courtice, '70; Miller and Osmond, '73; Osmond, Miller and Yoshida, '73). Bone marrow lymphoid cells smaller than 8.0μ nuclear diameter in smears with few borderline exceptions do not show DNA synthesis; they have a dense nucleus with prominent condensed chromatin and are morphologically similar to small lymphocytes in other tissues, closely resembling those in the thymus.

The foregoing criteria of morphology, nuclear diameter and proliferative activity define two basic bone marrow lymphoid cell populations for the purposes of this presentation. Small lymphocytes are non-dividing cells, though this in no way rules out the possibility that they may grow and enter cell cycle following suitable activation. Large lymphoid cells include essentially all the DNA- synthesising forms but, in addition, cells of near borderline size (9.0-9.0μ nuclear diameter) probably include some non-dividing cells (Miller and Osmond, '73; Osmond, Miller and Yoshida, '73).

Lymphoid cells form an especially prominent population in bone marrow of young, growing animals. Quantitative studies in C3H mice of various ages, combining differential and absolute cell counts with measurements of bone marrow volume, reveal a rapid postnatal increase in numbers of bone marrow lymphocytes (Miller and Osmond, '74). Small lymphocytes increase from 10% of nucleated bone marrow cells at birth to 30% at 2-4 weeks of age, at which time they total more than 600×10^{3} small lymphocytes per cu.mm. of femoral marrow. Subsequently, the cells fall to approximately half these numbers in young adult mice aged 12-16 weeks. The large lymphoid cells show parallel changes, comprising approximately 20% of the total lymphoid population in the bone marrow at all ages. This age-related sequence in normal animals raised under conventional conditions parallels the spontaneous growth and regression of thymus cell populations.

Continuous infusion of ^{3}H-thymidine to label all DNA-synthesising cells and their progeny in C3H mice of various ages has been used to characterise the proliferation and renewal of bone marrow lymphoid cells (Miller and Osmond, '75). Most large lymphoid cells are actively proliferating. In 2 month old C3H mice 58% of all large lymphoid cells are in DNA synthesis at any one time, the proportion of DNA-synthesising cells increasing with increasing cell size (50%, 8-10μ diameter; 85%, 10-12μ diameter; 100%, 12μ diameter) (S. C. Miller and D. G. Osmond, unpublished observations). During continuous ^{3}H-thymidine infusion the DNA- labeling of large lymphoid cells is virtually complete (97-99%) after 2 days (10.0-11.9μ diameter) or 3 days (8.0-9.9μ diameter), indicating that the bone marrow large lymphoid cells are rapidly turning over. In older mice (4 month) the labeling pattern is similar except that 5-10% of the 8.0-9.9μ diameter lymphoid cells remain unlabeled after 4-5 days ^{3}H-thymidine infusion, and are thus neither proliferating nor rapidly replaced by newly formed cells (S.C. Miller and D.G. Osmond, unpublished observation). These cells may represent either a subpopulation of "resting" large lymphoid cells or, alternatively, an overlap in size with some of the long-lived small lymphocytes, noted below.

The generally rapid turnover of large lymphoid cells in mouse bone marrow contrasts with the heterogeneity of such cells in young guinea pigs (Rosse and Yoffey, '67; Rosse, '70; Miller and Osmond, '73; Osmond, Miller and Yoshida, '73). Basophilic large lymphoid cells in guinea pig bone marrow show an increase in the incidence of DNA-synthesising cells with increasing cell size, together with a shortening of DNA synthesis and of cell cycle times. On the other hand, pale-staining cells, comprising approximately one-third of the large lymphoid cells, show low ^{3}H-thymidine labeling indices and appear to be predominantly resting cells. These species differences in morphology and proliferative properties of large lymphoid cells may reflect species differences in their potential. This presentation is concerned mainly with the potentials of large lymphoid cells as progenitors of small lymphocytes. We have demonstrated in cultures of guinea pig bone marrow fractions that some large lymphoid cells are the immediate progenitors of non-dividing small lymphocytes (Osmond, Miller and Yoshida, '73). However, the total proliferation rate of large lymphoid cells in guinea pig bone marrow exceeds the needs for the renewal of bone marrow small lymphocytes, suggesting that some large lymphoid cell proliferation may be concerned with the production of non-lymphoid cells (Osmond, Miller and Yoshida, '73). Several laboratories, using bone marrow fractionation techniques and repopulation assays, have produced strong evidence that some bone marrow cells with the morphology of large lymphoid cells function as hemopoietic progenitor cells for non-lymphoid marrow cell lines (Van Bekkum, van Noord, Maat and Dicke, '71; Moore,

Williams and Metcalf, '72) as suggested by Yoffey (Yoffey and Courtice, '70). The proportion of large lymphoid cells with such potential is not known but probably varies according to species and age. The early stages between the lymphomyeloid stem cell and the identified lymphocyte progenitors in the bone marrow remain poorly understood (Miller, this volume).

The small lymphocytes in mouse bone marrow, although not directly incorporating ^{3}H-thymidine, show a rapid increase in labeling index during the first three days of ^{3}H-thymidine infusion, indicating a rapid replacement by newly-formed cells (Ropke and Everett, '73; Miller and Osmond, '75). With continuing infusion a subsequent increase in proportion of labeled cells occurs at a slower rate. Computer analysis shows that a best-fit curve through the observed data can be represented by the summation of two exponential curves, indicating two corresponding cell populations. From such studies in C3H mice of various ages, together with quantitative marrow data, relative proportions and renewal rates may be defined for two populations of bone marrow small lymphocytes, a major population of rapidly renewed cells, accounting predominantly for the rapid early increase in labeling index during ^{3}H-thymidine infusion, and a minor population of slowly renewed cells which contribute a slow increment of labeling index throughout the infusion period (Miller and Osmond, '75).

The two populations of small lymphocytes in mouse bone marrow differ from one another in origin, life span and functional potential. The slowly-renewed lymphocytes in C3H mouse bone marrow actually increase in incidence (from 7.0% to 22% of marrow small lymphocytes) and numbers (from 45 to $60 \times 10^3/mm^3$) with increasing age (from 4 to 16 weeks) (Miller and Osmond, '75). They are functionally heterogeneous; some may be indigenous "resting" cells, comparable with the long-lived small lymphocytes detected in bone marrow of rats given prolonged ^{3}H-thymidine administration by Haas, Bohne and Fliedner ('71) and suggested as being hemopoietic stem cells because of their enlargement in rats treated with myelotoxic agents. On the other hand, parabiotic experiments show many of these cells to be immigrants (Rosse, '71; Ropke and Everett, '74), mainly recirculating T and B memory lymphocytes (Howard and Scott, '72; Benner and van Oudenaren, '75). It remains to be determined to what extent the increased numbers of long-lived small lymphocytes in bone marrow with age involve presumptive "resting progenitor cells" and/or the expanding pool of recirculating memory cells.

Our interest has centered around the major population of rapidly-renewed small lymphocytes in mouse bone marrow. In

4 week old C3H mice they account for 93% of all bone marrow small lymphocytes (600x10^3 per cu. mm.) and show a short exponential transit time in the bone marrow (50% renewal in 14 hours). With increasing age their numbers decline and their turnover becomes somewhat less rapid (16 weeks; 230x10^3 per cu. mm.; 50% renewal, 24 hours). Selective labeling with ^{3}H-thymidine in mice (Osmond, '75), as in other rodents (Osmond and Everett, '64; Everett and Caffrey, '67; Brahim and Osmond, '73), has demonstrated formally that these cells are produced locally in the bone marrow. As already noted, their precursors have been identified among the large lymphoid cells, while their subsequent migration has been traced from bone marrow to the spleen and lymph nodes (Everett and Caffrey, '67; Brahim and Osmond, '73, '76). Their production occurs continuously on a large scale, equivalent to a total production of approximately 10^8 cells per day in the entire bone marrow of young adult mice. Recent work has aimed to test whether this represents a primary genesis and differentiation of Ig-bearing (B) lymphocytes.

DEVELOPMENT OF SURFACE MEMBRANE RECEPTORS ON BONE MARROW LYMPHOID CELLS

Many mature B lymphocytes in peripheral lymphoid tissues, such as those in the spleen and lymph nodes, display a characteristic combination of surface membrane components, viz. readily-detectable Ig molecules and receptor molecules which bind complement (C'3) and the Fc portion of IgG molecules, respectively (Parish and Hayward, '74). These components have now been demonstrated on bone marrow lymphocytes by antiglobulin-binding and erythrocyte rosetting techniques.

Surface Ig molecules have been quantitated on bone marrow lymphocytes by the binding of ^{125}I-labeled antiglobulin, active against mouse μ and K chains. The percentage and intensity of labeled small lymphocytes were examined radioautographically after cell suspensions were exposed to a wide range of antiglobulin concentrations at 0°C for 30 mins (Osmond and Nossal, '74a). Small lymphocytes in the spleen and lymph nodes of CBA mice show similar labeling curves, increasing in their labeling index to plateau levels (spleen, 47%; lymph node, 23%) at antiglobulin concentrations of 1 μg/ml of more, under the conditions of the experiment. In the bone marrow of CBA mice some small lymphocutes are well labeled by low concentrations of ^{125}I-antiglobulin. However, the percentage of labeled small lymphocytes increases linearly with increase in antiglobulin concentration and shows no plateau effect. A similar labeling pattern is seen using bone marrow small lymphocytes from other mouse strains (C3H; DBA/2) and other antiglobulin preparations. These findings, together with the distribution of grain counts of labeled small lymphocytes at one given antiglobulin concentration, suggest that individual small lymphocytes in bone marrow show a wide range of antiglobulin-binding capacity relative to small lymphocytes in the spleen and lymph nodes. Thus, in 10 week old CBA mouse bone marrow three groups of small lymphocytes may be defined, somewhat empirically, by their density of surface Ig: 1) approximately 30% of the bone marrow small lymphocytes show strong antiglobulin binding, quantitatively equivalent to that of spleen lymphocytes, 2) a further 20% show weaker antiglobulin binding, detectable only by high antiglobulin concentrations, 3) approximately half have no detectable surface Ig; most of these are so-called "double-negative" cells, being also devoid of T lymphocyte characteristics, as shown by antiglobulin-binding studies in congenitally athymic (nude) mice and in CBA mouse cells treated with anti-Thy 1. serum and complement (Osmond and Nossal, '74a).

The class of surface Ig on mouse bone marrow small lymphocytes has been examined by the binding of ^{125}I-antisera specific for the various Ig heavy chains (D.G. Osmond, P.E. Wherry, M. Daeron and J. Gordon, unpublished observations). In 22 week old DBA/2 mice the largest proportion of bone marrow small lymphocytes (46%) show μ chains, fewer (27%) have surface γ_2 chains while relatively small numbers show γ_1 (5%) or α (5%) chains. The high incidence of μ- and γ_2-bearing cells indicates that at least some of them belong to the rapidly-renewing small lymphocyte population, while it has yet to be shown whether the small numbers of γ_1- and α-bearing lymphocytes are either indigenous or recirculating cells.

Rosetting techniques have been used to compare surface Ig, Fc and complement receptors on bone marrow small lymphocytes (Osmond and Yang, '75). The techniques were modified from those of Parish and Hayward ('74), the receptors being detected by the binding to the lymphocyte of either the appropriate antibody or antibody plus complement, previously coated on the surface of marker sheep erythrocytes. In stained cytocentrifuge preparations marrow lymphoid cells were measured and the proportions of cells with rosettes of four or more associated erythrocytes were counted, after correction for the small incidence of similar rosettes in control mixtures of small lymphocytes with uncoated erythrocytes. With this technique many bone marrow small lymphocytes in 10 week old C3H mice clearly do show receptors for

Fc (25) and complement (18%) as well as surface IgM (39%) (W.C. Yang and D.G. Osmond, unpublished observations). However, a higher proportion of Ig-bearing small lymphocytes appear to lack Fc and complement receptors in the bone marrow than elsewhere. The incidence of Fc or complement receptor bearing small lymphocytes is only 40-50% that of Ig-bearing cells in the bone marrow compared with 60-80% in the spleen and virtually 100% in the lymph nodes. Moreover, the density of Fc and complement receptors, as judged by rosette size, tends to be lower in the bone marrow than in spleen or lymph nodes.

Kinetic studies have been designed to relate surface receptor status with cell age and to identify bone marrow lymphoid cells having the potential to develop B lymphocyte surface features. The experiments involve a combination of markers, viz. repeated or continuous ^{3}H-thymidine administration together with exposure of the cells to either ^{131}I-labeled antiglobulin or a rosetting procedure. The ^{3}H-label in nuclear DNA and the surface markers are distinguishable from one another in radioautographs. The surface receptor bearing cells are small lymphocytes. At one hour after ^{3}H-thymidine administration the labeled DNA-synthesising large lymphoid cells with few borderline exceptions show no detectable surface Ig (Osmond and Nossal, '74b). Among the small lymphocytes in the bone marrow the first ^{3}H-labeled cells to be produced all lack the capacity to bind antiglobulin. The ^{3}H-labeling of such Ig-negative cells increases exponentially to a plateau of approximately 90% after 2 days (Osmond and Nossal, '74b). These cells are therefore predominantly younger than 2 days. Antiglobulin-binding small lymphocytes show ^{3}H-thymidine labeling only after a lag of approximately 1.5 days, but the labeling index increases rapidly thereafter. The ^{3}H-thymidine labeling index of the most strongly Ig-bearing cells increases only after a further 0.5 day lag. These experiments, confirmed in current ^{3}H-thymidine labeling studies of Ig rosettes, show that the appearance of detectable surface Ig and its subsequent increase in density represents a maturation sequence related to cell age. Similarly, complement receptor-bearing cells rapidly become labeled with ^{3}H-thymidine after a post-mitotic lag of approximately one day. They show 46% labeling at two days and 83% at four days compared with 84% and 88%, respectively, for bone marrow small lymphocytes lacking the complement receptor. Fc receptor bearing small lymphocytes show an even more rapid onset of ^{3}H-thymidine labeling (15%, 1 day; 65%; 2 days; compared with 31% and 71%, respectively, for Fc receptor-negative small lymphocytes) (W.C. Yang and D.G. Osmond, unpublished observations).

Thus, Fc and complement receptor-bearing cells form part of the rapidly renewed indigenous population of small lymphocytes and, like surface Ig, these receptors become readily detectable during maturation of the cells soon after their production from proliferating progenitors in the normal steady state.

Further evidence of a surface Ig density sequence in the development of bone marrow small lymphocytes is seen during perturbations of the steady state (K. Evoy and D. G. Osmond, unpublished observations). In 10 week old C3H mice given 150 rads whole body irradiation the absolute numbers of both large lymphoid cells and small lymphocytes per cu. mm of bone marrow rebound from an initial post-irradiation depletion to markedly supranormal levels at 10-14 days. Exposure to ^{125}I-antiglobulin at various times after irradiation reveals that the large lymphoid cells lack detectable surface Ig, even during the rebound peak when the cells form a conspicuous population in the bone marrow. Ig-bearing small lymphocytes recover rapidly but the shape of their labeling curve over a range of ^{125}I-antiglobulin concentrations changes with time. At three days after irradiation surface Ig is detectable only at high antiglobulin concentrations; subsequently, the bone marrow small lymphocytes show the normal linear increase in labeling index with increasing antiglobulin concentration, while at 21 days after irradiation the labeling curve resembles that of spleen small lymphocytes, which continues to show a well-marked plateau of labeling with antiglobulin concentrations of 1 μg/ml or more. Thus, the relative proportions of weak Ig-bearing and strong Ig-bearing small lymphocytes vary from time to time. Expressed as absolute numbers of cells per cu. mm of bone marrow these data reveal successive waves of Ig-negative and of weak Ig-bearing small lymphocytes, followed by a wave and overshoot of strong Ig-bearing small lymphocytes, increasing parallel to and approximately five days after the wave of Ig-negative small lymphocytes. Thus, both normal bone marrow lymphocytopoiesis and endogeneous post-irradiation regeneration exhibit a sequence in which large lymphoid progenitor cells and the most newly-formed small lymphocytes lack detectable surface Ig, while this becomes detectable with increasing ease during subsequent post-mitotic maturation. The other characteristic B lymphocyte receptors, i.e. for Fc and complement, also appear during the maturation sequence, possibly in a stepwise fashion. These findings, together with functional observations on the development of antibody-forming cells from bone marrow (Lafleur, Miller and Phillips, '72; Stocker, Osmond and Nossal, '74;

Miller, this volume), indicate that the major potential of mouse bone marrow lymphoid cells is the continuous production and differentiation of virgin B lymphocytes. This does not necessarily imply that all indigenous bone marrow lymphoid cells are so committed. Some "large lymphoid" cells, not clearly distinguishable from the B lymphocyte progenitors, may be progenitors for non-lymphoid cell lines, as already noted. Further, while "double negative" (Ig-ve; Thy.1-ve) small lymphocytes are the immediate precursors of mature, Ig-bearing B lymphocytes, it is possible that some of these cells may have other potentials, and some may die after a short life span, as suggested for certain cells in the thymus (Shortman and Jackson, '74).

Finally, the extent to which the various subpopulations of receptor-bearing small lymphocytes overlap one another in the bone marrow has yet to be determined. The simplest interpretation would be that initially "receptorless" small lymphocytes progressively display surface Ig of various heavy chain classes together with the other characteristic receptor molecules in an additive sequence. However, preliminary data suggest that this is not necessarily the case. Some young small lymphocytes appear to display Fc receptors before they would be expected to show surface Ig (W.C. Yang and D.G. Osmond, unpublished observations).

REGULATORY INFLUENCES ON BONE MARROW LYMPHOID CELLS

A high level of lymphocyte production in mouse bone marrow appears to occur in anticipation of immune needs, but the homeostatic mechanisms involved in regulating the process are not known. Experiments will now be described which have aimed to test the susceptibility of bone marrow lymphocytopoiesis to some possible regulatory influences and which bear upon the question of surface membrane differentiation of lymphoid progenitors.

The age-related changes in the magnitude of bone marrow lymphocyte production resemble those of thymic lymphocytopoiesis. However, they are apparently not dependent upon thymic influences. Early post natal development of the bone marrow lymphocyte population occurs normally in neonatally thymectomized C3H mice (Miller and Osmond, '76), while in "nude" mice with congenital thymic deficiency, even at 6 weeks after birth the renewal of lymphocytes in the bone marrow, as shown by repeated labeling with ^{3}H-thymidine, proceeds at the same high rate as that in normal littermates (D.G. Osmond, unpublished). On the other hand, either altered antigen levels or the administration of anti-IgM antibodies *in vivo* do produce marked effects on bone marrow lymphocytes, as follows.

The bone marrow of germfree mice from whom microorganisms have been excluded from birth until 10 weeks of age shows normal numbers of total lymphocytes (Claesson, Ropke and Hougen, '74; Osmond, '75) and of Ig-bearing lymphocytes (Osmond and Nossal, '74a). However, ^{3}H-thymidine labeling indicates that the turnover of the bone marrow lymphocyte population is prolonged 4-fold when compared with conventionally-reared mice (half-renewal times; 125 hours and 32 hours, respectively) (Osmond, '75 and unpublished observations). No such effect occurs in the thymus. Thus, in the bone marrow, though not in the thymus, a substantial basic proliferation of lymphoid progenitors appears to be amplified by bacterial antigens or mitogens in normal life. Possibly, some antigen-driven amplification of lymphoid proliferation may be involved in generating the normally diverse repertoire of virgin B lymphocytes.

The influence of anti-Ig antibodies on bone marrow lymphoid cells *in vivo* has been tested in 2 groups of DBA/2 mice, viz. newborn mice and adult animals given whole body X-irradiation (950r) plus bone marrow transfusion (D.G. Osmond, P.E. Wherri, M. Daeron and J. Gordon, unpublished observations). Each group was treated for 3 months by Dr. J. Gordon with repeated injections of an antiserum raised in rabbits against mouse IgM. In both groups of anti-IgM treated mice the incidence of small lymphocytes in the bone marrow was reduced severely to approximately 3% compared with 17-25% in control animals given saline injections, while the incidence of large lymphoid cells was also reduced to 0.2% from 2.0%. Radioautography after exposure of the cells to ^{125}I-labeled monospecific anti-Ig antisera shows that in the anti-IgM treated mice lymphocytes binding antiglobulins specific for μ, γ_1, γ_2 and α chains are all virtually absent from both the bone marrow and the spleen. The few Ig-bearing small lymphocytes which can be found tend to be weakly labeled and pycnotic.

Two important conclusions can be drawn from these results. Firstly, anti-IgM serum prevents the development of lymphocytes bearing *all* Ig heavy chains in the bone marrow, suggesting that IgM-bearing cells give rise to lymphocytes carrying the other classes of Ig, and that murine bone marrow, like the avian bursa of Fabricius, is the site of an IgM to IgG sequence in lymphocyte production. Secondly, anti-IgM treatment eliminates from the bone marrow virtually all the "double negative" small lymphocytes and many large lymphoid cells as well as the Ig-bearing lymphocytes. Apparently, the anti-IgM acts at, or even before, the stage of the identifiable large lymphoid cell, even though sensitive radioautographic techniques fail to detect surface IgM at this stage. The solution to this apparent paradox may be provided by recent work of Melchers *et al.* ('75). They describe a population of cells (type I) in mouse bone marrow which have electrophoretic mobilities and sedimentation velocities comparable with those of large lymphoid cells and which synthesise 7-8s IgM with a considerably more rapid turnover rate than that of small lymphocytes. Such cells might be sensitive to anti-IgM antibody yet be apparently Ig-negative by conventional antiglobulin-binding techniques, due to the rapidity of shedding of Ig molecules from the surface membrane. From current data it seems probable that the large lymphoid cells in bone marrow are identical with both the type I cells of Melchers *et al.* ('75) and the anti-IgM sensitive "pre-B" cells assayed by Lafleur *et al.* ('72) and Miller (this volume).

CONCLUSION

Lymphoid cells in mammalian bone marrow have a variety of developmental potentials, summarised in Fig. 1. In mice, as shown previously in guinea pigs and rats, the bone marrow is a major site of continuous production of small lymphocytes. These newly-formed cells show a rapid exponential renewal and a partial maturation in the bone marrow; after an initial delay, they display a sequential appearance and increase in density of characteristic surface membrane components (IgM and IgG_2 molecules; Fc and complement receptors) indicating a primary, large-scale genesis of Ig-bearing (B) lymphocytes. Their immediate progenitors are included in a population of large lymphoid cells whose normally active proliferation is modified by age-related factors, including environmental microorganisms, but not by thymic influences. Although they are apparently Ig-negative by sensitive surface antiglobulin-binding studies, the large lymphoid cells in mouse bone marrow are susceptible *in vivo* to anti-IgM antibodies which prevent the subsequent development of cells bearing Ig of all heavy chain classes, suggesting an IgM to IgG sequence during differentiation. Further work is required to determine the proliferative sequence which maintains the large lymphoid cell population and the steps involved in initiating differentiation and Ig synthesis.

Finally, in considering the potentials of lymphoid cells in the bone marrow the presence of a minor population of long-lived small lymphocytes must be borne in mind; functionally heterogeneous, they include "resting" putative progenitor cells, as well as T and B memory small lymphocytes, immigrants from the recirculating lymphocyte pool.

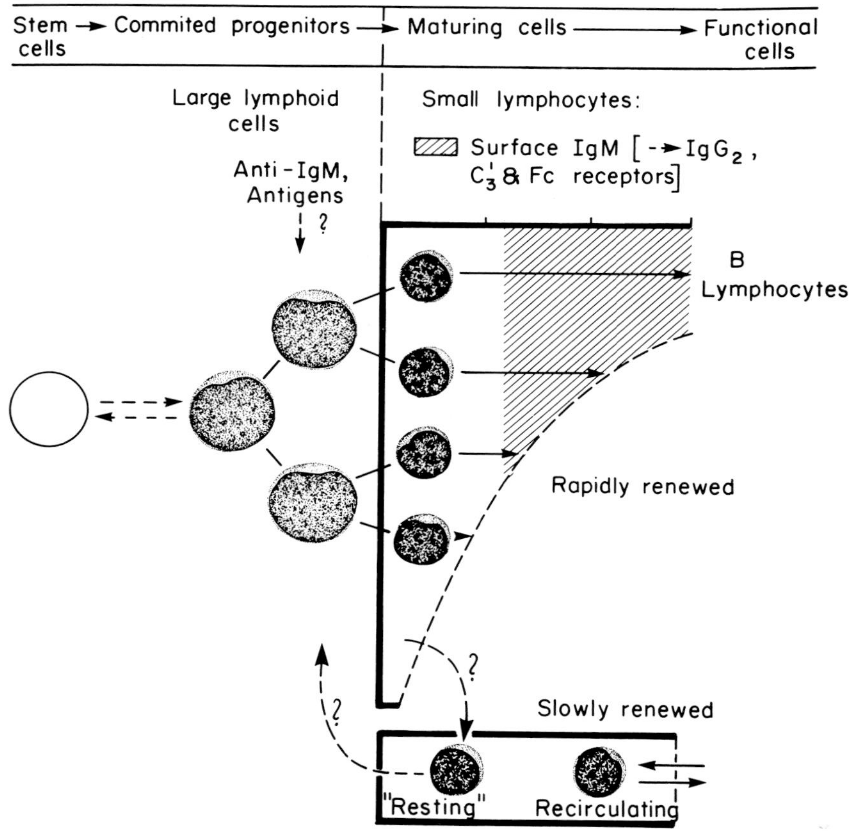

Fig. 1. Scheme of lymphoid cells in mouse bone marrow. Proliferating large lymphoid cells give rise to a population of rapidly-renewed small lymphocytes which show an exponential disappearance from the bone marrow with respect to time. These small lymphocytes show a sequential appearance of surface Ig molecules, Fc and C'3 receptors, characteristic of "B" lymphocytes, after a variable post-mitotic lage period during which the receptors are not detectable by surface binding techniques. Although the large lymphoid progenitor cells are also Ig-negative by these techniques, their proliferation appears to be effected by administration of anti-IgM anti-

ACKNOWLEDGEMENT

This work was supported by the Medical Research Council of Canada.

LITERATURE CITED

Benner, R. and A. van Oudenaren. 1975. Antibody formation in mouse bone marrow. IV. The influence of splenectomy on the bone marrow plaque-forming cell response to sheep red blood cells. *Cell Immunol. 19:* 167-182.

Brahim, F. and D. G. Osmond. 1973. The migration of lymphocytes from bone marrow to popliteal lymph nodes demonstrated by selective bone marrow labeling with ^{3}H-thymidine *in vivo*. *Anat. Rec. 175:* 737-746.

Brahim, F. and D. G. Osmond. 1976. Migration of newly-formed small lymphocytes from bone marrow to lymph nodes during primary immune responses. *Clin. Exp. Immunol.* In press.

Claesson, M. H., C. Ropke, and H. P. Hougen. 1974. Distribution of short-lived and long-lived small lymphocytes in the lymphomyeloid tissues of germ free NMRl mice. *Scand. J. Immunol. 3:* 597-606.

Everett, N. B. and R. Caffrey. 1967. Radioautographic studies of bone marrow small lymphocytes. In *The Lymphocyte in Immunology and Haemopoiesis*. J. M. Yoffey, ed. p. 108. Edward Arnold Ltd., London.

Haas, R. J., F. Bohne and T. M. Fliedner. 1971. Cytokinetic analysis of slowly proliferating bone marrow cells during recovery from radiation injury. *Cell Tissue Kinet. 4:* 31-45.

Howard, J. C. and D. W. Scott. 1972. The role of recirculating lymphocytes in the immunological competence of rat bone marrow cells. *Cell Immunol. 3:* 421-429.

Hudson, G., D. G. Osmond and P. J. Roylance. 1963. Cell-populations in the bone marrow of the normal guinea pig. *Acta Anat. 52:* 234-239.

Lafleur, L., R. G. Miller and R. A. Phillips. 1972. A quantitative assay for the progenitors of bone marrow-associated lymphocytes. *J. Exp. Med. 135:* 1363-1374.

bodies and by the presence of normal environmental micro-organisms in vivo. A second population of slowly-renewed, long-lived, small lymphocytes includes recirculating "B" and "T" memory cells and, possibly, resting progenitor cells.

Melchers, F., H. von Boehmer and R. A. Phillips. 1975. B-lymphocyte subpopulations in the mouse. Organ distribution and ontogeny of immunoglobulin-synthesizing and of mitogen-sensitive cells. *Transplantation Rev.* In press.

Miller, S. C. and D. G. Osmond. 1973. The proliferation of lymphoid cells in guinea pig bone marrow. *Cell Tissue Kinet. 6:* 259-269.

Miller, S. C. and D. G. Osmond. 1974. Quantitative changes with age in bone marrow cell populations of C3H mice. *Exp. Haematol. 2:* 227-236.

Miller, S. C. and D. G. Osmond. 1975. Lymphocyte populations in mouse bone marrow. Quantitative kinetic studies in young, pubertal and adult C3H mice. *Cell Tissue Kinet. 8:* 97-110.

Miller, S. C. and D. G. Osmond. 1976. Quantitative studies of lymphocytes and other cell populations in the bone marrow of neonatally thymectomized C3H mice. *Anat. Rec.* In press.

Moore, M.A.S., N. Williams and D. Metcalf. 1972. Purification and characterisation of the *in vitro* colony forming cell in monkey hemopoietic tissue. *J. Cell. Physiol. 79:* 283-292.

Osmond, D. G. 1975. Formation and maturation of bone marrow lymphocytes. *J. Reticuloendothelial Soc. 17(2):* 99-114.

Osmond, D. G. and N. B. Everett. 1965. Radioautographic studies of bone marrow lymphocytes *in vivo* and in diffusion chamber cultures. *Blood 23:* 1-17.

Osmond, D. G. and W. C. Yang. 1975. Differentiation of mouse bone marrow lymphocytes: complement (C'3) and Fc receptors. *Proc. Can. Fed. Biol. Soc. 18:* 84.

Osmond, D. G., S. C. Miller and Y. Yoshida. 1973. Kinetic and haemopoietic properties of lymphoid cells in the bone marrow. In *Haemopoietic Stem Cells. CIBA Foundation Symposium,* vol. 13. Associated Scientific Publishers, Amsterdam. p. 131-156.

Osmond, D. G. and G.J.V. Nossal. 1974a. Differentiation of lymphocytes in mouse bone marrow. I. Quantitative radioautographic studies of antiglobulin-binding by lymphocytes in bone marrow and lymphoid tissue. *Cell Immunol. 13:* 117-131.

Osmond, D. G. and G.J.V. Nossal. 1974b. Differentiation of lymphocytes in mouse bone marrow. II. Kinetics of maturation and renewal of antiglobulin-binding cells studied by double labeling. *Cell. Immunol. 13:* 132-145.

Parish, C. R. and J. A. Hayward. 1974. The lymphocyte surface. II. Separation of Fc receptor, C3 receptor and surface immunoglobulin-bearing lymphocytes. *Proc. R. Soc. Lond. (Biol.) 187:* 65-81.

Ropke, C. and N. B. Everett. 1973. Small lymphocyte populations in the mouse bone marrow. *Cell Tissue Kinet. 6:* 499-507.

Ropke, C. and N. B. Everett. 1974. Migration of small lymphocytes in adult mice demonstrated by parabiosis. *Cell Tissue Kinet. 7:* 137-150.

Rosse, C. 1970. Two morphologically and kinetically distinct populations of lymphoid cells in the bone marrow. *Nature 227:* 73-75.

Rosse, C. 1971. Lymphocyte production and lifespan in the bone marrow of the guinea pig. *Blood 38:* 372-377.

Rosse, C. and J. M. Yoffey. 1967. The morphology of the transitional lymphocyte in guinea pig bone marrow. *J. Anat. 102:* 113-124.

Shortman, K. and H. Jackson. 1974. The differentiation of T lymphocytes. I. Proliferation kinetics and interrelationships of subpopulations of mouse thymus cells. *Cell Immunol. 12:* 230-246.

Stocker, J. W., D. G. Osmond and G.J.V. Nossal. 1974. Differentiation of lymphocytes in mouse bone marrow. III. The adoptive response of bone marrow cells to a thymus cell-independent antigen. *Immunology 27:* 795-806.

van Bekkum, D. W., M. J. van Noord, B. Maat and K. A. Dicke. 1971. Attempts at identification of hemopoietic stem cell in mouse. *Blood 38:* 547-558.

Yoffey, J. M. and F. C. Courtice. 1970. *Lymphatics, Lymph and the Lymphomyeloid Complex*. Academic Press, London.

Yoffey, J. M., G. Hudson and D. G. Osmond. 1965. The lymphocyte in guinea pig bone marrow. *J. Anat. 99:* 841-960.

Some Techniques for Analyzing B Lymphocyte Development

R. G. MILLER

Department of Medical Biophysics,
University of Toronto,
Toronto, Ontario, Canada

The cells involved in the humoral immune response can be conveniently organized into three compartments: the effector cell compartment, which contains mature antibody-producing cells, the "antigen-sensitive unit" which contains B lymphocytes and other cells required for their development into antibody-producing cells, and the stem-cell compartment which can replace cells in the antigen-sensitive unit as required. The final effector molecules of the response, antibody molecules, react specifically with the antigen which induced their production, the specificity being conferred by the amino acid sequence of the combining site of the antibody molecule. The ability to produce a specific antibody molecule precedes exposure to the antigen. It is expressed by surface receptors on the B lymphocyte which have combining sites that appear to be identical to the combining site of the antibody molecule ultimately produced by the progeny antibody-producing cells. A given B cell appears to have receptors of only one, or a limited number of specificities.

In an adult animal, B cells have relatively short life spans and are being continually replaced by new cells differentiating from more primitive precursors. The ultimate stem cell is not specificity restricted in the sense that a single stem cell appears to be capable of giving rise to B cells of all possible specificities. A central problem of immunology today is to explain how this specificity is acquired. There are two major opposing points of view. The first (germ-line theory) would say that the genetic coding for all possible specificities already exists in the stem cell and is differentially expressed during development. The major difficulties with this kind of theory involve the apparently large number of specificities which can be expressed and how the coding for all these specificities could be preserved in the germ line. The second theory (somatic mutation theory) would say that all specificities are generated by somatic mutation of one (or a few) genes at some point along the development pathway between the stem cell and the B cell. The major problem with this theory is that no completely satisfactory mechanism for generating the necessary somatic mutations has been documented.

If one could identify that stage in the B cell development pathway at which specificity restriction occurs, it might be possible to determine the mechanism of this specificity restriction. This objective has not yet been achieved although it has been and remains a long-term goal of our laboratory. In studies with mice, some progress has been made. Thus, from the work of Wu et al. ('63) and Edwards et al. ('70), the ultimate stem cell appears to be identical to or at least very closely related to the CFU (hemopoietic stem cell). A discrete development stage between the stem cell and the B cell has been identified (Lafleur et al., '72a). This cell, called the PB cell, is already specificity restricted but is still incapable of being induced by antigen (Lafleur et al., '73). The PB cell differentiates into a B_1 cell which in turn differentiates into a B_2 cell (Lafleur et al., '72b; Miller et al., '75). B_1 and B_2 cells have only been distinguished on the basis of differences in their sedimentation profile. The stage at which specificity is acquired lies between the stem cell and the PB cell. In birds, the specificity restriction appears to occur in a special anatomical location, the bursa of Fabricius. We find no evidence for a similar privileged anatomical location in mice (Phillips and Miller, '74). See Miller and Phillips ('75) and Miller et al. ('75) for reviews.

Progress in understanding B cell development has been almost entirely dependent upon the development of new technology. Rather than review the biological data, this article will concentrate on the technological advances made, the kinds of questions each allowed to be answered, and the kinds of questions each cannot answer. Current and possible future technological developments will also be discussed. Emphasis throughout is placed on those techniques we have employed ourselves or intend to employ in the future.

1. T6 chromosome marker. It has been known for some time that lethally irradiated animals can be protected from death by syngeneic bone marrow transplantation. What was not clear, until the work of Ford et al. ('56), was whether the grafted cells actually grew in the animal to form a chimera or whether they merely in some way stimulated the animal to

reconstitute itself. Ford et al. developed a substrain of CBA mice with a distinctive chromosomal abnormality present in all cells, the T6 marker. Irradiated CBA mice transplanted with T6-CBA bone marrow cells survived the irradiation and later showed T6 marker cells in bone marrow, thymus, spleen, lymph node, and Peyer's patches (Evans et al., '67). It is strongly implied that stem cells in the grafted bone marrow have reconstituted both the hemopoietic and immune systems. The major problem with this technique in developing a further understanding of the system is that the T6 marker only allows one to distinguish between graft and host. Thus, for example, B lymphocytes, granulocytes and red cells may have a common stem cell or may each have a separate stem cell. The two possibilities cannot be distinguished.

2. Spleen colony assay. Large, discrete cell colonies can be found in the spleens of lethally irradiated mice transplanted with syngeneic bone marrow (Till and McCulloch, '61). They found that each of these colonies is comprised of the descendants of a single hemopoietic stem cell: immature erythrocytes, immature granulocytes, megakaryocytes, additional stem cells and other cells that cannot be identified. The number of colonies obtained can be used to obtain a quantitative estimate of the number of stem cells present in the cell suspension transplanted. The composition of the colony can be used to infer something about the different types of progeny the stem cell can produce. However, there are two problems with this last method of analysis. First, the environment provided by the developing colony may be suitable for the growth of unrelated cells, either from the graft or the host. Thus, one cannot conclude that all cells found in the colony are in fact descendants of the stem cell that initiated it, or even of the grafted cell population. Secondly, not all of the possible differentiated descendants of the stem cell may be demonstrable in the colony, either because intermediate development stages of a particular cell line leave the colony before they are recognizable or because the development of a recognizable descendant takes longer than the time for which a discrete colony can be recognized (about 14 days).

3. Chromosome markers unique to one clone. Wu et al. ('67) have developed a procedure for repopulating a mouse with the progeny of a single CFU bearing a unique chromosome marker. Any cell in that mouse with the chromosome marker must necessarily have derived from the original CFU. By doing karyotype analysis on the cells of spleen colonies formed from the marrow of such a mouse, coupled with tests

for immature erythrocytes and granulocytes, they were able to demonstrate beyond any doubt that a single CFU can give rise to more CFU, red cells, and granulocytes. They were also able to demonstrate the presence of marked cells in lymph node and thymus (Wu et al., '68) from which one can conclude that T cells are almost certainly descended from CFU. This analysis, however, provided no definitive data on the origin of B cells (Miller and Phillips, '75). The major problem arises from the fact that several independent measurements must be made to show that a cell type is in the clone. Suppose one wants to know if cell type X is in the clone. Consider a cell suspension which is 10% cell type X by either morphological or functional criteria. One has determined independently that 1% of the cells in the suspension and 5% of cell type X are undergoing mitosis. Therefore half the mitotic cells are type X. On doing a karyotype analysis, one finds that 50% of the mitoses are marked. Since at present one cannot simultaneously do a karyotype analysis and a functional or morphological analysis, it is impossible to conclude whether X is in the clone or not. One sees that this problem becomes much more severe if cell X comprises a smaller fraction of the mitotic cells. Even if one can establish that cell X is in the clone, it is impossible to deduce how cell X is related to other cells in the clone. Thus two cells may develop along completely different pathways directly from the CFU as do, apparently, red cells and neutrophils or they may be closely related as are, for example, B cells and antibody-producing cells.

4. Physical cell separation-enrichment. One problem raised in the last section - determining whether a minority cell type is a member of a marked clone - could be solved if one could obtain the cell type in sufficiently high purity before karyotype analysis. Cell separations on the basis of differences in physical properties, e.g. density (Shortman, '60), electrophoretic mobility (Hannig, '71), sedimentation rate (Miller and Phillips, '69) are the simplest and most reproducible procedures available. The problem is to find a procedure which yields the cell type of interest in sufficiently high purity to be useful. Consider, for example, the problem of separating B lymphocytes from T lymphocytes. In density separation, each population is split up into multiple subpopulations of largely unknown significance and little overall separation is achieved (Shortman, '74). In sedimentation separation, T lymphocytes sediment slightly more rapidly than B lymphocytes, but the two populations largely overlap (Miller et al., '75). Only electrophoresis produces good separation (Zeiller et al., '74), but the procedure requires sophisticated and expensive special

equipment. When we wanted to establish whether B cells were descended from CFU, by chromosome marker analysis, we coupled physical cell separation with a biological property of B cells and their descendants: their ability to form rosettes. When a suspension of lymphoid cells is incubated with SRBC (sheep erythrocytes) under appropriate conditions, the SRBC will bind to those cells carrying receptors specific for SRBC to form an object called a rosette. Both B and T cells can form rosettes, but unless special precautions are taken to stabilize them, the T cell rosettes will rapidly dissociate (Elliott and Haskill, '73). Sedimentation separation separates cells primarily on the basis of differences in size and allows rosettes to be obtained in relatively high purity. In this way we were able to show that some, and perhaps all, rosette-forming cells in a mouse undergoing a secondary immune response to SRBC were descended from the CFU (Edwards et al., '70). If all (or nearly all) the rosette-forming cells analyzed were in fact in the B cell lineage, then the B cell is descended from the CFU. This problem is currently under reinvestigation (S. Abramson, private communication).

5. Physical cell separation-analysis. Using physical cell separation, it is often difficult to obtain the cell population of interest in high purity. However, provided one has a good quantitative assay for the cell of interest, purity may not be necessary. This is demonstrated by the sedimentation studies of Lafleur et al. ('72a, '72b, '73) defining the PB cell. The relative frequency of B cells in different cell suspensions can be determined in a quantitative transplantation assay through their ability to generate antibody-producing cells against SRBC. The test cell suspension, SRBC, and an excess of the other cells required to initiate B cells (T helper cells and A cells) are injected into a lethally irradiated mouse. An appropriate time later, the spleen of this mouse is assayed for its content of antibody-producing cells. This number is taken as a relative measure of the number of B cells injected. When this assay is used to measure the sedimentation profile of B cells, they are found to sediment in a narrow band with a characteristic sedimentation velocity of 3.0 mm/hr under the conditions used. The same profile is found for B cells in bone marrow, spleen or lymph node. Different results, however, are obtained if the transplantation assay is slightly modified. If the SRBC are injected 7 days after transplantation of the other cells, instead of with the other cells, a second band of activity, at larger sedimentation velocity (around 5 mm/hr) is found in bone marrow and spleen, but not in lymph node. This activity is due to the PB cell, whose properties were outlined in the Introduction. If one delays injecting the SRBC for a much

longer time, the activity per fraction increases substantially and, in bone marrow, the activity profile ultimately resembles that of the CFU.

The major drawback to this type of analysis is the assay system. It is indirect in the sense that one is measuring the frequency of the cell of interest (B cell, PB cell or CFU) by counting antibody-producing cells, the differentiated progeny. A large number of assumptions about cell proliferation rates, life spans, etc., all difficult to establish with certainty, must be made before conclusions can be reached. Much progress would be made by developing reliable, quantitative assays for a stage in the B cell pathway earlier than the final stage, the antibody-producing cell. Another major problem involves the necessity of using an *in vivo* transplantation assay in which one ends up assaying only what happens in the spleen. Unrealized problems concerning spleen trapping efficiency and cell migration could have quantitative, and perhaps even qualitative effects on the results. It is probably true that completely unambiguous precursor-progeny relationships will be established only after quantitative *in vitro* assays have been developed.

Despite these limitations, five discrete stages in B cell development have been characterized by sedimentation separation. The sedimentation profiles of each are shown in Figure 1.

6. Physical cell separation using biological markers. All B lymphocytes carry internally-produced immunoglobulin surface receptors. These can be reliably used to distinguish cells in the B cell lineage from all other cells. In addition, the B cell lineage can be divided into subpopulations according to the presence or absence of other surface markers such as a receptor for the C3 component of complement or a receptor for the Fc end of certain antibody molecules. Quantitative B cell assays have been developed by using radiolabelled or fluorescent antibody against these receptors. See, for example, the review of Osmond ('75). These assays tend to be tedious and are perhaps not too useful for studying minor subpopulations because of the time required to assay large numbers of cells. (Note that one would expect that the more primitive a cell, the less frequent it will be because of the clonal expansion usually involved in a differentiating system). One would also like to be able to separate the marked cells from all other cells so that one could follow their subsequent fate either *in vitro* or *in vivo*.

Methods capable of meeting the above criteria, i.e. rapid analysis of large numbers of cells plus the capability to

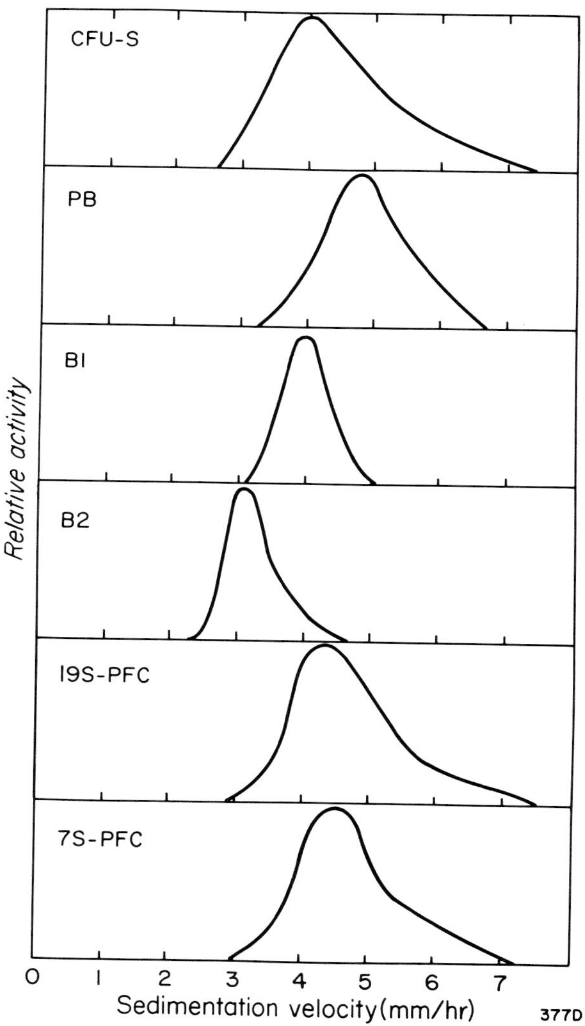

Fig. 1. Discrete stages in the B-cell development pathway as characterized by velocity sedimentation cell separation.

separate marked cells from all other cells, now exist. In practice, the most useful system has been that developed by Hulett et al. ('73). The principle is as follows. First one fluoresceinates an antibody directed against a surface marker on the cell of interest and reacts the antibody with the cell suspension containing this cell. The cell suspension is then squirted through a small orifice to form a continuous jet of fluid in free air. An intense monochromatic light beam is focussed on the jet and will excite fluorescence in the labelled cells. This can be detected in detectors equipped with filters to exclude scattered light from the incident beam. The jet of fluid does not persist for long as a continuous jet, but breaks up into droplets. It is easily arranged so that the jet breaks up into droplets of uniform size at a fixed point in space. Droplets form at the rate of about 50,000 per second. When a cell of interest is detected, the jet is momentarily electrically charged just as the cell reaches the break-off point. Thus the droplet carrying the cell of interest will have an electrical charge and can be electrically deflected away from the other uncharged droplets. Cells can be satisfactorily processed at a rate of 10,000 per second and retain full viability.

We have just built such a system and are in the final stages of testing it. We anticipate that it will be extremely useful in further studies of B cell differentiation.

LITERATED CITED

Edwards, G. E., R. G. Miller and R. A. Phillips. 1970. Differentiation of rosette-forming cells from myeloid stem cells. *J. Immunol. 105:* 719-729.

Elliott, B. E. and J. S. Haskill. 1973. Characterization of thymus-derived and bone marrow-derived rosette-forming lymphocytes. *Europ. J. Immunol. 3:* 68-74.

Evans, E. P., C. A. Ogden, C. E. Ford and H. S. Micklem. 1967. Repopulation of Peyer's patches in mice. *Nature 216:* 36-38.

Ford, C. E., J. L. Hamerton, D.W.H. Barnes and J. F. Loutit. 1956. Cytological identification of radiation-chimaeras. *Nature 177:* 452-454.

Hannig, K. 1971. Free-Flow Electrophoresis (a technique for continuous preparative and analytical separation). In *Methods in Microbiology.* J. R. Norris and D. W. Ribbons eds. Academic Press, London, vol. 5B, pp. 513-548.

Hulett, H. R., W. A. Bonner, R. G. Sweet and L. A. Herzenberg. 1973. Development and application of a rapid cell sorter. *Clin. Chem. 19:* 813-816.

Lafleur, L., R. G. Miller and R. A. Phillips. 1972. A quantitative assay for the progenitors of bone marrow-associated lymphocytes. *J. Exp. Med. 135:* 1363-1374.

Lafleur, L., B. J. Underdown, R. G. Miller and R. A. Phillips. 1972. Differentiation of lymphocytes: Characterization of early precursors of B lymphocytes. *Ser. Haematol. V:* 50-63.

Lafleur, L., R. G. Miller and R. A. Phillips. 1973. Restriction of specificity in the precursors of bone or marrow-associated lymphocytes. *J. Exp. Med. 137:* 954-966.

Miller, R. G. and R. A. Phillips. 1969. Separation of cells by velocity sedimentation. *J. Cell Physiol. 73:* 191-201.

Miller, R. G. and R. A. Phillips. 1975. Development of B lymphocytes. *Fed. Proc. 34:* 145-150.

Miller, R. G., R. M. Gorczynski, L. Lafleur, H. R. MacDonald and R. A. Phillips. 1975. Cell separation analysis of B and T lymphocyte differentiation. *Transpl. Rev. 25:* 59-97.

Osmond, D. G. 1975. Formation and maturation of bone marrow lymphocytes. *J. Reticuloendothelial Soc. 2:* 97-112.

Phillips, R. A. and R. G. Miller. 1974. Marrow environment not required for differentiation of B lymphocytes. *Nature 251:* 444-446.

Shortman, K. 1963. The separation of different cell classes from lymphoid organs. II. The purification and analysis of lymphocyte populations by equilibrium density gradient centrifugation. *Aust. J. Exp. Biol. Med. Sci. 46:* 375-396.

Shortman, K. 1974. Separation methods for lymphocyte populations. *Contemporary Topics in Molec. Immunol. 3:* 161.

Till, J. E. and E. A. McCulloch. 1961. A direct measurement of the radiation sensitivity of normal mouse bone marrow cells. *Rad. Res. 14:* 213-222.

Wu, A. M., J. E. Till, L. Siminovitch and E. A. McCulloch. 1967. A cytological study of the capacity for differentiation of normal hemopoietic colony-forming cells. *J. Cell. Physiol. 69:* 177-184.

Wu, A. M., J. E. Till, L. Siminovitch and E. A. McCulloch. 1968. Cytological evidence for a relationship between normal hematopoietic colony-forming cells and cells of the lymphoid system. *J. Exp. Med. 127:* 455-464.

Zeiller, K., G. Pascher, G. Wagner, H. G. Liebich, E. Holzberg and K. Hannig. 1974. Distinct subpopulations of thymus-dependent lymphocytes. Tracing of the differentiation pathway of T cells by use of preparatively electrophoretically separated mouse lymphocytes. *Immunol. 26:* 995-1012.

Hemopoietic Stem Cell Migration

N. B. EVERETT AND W. D. PERKINS

Department of Biological Structure
University of Washington
School of Medicine
Seattle, Washington 98195

Appropriately the stem cell has been functionally defined as a cell having the capacity for extensive proliferation resulting in renewal of its own kind as well as giving rise to fully differentiated cells (Becker, McCulloch and Till '63; Siminovitch, McCulloch and Till, '63; Tyler and Everett, '66). Although there are no uniformly acceptable morphological criteria for characterizing hemopoietic stem cells, the problem of identification is believed to have been simplified by the convincing evidence provided during recent years that these cells have pluripotential capacity (Becker et al., '63; Siminovitch et al., '63; Tyler and Everett, '66; Murphy, Bertles and Gordon, '71; Trentin, '71; Nowell et al., '70, Micklem et al., '66; Wolf and Trentin, '68). The problem has also been narrowed by having essentially eliminated reticular cells as candidates for stem cells or as stem cell precursors. Although we recognize that some investigators support the more traditional view that reticular cells have stem cell capacity, the results of our own studies as well as of others have led us to conclude that the view is no longer tenable Everett and Tyler, '67a).

The strongest evidence for this comes from kinetic studies of cell proliferation employing ^{3}H-thymidine and radioautography in growing animals and, more particularly, from studying hemopoietic recovery in animals subjected to lethal and sublethal doses of irradiation subsequent to labeling a high percentage of reticular cells with ^{3}H-thymidine during the period of rapid growth. The results have shown that reticular cells are slowly proliferating and their rate of formation parallels the growth rate of the animal. Thus they are long lived. Numerous distinctly labeled cells have, in fact, been noted in hemopoietic tissues for as long as one year after labeling. In addition, these cells have been shown to be very radioresistant. In rats having a high percentage of labeled reticular cells there was no detectable decrease in number following sublethal or lethal doses of Co^{60} irradiation. Further, during hemopoietic recovery, no labeled precursors for any of the blood cell lines have been observed.

Although the small lymphocyte has been a strong candidate for the pluripotent stem cell, most investigators would agree that this is no longer the case, particularly in view of the failure to obtain hemopoietic recovery following radiation depletion of hemopoietic tissues by transfusion of relatively pure populations of lymphocytes, e.g., from lymph node or thymus (Micklem and Ford, '60; Curry, Trentin and Cheng, '67). There is, however, general agreement that the pluripotent stem cell of the adult is marrow-derived and falls within a class of mononuclear cells which includes lymphoid cells (Murphy et al., '71; Cudkowicz et al., '64; Osmond, Miller and Yoshida, '73; Yoffey, '73). Yoffey ('73) designates this category of marrow mononuclear cells as the lymphocyte-transitional cell compartment.

In attempts to more precisely identify the marrow-derived cells which migrate to hemopoietically depleted tissues and promote recovery of the respective blood cell lines, we have made rather extensive use of radioautography in irradiated parabiotic animals (Tyler and Everett, '66; '72). Pairs of rats from an inbred Lewis colony were joined in parabiosis and two weeks later were given 1000 R (Co^{60}) while the hind limbs of one member were shielded. (This was to restrict the source of any potential source of stem cells to a limited bone marrow compartment and to increase the possibility of identifying migrant precursor cells.) Twelve hours after irradiation the shielded animals were given ^{3}H-thymidine (1 μCi/g body weight) while the cross circulation was arrested by a rubber compression clamp. After 15 minutes the non-shielded rat was given 5 mg of non-radioactive thymidine and the compression clamp was released, restoring the cross circulation. One pair was sacrificed 1 1/2 days postinjection and at daily intervals thereafter for 8 days. The ^{3}H-thymidine injections were repeated daily in the same way for the animals remaining after each sacrifice. Thus the number of injections per animal ranged from 1 to 8. Radioautographs were made of blood, and of tissue smears and sections of bone marrow, thymus, spleen and mesenteric lymph nodes of each pair.

A detailed cellular analysis of the recovery which ensued in the respective hemopoietic compartments of the parabionts established that cells responsible for recovery had their origin in the shielded marrow. Further, it was concluded that it was cells of a category we designated as "monocytoid" which served as stem cells for both the erythrocytic and granulocytic cells in the irradiated bone marrow. These studies also implicated monocytoid cells as stem cell precursors in spleen, mesenteric lymph nodes and thymus. We have applied the term monocytoid to these cells since they have morphological features in common with the earlier stages of the monocyte category and no distinct morphological boundary could be established to distinguish them from cells of the monocyte cell line. They compare to the medium lymphocyte in size with a nuclear diameter of 8-10 μ. At the light microscope level the nucleus, generally centrally located, is irregularly shaped and evidences many invaginations and folds. The chromatin is typically leptochromatic and the cytoplasm is quite abundant and pale staining, only slightly basophilic. It is recognized that some investigators would include them as lymphoid cells but we feel that they are morphologically as well as functionally distinct from the typical medium to large lymphocytes of lymphoid tissues, including thoracic duct lymph. More convincing evidence that monocytoid cells serve as stem cells for thymocytes came from a later and more extensive series of studies (Everett and Tyler, '69) employing parabiosis and shielding as described above as well as the transfusion of labeled bone marrow cells into irradiated recipients. The latter provided for following heavily labeled cells, administered as a single pulse, into the depleted thymus and allowed for determining rather precisely a time sequence of their development during thymic recovery.

As implied, the only suspect immigrant cells found in the thymus which could have served as thymic cell precursors were monocytoid cells. In the parabiotic series, labeled monocytoid cells were noted within the thymus of the non-shielded members as early as 2 1/2 days post-irradiation and mostly within the connective tissue septa, within or adjacent to blood vessels. From 2 1/2 - 4 1/2 days post-irradiation labeled immigrant cells, many in mitosis, and their progeny were noted within the lobules near blood vessels at the outer edge of the thymic cortex. Essentially no labeled lymphocytes were noted in blood until 5 1/2 days post-irradiation. The results of the bone marrow transfusion experiments provided further evidence that the labeled dividing blasts noted at 3 1/2 days within the thymic cortex of nonshielded parabionts were descendants of the labeled immigrant monocytoid cells observed within the blood vessels and connective tissue at 2 1/2 days. At 16 hours post-transfusion, numerous labeled monocytoid cells were noted in the thymus. Approximately 20% were intravascular, about 50% were in the connective tissue septa near vessels and the remainder were within the outer rim of the thymic cortex. At 24 and 40 hours post-transfusion, small clusters of labeled blast cells were observed in the outer cortex which were more weakly labeled than the immigrant cells. It may be noted that at 40 hours post-injection significant numbers of distinctly labeled fibrocyte-like cells were noted within the septa. These were interpreted as distorted migrating forms of monocytoid cells. Heavily labeled macrophages were also noted within the septa and within the medulla at this interval. No labeled monocytoid or blast cells were noted 64 hours post-transfusion, although heavily labeled macrophages as well as labeled fibrocyte-like cells were present at this time in about the same numbers as at the 40-hour interval. These results were interpreted as indicating that from one morphological cell type migrating into the thymus (monocytoid cell) three morphological cell types were derived: blast cells, fibrocyte-like cells and macrophages.

Monocytoid cells were very conspicuous in blood of parabiotic rats following irradiation and, in fact, after 2 1/2 days comprised from 50-90% of the nucleated cells in blood. Further, between 3 1/2 and 5 1/2 days postirradiation, leukocytes in the blood of nonshielded parabionts were of the monocytoid type. Only small numbers of lymphocytes were observed in the blood of either the shielded or nonshielded parabionts before 6 1/2 days postirradiation.

Taking note of the observations from the parabiotic studies that the hemopoietic recovery of the lethally irradiated bone marrow resulted from bone marrow cells of the protected marrow which crossed via the blood, and of the great increase in blood monocytoid cells of the partially shielded animals, Lord ('67) compared the effectiveness of blood buffy coat cells with that of bone marrow cells in promoting erythropoietic recovery of lethally irradiated rats. Animals with both hind limbs shielded were given 870 R whole body Co^{60} irradiation. Five and one half days later the buffy coat cells were recovered from the blood of these animals and transfused in the dose range of 4,500 to 30,000 cells into lethally irradiated rats. The erythropoietic recovery was measured in these and in similarly irradiated rats transfused with normal bone marrow cells by determining the numbers of reticulocytes appearing in the circulation. It was found that the buffy coat recovery curve paralleled that of the bone marrow and indicated that the stem cells of the buffy coat were probably identical to the stem cells of the marrow. Further, the calculations revealed that the buffy coat was approximately 9-10 times richer in stem cells than the bone marrow.

Earlier studies have provided evidence for circulating stem cells in blood (Popp, R.A.,'60; Goodman and Hodgson, '62) and it has been reported that normal buffy coat is approximately 100 times less rich in stem cells than normal bone marrow (Trobaugh and Lewis, '64). Lord rarely obtained measurable recovery following the transfusion of normal buffy coat cells since with the numbers used, up to 400,000, death usually ensued at an early stage.

In autologous cultures of blood buffy coat cells of rabbit blood within Millipore diffusion chambers placed intraperitoneally in severely bled donors, Grigoriu, Antonescu, and Iercan ('71) noted a substantial formation of erythroblasts of all stages. These reached a peak, comprising approximately 45% of the nucleated cells present in the chambers by day 6 or 7 of culture. In many of the chambers basophilic mononuclear cells were noted from the fourth to the eighth day of culture which were interpreted as transitional cells in accord with Yoffey's designation. A direct correlation was reported between the numbers of these cells and basophilic erythroblasts.

Another method which we have used to promote the appearance of monocytoid cells in large numbers into the peripheral circulation and one that is simply executed is that of blood withdrawal. We have found that by 3 days after withdrawing approximately 1/3 the blood volume of a young adult rat, usually by heart puncture, a high percentage of the mononuclear cells in the blood are of the monocytoid type. For light microscope study these have been conveniently recovered from the buffy coat after slow speed centrifugation. Light microscope photomicrographs of smear preparations of representative monocytoid cells of these buffy coat preparations, stained with MacNeil's tetrachrome, are shown in Figure 1. It may be noted that these cells vary somewhat in morphology but several show the typical features as described for the monocytoid cells which appeared in the blood and other tissues of the irradiated parabionts and evidenced pluripotential stem cell function.

Ultrastructural studies were also carried out 3 days after partial exsanguination to more specifically determine the kinds of cells present and their morphological characteristics. Since anticoagulants could alter the morphological features of cells in the peripheral blood, the cells were prepared for electron microscopy without their use. To do this, peripheral blood was removed from the tail veins of rats by means of small capillary tubes and immediately separated by centrifugation while the fibrin clot was being formed. In this manner the erythrocytes sedimented faster than the fibrin. As a result, the clot remained in the serum portion of the separated blood and entrapped the slower sedimenting leukocytes. The leukocyte-fibrin clot was then removed from the capillary tube, cut into small pieces, fixed in formaldehyde-glutaraldehyde

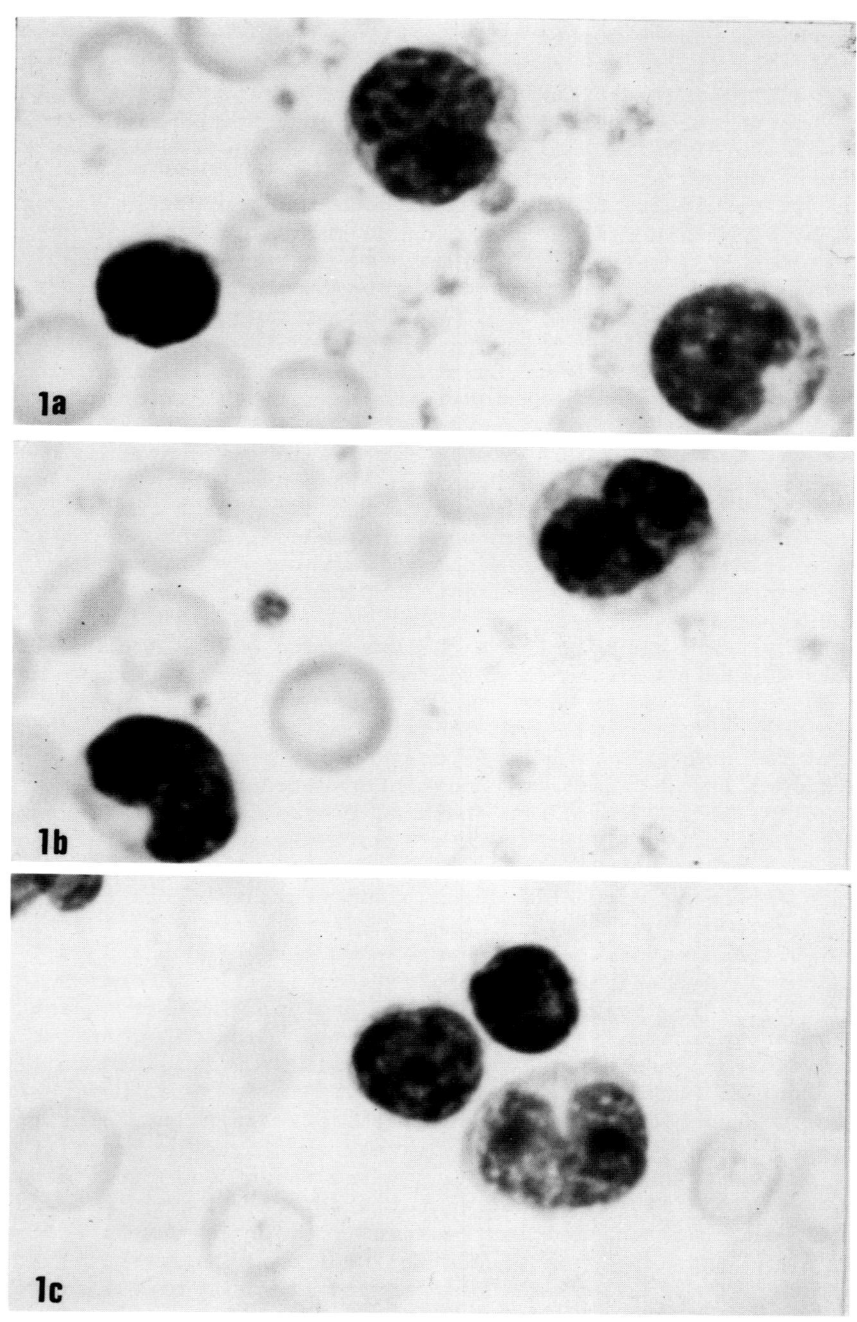
1a
1b
1c

(Karnovsky, '65) and prepared for conventional electron microscopy. The ultrastructural findings of these experiments corroborated the light microscopic observations. A variety of mononuclear cells was seen and included small lymphocytes, medium lymphocytes, monocytes and cells of the monocytoid category as cells previously described. The ultrastructural features of these cells are shown in Figures 2-6.

During the course of studies designed to determine the origin of mononuclear cells which appear early in inflammatory exudates and differentiate into monocytes and macrophages (Everett and Tyler, '68) by employing a modified form of the Rebuck skin window technique (Rebuck and Crowley, '55), it was noted that the cells which first appeared were similar in morphology to cells which we proposed as stem cells for the erythroid and granulocytic series of bone marrow and for lymphocytes of the thymus in irradiated rats. Thus it seemed propitious to determine if these mononuclear cells might have the potential for promoting erythropoietic recovery in lethally irradiated animals. Accordingly, experiments were carried out using C57Bl mice (Tyler, Rosse and Everett, '72). Mice were given 800 R total body irradiation and 18 hours later each of 16 animals was transfused with 1×10^6 subcutaneous exudate cells which were collected at 18 hours after inserting glass coverslips to produce the inflammatory responses. For comparison, three groups of other irradiated mice were injected respectively with bone marrow cells, thoracic duct lymphocytes and blood leukocytes from nontreated mice.

Erythroid recovery was assessed by the method of measuring the uptake of radioactive iron in the spleen (Hodgson, '62; Smith, '64) 18 hours after intraperitoneal injections of 5 µCi of ^{59}Fe ferrous sulfate administered to each mouse 7 days after the cellular transfusions.

It was found that the serum exudate cells promoted erythroid recovery but were only 1/10 as effective as comparable numbers of bone marrow cells. They were at least twice as effective as an equal number of blood leukocytes. The ^{59}Fe up-

---

Fig. 1. Light microscope photomicrographs of mononuclear cells recovered from the buffy coat of rat blood 3 days after withdrawing approximately 1/3 of the animal's blood volume which illustrate the range in the morphology of these cells. 1A shows two monocytoid cells, and a small lymphocyte at the lower left. 1B shows two monocytoid cells. 1C shows two lymphocytes above, medium and small, and a cell considered to have typical monocyte morphology.

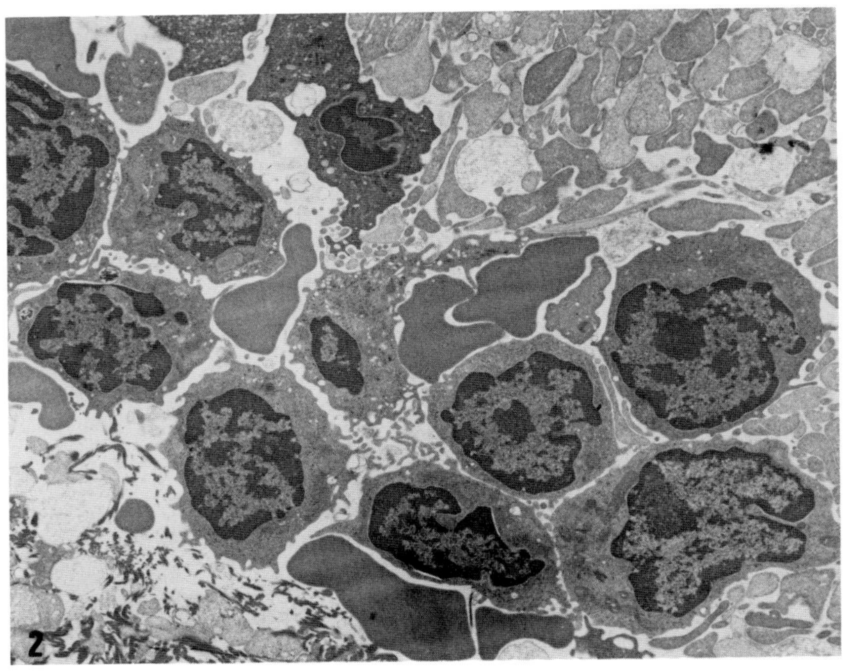
2

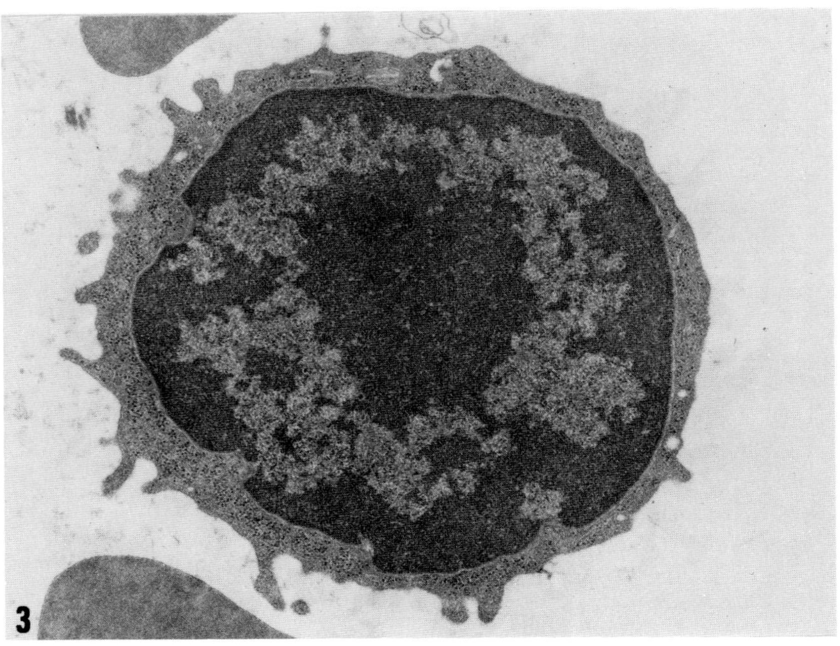
3

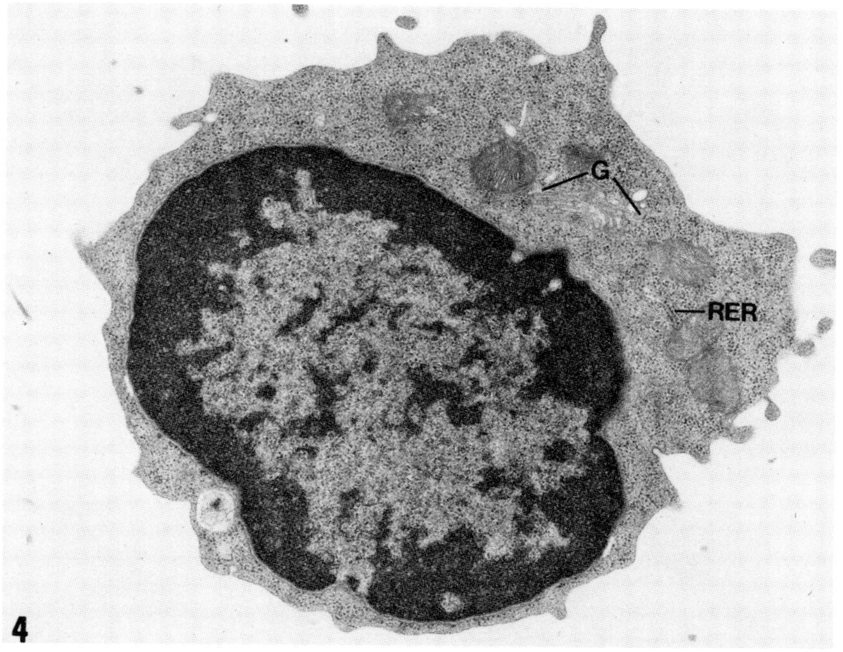

Figs. 2-6. Electron micrographs of mononuclear cells recovered from the buffy coat of rat blood 3 days after withdrawing approximately 1/3 of the animal's blood volume.
Fig. 2. Low power electron micrograph of peripheral blood monocytoid cells which illustrates the range in morphology of these cells.
Fig. 3. Electron micrograph of a typical small lymphocyte with the characteristic high nuclear to cytoplasmic ratio. The nucleus is spherical and contains patchy heterochromatin. The cytoplasm is scant and there are few organelles.
Fig. 4. Electron micrograph of a medium lymphocyte. The nuclear to cytoplasmic ratio is about one. Cytoplasmic organelles include strands of rough endoplasmic reticulum (RER) and a small Golgi apparatus (G).

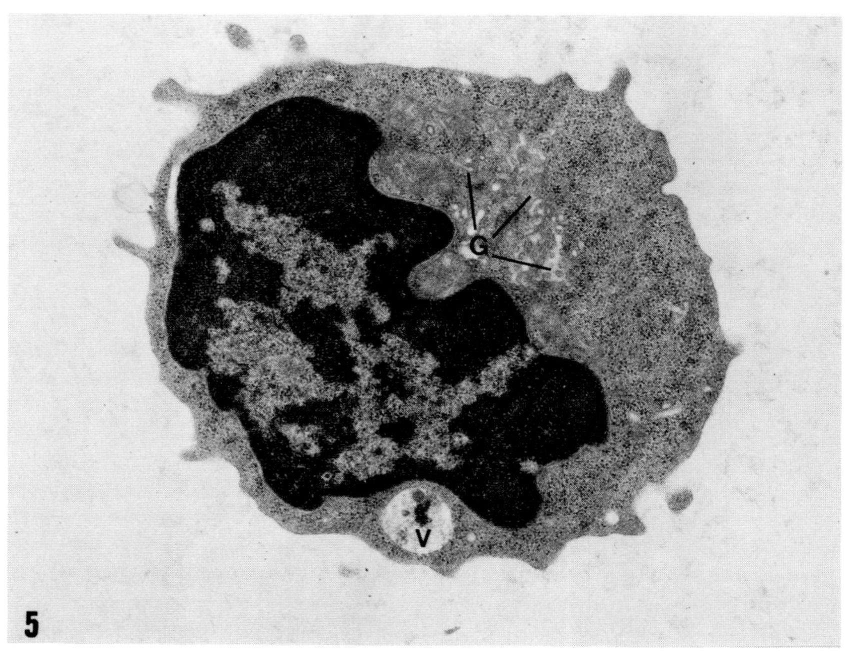
G
V
5

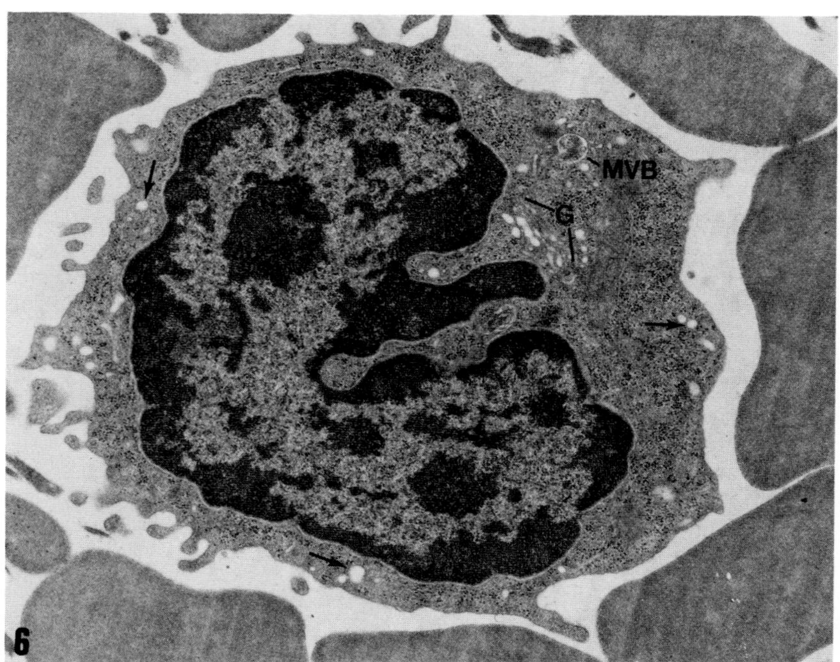
MVB
G
6

take in spleens of thoracic duct lymph recipients was no greater than in the controls.

The effectiveness of the exudate cells was destroyed by giving 1000 R *in vitro* irradiation before transfusion.

Differential counts made on supravitally stained exudate cells showed that approximately 54% were neutrophils, 7% monocytes with distinct rosettes of neutral red vacuoles, 23% monocytoid cells with neutral red vacuoles but no rosettes, 10% macrophages, 3% small lymphocytes and 3% damaged and nonclassifiable cells.

Until recently at least, the reports concerning the identity and source of mononuclear cells which appear in inflammatory exudates and from which macrophages are derived have been somewhat controversial. Rebuck and associates (Sieracki and Rebuck, '60) implicated small lymphocytes as the precursor cells and Kosunen et al. ('63) suggested that large or medium lymphocytes were primarily involved. Volkman and Gowans ('65a,b), however, using the subcutaneous coverslip technique, provided strong evidence for eliminating lymphocytes as macrophage precursors and concluded that monocytes from bone marrow and spleen were the cells of origin. In accord with the latter observations are those of Spector and associates (Spector, '67; Spector and Coote, '65; Spector, Walter and Willoughby, '65) who found that essentially all mononuclear cells which appeared in inflammatory exudates were from circulating monocytes. In view of these conflicting results, we approached the problem in rats and as indicated in a preliminary report, our findings were essentially in accord with those of Volkman and Gowans that mononuclear cells which appeared in subcutaneous exudates and gave rise to macrophages were of bone marrow origin (Everett and Tyler, '68). The results of recent fine structural studies of the mononuclear cells which appear in the subcutaneous tissue within the first few hours after coverslip placement include a high percentage of cells of the monocytoid category as well as cells of definite monocyte morphol-

---

Fig. 5. Electron micrograph of a monocytoid cell having an eccentric nucleus that is irregular in outline. The cytoplasm contains a vacuole with inclusions (V) and an extensive Golgi apparatus (G).

Fig. 6. Electron micrograph of a monocyte that has an eccentric nucleus which is bean shaped. The cytoplasm contains a Golgi apparatus (G), multivesicular body (MVB) and a number of small vesicles (arrows).

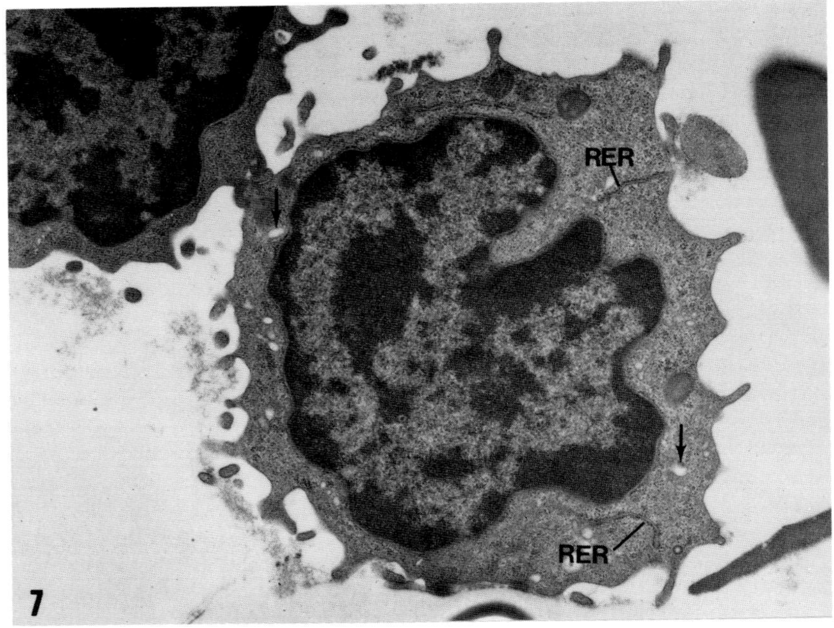
RER
RER
7

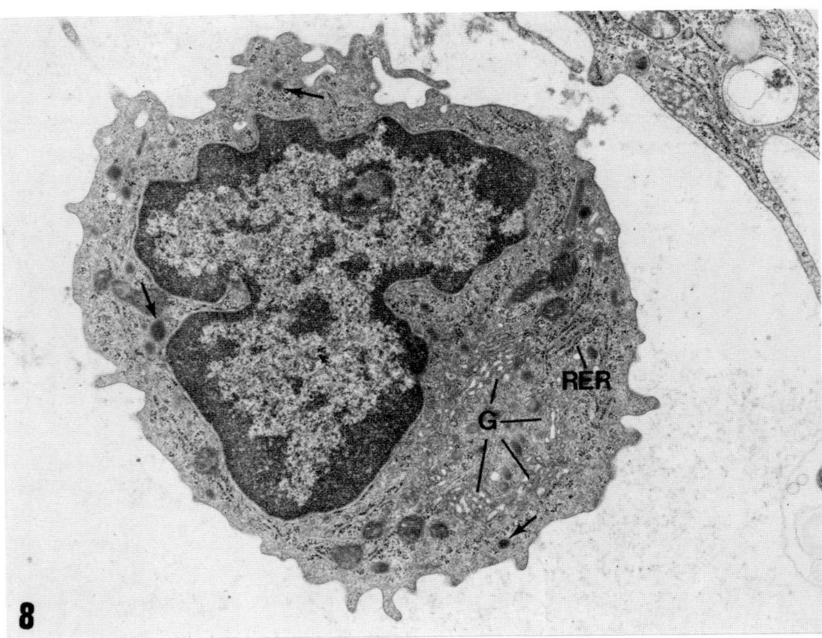
RER
G
8

ogy. With time, a higher percentage of the cells evidence characteristics of the monocyte-macrophage cell line. These studies were made from sections of subcutaneous tissues removed at intervals from 6 to 36 hours after inserting the coverslips, fixed in formaldehyde-glutaraldehyde and prepared for conventional electron microscopy. The fine structural features of cells in this developmental sequence are shown in Figures 7-9.

An extension of the radioautographic studies for determining the tissue origin of the monocytoid cells in the exudates has provided definite evidence that the bone marrow is the primary source. The procedure has been that of injecting ^{3}H-thymidine into the tibial marrow as described previously (Everett and Caffrey, '67), with appropriate measures to preclude labeling cells in other compartments, 48 hours before inserting coverslips. At intervals thereafter, smear preparations were made of the exudate fluid and processed by conventional radioautography. The labeled monocytoid cells shown in Figure 10 from exudate fluid 24 hours after inserting coverslips are typical of the primitive bone marrow-derived cells which appeared in the exudates.

In summary it may be stated that mononuclear cells which appear in large numbers within peripheral blood following irradiation or massive bleeding, and in subcutaneous exudates in response to an inflammatory stimulus, include cells having stem cell capacity. The fine structural observations have in particular provided convincing evidence that this broad category of cells which includes lymphocytes as well as cells of the monocyte-macrophage series also includes cells which are morphologically distinct from cells of these well recognized

---

Figs. 7-9. Electron micrographs of cells recovered from subcutaneous exudates of rats from 6 to 36 hours after inserting glass coverslips.
Fig. 7. Electron micrograph of a monocytoid cell having a nucleus that is irregular in outline and contains patchy heterochromatin. The cytoplasm contains a few strands of rough endoplasmic reticulum (RER) and a number of vacuoles (arrow).
Fig. 8. Electron micrograph of a monocyte having a nucleus that is irregular in outline and indented, and contains marginated heterochromatin. The cytoplasm contains rough endoplasmic reticulum (RER), an extensive Golgi apparatus (G), and many small granules (arrows).

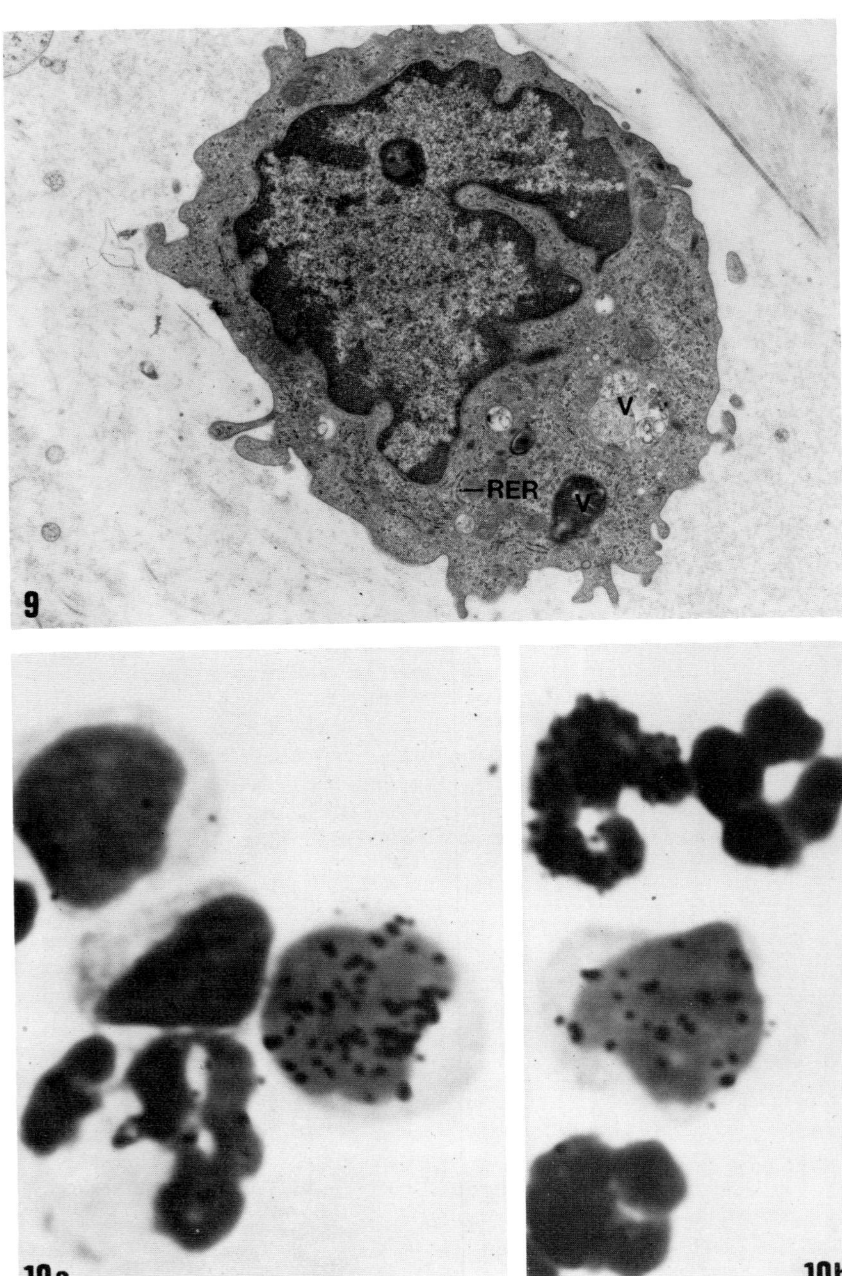
9
RER
V
V
10a
10b

and definable cell lines. This latter group of cells, designated as monocytoid, is of bone marrow origin and is believed to include the pluripotential stem cells.

It would appear that there are certain similarities in the ultrastructural features of the monocytoid cells with those reported by others (Murphy et al., '71; Rubinstein and Trobaugh, '73; van Bekkum et al., '71) for the presumptive stem cells isolated from bone marrow.

This work supported by US ERDA Contract AT(45-1)-2225 and USPHS Grant AI-08910 from the National Institutes of Health.

Fig. 9. Electron micrograph of a macrophage having a nucleus that is eccentric and irregular in outline and has marginated heterochromatin. The cytoplasm contains rough endoplasmic reticulum (RER). Typical of macrophages are a number of phagocytic vacuoles (V).

Fig. 10. Radioautographs of cells recovered from subcutaneous exudates of rats 24 hours after insertion of glass coverslips, and 72 hours after localized injections of ^{3}H-thymidine into the tibial marrow parenchyma. 10A shows a labeled monocytoid cell, two nonlabeled large mononuclear cells, and a weakly labeled neutrophil. 10B shows a labeled mononuclear cell, considered to be within the monocytoid category and three neutrophils, one of which is heavily labeled.

LITERATURE CITED

Becker, A. J., E. A. McCulloch and J. E. Till. 1963. Cytological demonstration of the clonal nature of spleen colonies derived from transplanted mouse marrow cells. *Nature (Lond.) 197:* 452-454.

Cudkowicz, G., A. C. Upton, M. Shearer, and W. L. Hughes. 1964. Lymphocyte content and proliferative capacity of serially transplanted mouse bone marrow. *Nature (Lond.) 201:* 165-167.

Curry, J. L., J. J. Trentin, and V. Cheng. 1967. Hemopoietic spleen colony studies. III. Hemopoietic nature of spleen colonies induced by lymph node or thymus cells with or without phytohemagglutinin. *J. Immunol. 99:* 907-916.

Everett, N. B. and R.W.C. Tyler. 1967a. Radioautographic studies of reticular and lymphoid cells in germinal centers of lymph nodes. In: *Germinal Centers in Immune Responses*. H. Cottier, N. Odartchenko, R. Schindler and C. C. Congdon, eds. Springer-Verlag, Berlin, pp. 145-151.

Everett, N. B. and R. W. Caffrey. 1967b. Radioautographic studies of bone marrow small lymphocytes. In: *Lymphocyte in Immunology and Haemopoiesis*. J. M. Yoffey, ed. Edward Arnold Ltd., London, pp. 108-119.

Everett, N. B. and R. W. Tyler. 1968. Studies of lymphocytes: relationship to mononuclear cells of inflammatory exudates. *Biochem. Pharmacol. Suppl:* 185-196.

Everett, N. B. and R. W. Tyler. 1969. Radioautographic studies of the stem cell in the thymus of the irradiated rat. *Cell Tissue Kinet. 2:* 347-362.

Goodman, J. W. and G. S. Hodgson. 1962. Evidence for stem cells in the peripheral blood of mice. *Blood 19:* 702-714.

Grigoriu, G., M. Antonescu, and E. Iercan. 1971. Evidence for a circulating stem cell: newly formed erythroblasts found in autologous leukocyte-filled diffusion chambers inserted into bled rabbits. *Blood 37:* 187-195.

Hodgson, G. S. 1962. Erythrocyte Fe^{59} uptake as a function of bone marrow dose injected in lethally irradiated mice. *Blood 19:* 460-467.

Karnovsky, M. J. 1965. A formaldehyde-glutaraldehyde fixative of high osmolarity for use in electron microscopy. *J. Cell Biol. 27:* 137A.

Kosunen, T. U., B. H. Waksman, M. H. Flax and W. S. Tihen. 1963. Radioautographic study of cellular mechanisms in delayed hypersensitivity. I. Relayed reactions to tuberculin and purified proteins in the rat and guinea pig.

Immunology 6: 276-290.

Lord, B. I. 1967. Improved erythropoietic recovery in lethally irradiated rats after transfusion of buffy coat cells from the blood of partially shielded, heavily irradiated donors. *Nature (Lond.) 214:* 924-925.

Micklem, H. S. and C. Ford. 1960. Proliferation of injected lymph node and thymus cells in lethally irradiated mice. *Plastic Reconstruc. Surg. 26:* 436-441.

Micklem, H. S., C. E. Ford, E. P. Evans and J. Gray. 1966. Interrelationships of myeloid and lymphoid cells: studies with chromosome-marked cells transfused into lethally irradiated mice. *Proc. Roy. Soc. B 165:* 78-102.

Murphy, M. J., J. F. Bertles, and A. S. Gordon. 1971. Identifying characteristics of the haematopoietic precursor cell. *J. Cell Sci. 9:* 23-48.

Nowell, P. C., B. E. Hirsch, D. H. Fox and D. B. Wilson. 1970. Evidence for the existence of multipotential lympho-hematopoietic stem cells in the adult rat. *J. Cell Physiol. 75:* 151-158.

Osmond, D. G., S. C. Miller and Y. Yoshida. 1973. Kinetic and haemopoietic properties of lymphoid cells in the bone marrow. In: *Haemopoietic Stem Cells.* Ciba Foundation Symposium 13, New Series. Elsevier, North-Holland, Amsterdam, pp. 131-156.

Popp, R. A. 1960. Erythrocyte repopulation in X-irradiated recipients of nucleated peripheral blood cells of normal mice. *Proc. Soc. Exp. Biol. Med. 104:* 722-724.

Rebuck, J. W. and J. A. Crowley. 1955. A method of studying leukocytic functions *in vivo*. *Ann. N.Y. Acad. Sci. 59:* 757-805.

Rubinstein, A. S. and F. E. Trobaugh, Sr. 1973. Ultrastructure of presumptive hematopoietic stem cells. *Blood 42:* 61-80.

Sieracki, J. C. and J. W. Rebuck. 1960. Role of the lymphocyte in inflammation. In: *The Lymphocyte and Lymphocytic Tissue.* International Academy of Pathology Monograph No. 1. J. W. Rebuck, ed. Hoeber, New York, pp. 71-81.

Siminovitch, L., E. A. McCulloch and J. E. Till. 1963. The distribution of colony forming cells among spleen colonies. *J. Cell. Comp. Physiol. 62:* 327-336.

Smith, L. H. 1964. Marrow transplantation measured by uptake of Fe^{59} by spleen. *Am. J. Physiol. 206:* 1244-1250.

Spector, W. G. 1967. Histology of allergic inflammation. *Brit. Med. Bull. 23:* 35-38.

Spector, W. G. and E. Coote. 1965. Differentially labelled blood cells in the reaction to paraffin oil. *J. Path.*

Bact. 90: 589-598.

Spector, W. G., M.N.I. Walter and D. A. Willoughby. 1965. The origin of the mononuclear cells in inflammatory exudates induced by fibrinogen. *J. Path. Bact. 90:* 181-192.

Trentin, J. J. 1971. Determination of bone marrow stem cell differentiation by stromal hemopoietic inductive microenvironments (HIM). *Am. J. Path. 65:* 621-628.

Trobaugh, F. E. and J. P. Lewis. 1964. Repopulation potential of blood and marrow. *J. Clin. Invest. 43:* 1306. (Abstract).

Tyler, R.W.C. and N. B. Everett. 1966. A radioautographic study of hemopoietic repopulation using irradiated parabiotic rats. Relation to the stem cell problem. *Blood 28:* 873-890.

Tyler, R. W. and N. B. Everett. 1972. A radioautographic study of cellular migration using parabiotic rats. *Blood 39:* 249-266.

Tyler, R. W., C. Rosse, and N. B. Everett. 1972. The hemopoietic repopulating potential of inflammatory exudate cells. *J. Reticuloendothel. Soc. 11:* 617-626.

van Bekkum, D., M. van Noord, B. Maat and K. Dicke. 1971. Attempts at identification of hematopoietic stem cells in mouse. *Blood 38:* 547-558.

Volkman, A. and J. L. Gowans. 1965a. The production of macrophages in the rat. *Brit. J. Exp. Path. 46:* 50-61.

Volkman, A. and J. L. Gowans. 1965b. The origin of macrophages from bone marrow in the rat. *Brit. J. Exp. Path. 46:* 62-70.

Wolf, N. S. and J. J. Trentin. 1968. Hemopoietic colony studies. V. Effect of hemopoietic organ stroma on differentiation of pluripotent stem cells. *J. Exp. Med. 127:* 205-214.

Yoffey, J. M. 1973. Stem cell role of the lymphocyte-transitional cell (LT) compartment. In: *Haemopoietic Stem Cells.* Ciba Foundation Symposium 13, New Series. Elsevier, North-Holland, Amsterdam, pp. 5-39.

Regulation of Stem Cells After Local Bone Marrow Injury: The role of an osseous environment

HARVEY M. PATT AND MARY A. MALONEY

Laboratory of Radiobiology
University of California
San Francisco, California 94143

Comparison of various vertebrates reveals a progressive confinement of blood formation, from widely scattered regions of the body in fishes to restricted foci in amphibians and reptiles and finally to a primary focus in birds and mammals (Siegel, '70). Since the evolution of bone marrow as the major site of blood formation in terrestrial vertebrates occurred *pari passu* with the development of an internal framework of cellular bone, the question naturally arises about the significance of the connection between bone and its hematopoietic marrow. Did the eventual confinement of blood formation (excluding of course lymphoid cells) to a bony setting merely reflect an efficient utilization of open space or did the intimacy of osseous tissue provide a more advantageous environment and perhaps a reservoir of essential cellular elements? It is clear that there is no obligatory functional relationship between bone and blood-forming tissue even in higher vertebrates as indicated by the independent occurrence of hematopoiesis in yolk sac, fetal liver and spleen, and in various tissues after bone marrow damage. Yet, it is also clear that bone imparts a special quality for hematopoiesis, since wherever true bone is formed, irrespective of the initiating mechanism, it generally leads to a new hematopoietic marrow (Trueta, '68; Urist et al., '69). Unlike the surge of compensatory blood formation that may occur in an adult tissue such as spleen, hematopoiesis associated with ectopic bone takes place in the absence of a compensatory requirement. Significantly, the destruction, removal or heterotopic implantation of a discrete segment of bone marrow leads to a strikingly similar cascade of events (Amsel et al.,

'69; Brånemark et al., '64; Patt and Maloney, '70; Röhlich, '41; Tavassoli and Crosby, '68). In each case, blood-forming tissue emerges in intimate association with bone. In this paper we will consider the possible nature of bone-bone marrow relationships within the context of stem cell regulation in response to local injury. For our purpose, local injury will refer to those circumstances in which damage includes the marrow matrix and thus in which it is necessary to restore or regenerate marrow as an organized tissue. This is in contrast to other types of injury where restoration of normal marrow function may depend merely upon the availability of comparatively few hematopoietic stem cells as in the case, for example, of moderate radiation damage.

CELLS OF BONE AND MARROW

Viewed in developmental perspective, it is not surprising that bone and its associated marrow contain cells in common. Precursors of the specialized cells of bone cover all bone surfaces and it is particularly important to recall that marrow is surrounded by mesenchymal cells which line the external endosteal envelope of trabecular bone as well as the internal endosteal layer of cortical bone (Bloom et al., '41; Rasmussen and Bordier, '74). The endosteal covering can be regarded as a condensed peripheral layer of the stroma of bone marrow (McLean and Urist, '68). Thus, it is understandable that cells characteristic of bone can be incorporated into the marrow stroma as reticular cells and, moreover, that reticular cells of the marrow stroma can display osteogenic activity and transform into cells of bone.

Some insight about the connection between the cells of bone and marrow is provided by the sequence of events in a heterotopic marrow implant or in an evacuated medullary cavity. When an autologous marrow fragment is placed in a subcutaneous site, frankly hemic cells disappear and persisting reticular cells form an ossicle that eventually contains an active marrow which may remain for many months (Tavassoli and Crosby, '68). The nature of the bone induction mechanism is unknown but it is known from immunogenetic studies that the new ectopic bone is derived from stromal cells in the implanted marrow (Friedenstein et al., '68). This property of normal marrow stroma reappears in the connective tissue that forms a few days after evacuation of a medullary cavity (Patt and Maloney, '72a). It coincides with an influx of mesenchymal cells from nearby enlarging haversian canals as a prelude to the transient trabeculization of the medullary cavity (Patt and Maloney, '70). Whether marrow is implanted, removed or disrupted *in situ*, new hematopoietic foci tend to appear in sites of active bone resorption, perhaps in part at least because such sites provide a rich source of stromal cells.

While it is conceivable that mesenchymal cells associated with bone can generate hematopoietic elements as well as stromal cells, there is little, if any, evidence of this. Rather, the prevailing evidence indicates that hematopoietic and stromal stem cells are disparate forms. Thus, the existence of independent lines of hematopoietic and stromal precursors is supported by their differential radiosensitivity (Amsel and Dell, '71a) as well as by karyotypic and immunogenetic analyses of marrow implants (Friedenstein et al., '68; Friedenstein and Kuralesova, '71). In established heterotopic implants, hematopoietic cells generally represent the genotype of the recipient while stromal cells retain the genotype of the donor. These findings do not necessarily exclude the possibility that some hematopoietic stem cells may exist within the confines of osseous tissue, a point to which we will return later.

THE BONE-BONE MARROW INTERFACE

The early observation that hematopoiesis tends to be concentrated near the periphery of the medullary cavity (Drinker et al., '22) has been elaborated upon in recent studies which reveal that hematopoietic stem cells (CFU_S) are not distributed uniformly throughout the marrow. There are about twice as many CFU_S near the bone surface as in the center of the shaft (Lord and Hendry, '72) and those that are adjacent to bone appear to proliferate much more rapidly than those more distant from bone (Lord et al., '75). The distribution of presumptive granulocyte precursors (CFU_C) is different with a peak inside the medullary cavity and lower concentrations at periphery and center. As CFU_S differentiate and move away from bone, one would expect to find higher concentrations of differentiated progeny such as CFU_C. Thus, the spatial organization of stem cells points to the possible importance of endosteal surfaces in their maintenance. A further indication of this is perhaps provided by comparison of the origin of blood-forming cells in heterotopic marrow and femur shaft implants. In contrast to an exclusively host origin in the former (Friedenstein et al., '68), the new hematopoietic tissue can be of both donor and host origin in the latter (Amsel and Dell, '71b; Fried et al., '73). This difference may be reconciled by assuming a different post-implantation survival for CFU within the marrow and CFU in a more intimate relationship with bone surfaces.

The fact that CFU_S tend to concentrate near bone is germane to the question of the origin of hematopoietic stem cells in the repopulation of an evacuated medullary cavity. Several lines of evidence discussed elsewhere (Patt and Maloney, '75) indicate that immigrant stem cells are not essential for the repopulation, although they can contribute to later stages of regeneration with the emergence of a sinusoidal system (Knospe et al., '72). The reason for this is now clear. As seen in our studies with +/+ marrow engrafted W/W^v mice, there are residual stem cells in the evacuated medullary cavity and it is likely that these cells are closely associated with endosteum and surface haversian canals (Maloney and Patt, '75).

In recent work, we determined the capacity of subcutaneous implants of femur shafts depopulated *in vivo* to elevate the hematocrit of W/W^v mice. An increased hematocrit after implantation of +/+ marrow is correlated with the appearance of +/+ CFU in the recipient W/W^v. It turned out that +/+ femur shaft implants could still increase the hematocrit of recipient W/W^v after *in vivo* marrow removal. On the other hand, *in vivo* depopulated femur shafts from W/W^v (+/+) chimeras were ineffectual as judged by failure of the implants to elevate the hematocrit of W/W^v recipients even after intervals of 90 days or more (Maloney et al., unpublished data). These results might signify that more stem cells are in endosteal crevices or otherwise bone-associated, and therefore more difficult to remove, in +/+ mice than in W/W^v chimeras produced by intravenous injection of +/+ marrow. Although this facile interpretation would seem to be inconsistent with our earlier finding that residual +/+ stem cells were responsible for nearly all of the CFU repopulation in a locally evacuated medullary cavity of a W/W^v (+/+) chimera, there is a plausible explanation for the apparent discrepancy. It might simply reflect differences in the number of residual stem cells required for *in situ* repopulation of a given medullary cavity and for elevation of the hematocrit after subcutaneous implantation of a femur shaft. It is also expected that fewer residual cells would survive in the immediate environment of a depopulated femur implant than of a depopulated medullary cavity. Thus far, our studies with femur shafts and bone cell suspensions after complete marrow removal *in vitro* are inconclusive and we are, therefore, unable to say whether hematopoietic stem cells are also present within the bone proper. Conceivably, bone CFU could be equilibrated with marrow CFU and arise simply by migration from the contiguous marrow. But, if CFU indeed exist within bone, such cells could also constitute a slowly turning-over residue derived during the formation of embryonic marrow. The existence of a small pool of quiescent ancestral CFU somewhat removed from endosteal surfaces would be consistent with the apparent requirement for a form of clonal succession in stem cell generation (Kay, '65; Metcalf and Moore, '71).

There seems little doubt that bone contributes to the establishment and maintenance of a matrix for hematopoiesis. On the one hand, mesenchymal cells of bone provide a source of stromal elements (Patt and Maloney, '70) and on the other, there is a close functional relationship between bone formation and resorption and restoration of the marrow microvasculature (Amsel et al., '69; Brånemark et al., '64; McClugage et al., '71). Trabeculization appears to be a necessary step in the formation of a new marrow matrix *in situ* and similarly the formation of ectopic cancellous bone signals the impending formation of an associated hematopoietic marrow. It is significant that spleen stroma unlike marrow stroma does not have osteogenic potential and that heterotopic implants of spleen do not lead to persisting hematopoietic tissue under normal conditions (Haley et al., '75; Tavassoli et al., '73). In this connection, it is of interest to recall the recurring transient trabeculization of the medullary cavity associated with the ovulatory cycle in birds (Bloom et al., '41). Hematopoietic tissue shrinks periodically as cancellous bone encroaches on marrow space and is rapidly reformed with the onset of bone resorption. An extreme picture may be seen in the osteopetrotic mouse where essentially all medullary space is occupied by bone because of a defect in bone resorption, and hematopoiesis is centered in the spleen. When the bone resorption defect is corrected, for example, by injection of normal bone marrow cells which provide a source of osteoclastic precursors, there is a rapid development of medullary hematopoiesis in resorption sites (Walker, '75).

In view of the quite small number of circulating stem cells (ca. 1 per 50 mm^3 blood), it is remarkable that some appear at sites of ectopic bone formation even though the target volume may be less that 1 mm^3 (Patt and Maloney, '72b). Irrespective of the stem cell delivery mode, i.e., via the circulation or by direct migration from neighboring tissue, the evolving bony matrix must provide a particularly fruitful soil for stem cell activation, especially since there is no organismal demand for additional or compensatory hematopoiesis. The picture with heterotopic implants is not unlike that during embryonic life when migratory hematopoietic stem cells derived mainly from fetal liver settle in newly developed bone (Metcalf and Moore, '71). Apropos the seeding of hematopoietic stem cells in sites of new bone formation, there is apparently minimal traffic from one extant marrow region to another in the absence of perturbed hematopoiesis (Ford et al., '66). Unfortunately, it has not yet been possible to study stem cell migration streams under normal physiological conditions and little is known about the life history of circulating hematopoietic stem cells. It would seem, however, from chromosome marker studies with intravenously injected bone marrow cells in normal recipients that only a few donor stem cells are activated in an existing marrow, most likely because few proliferative sites are available in a fully populated tissue (Micklem et al., '75). On the other hand, W/W^v marrow, which has about two-thirds the cellularity of normal marrow, is readily colonized and over-run by +/+ stem cells because of their proliferative advantage (Seller, '68). From these and the many studies in irradiated animals, it appears that hematopoietic stem cells are not endowed with a specific homing instinct but rather that their initial distribution is determined by blood flow patterns and their potential as hematopoietic progenitors is determined by the stromal environment in which they settle. That hematopoietic tissue reflects the functional potential of its stromal bed is suggested, for example, from comparison of blood formation patterns in red and yellow marrow (Tavassoli and Crosby, '70), in marrow and spleen (Wolf and Trentin, '68), and in normal and Sl/Sl^d mice (McCulloch et al., '65).

Stromal cells as such manifest little, if any, traffic. There is no indication of their migration from the epiphyses to a depopulated shaft (Knospe et al., '66; Patt and Maloney, '70), even though stromal cells adjacent to a depopulated region may show considerable proliferative activity (Meyer-Hamme et al., '71). It is known, moreover, that stromal cells must be placed directly into an aplastic medullary cavity induced by heavy local irradiation of a limb for hematopoiesis to recur after intravenous injection of bone marrow cells (Knospe et al., '68). Despite the apparent absence of significant stromal cell migration, there is a fairly ubiquitous source of their progenitor elements. Thus, mesenchymal cells of many connective tissues can modulate in a proper environment, such as that provided by certain epithelia or by bone matrix, and produce new bone that forms the requisite stromal matrix which eventually contains new marrow (Reddi and Huggins, '75; Urist et al., '69).

It should be noted that the apparent requirement for bone formation in the genesis of hematopoietic tissue when marrow fragments are implanted in a foreign setting is not absolute. Hematopoietic tissue of donor origin can persist for several weeks in the absence of bone when a piece of intact marrow or, for that matter, spleen is placed in the omentum of irradiated, but not of unirradiated, mice (Haley et al., '75; Meck et al., '73). Apparently, sufficient reticular elements survive for a time in an irradiated site to influence proliferation and differentiation of the implanted hematopoietic stem cells in the face of the enormous pressure for blood formation consequent to irradiation. The failure of the marrow reticulum to express its osteogenic potential under these conditions may be related to a variety of radiation-induced effects, e.g., suppressed vascularization of the implant. The local nature of these effects is indicated by the significant observation that marrow fragments implanted in a shielded omentum of an otherwise completely irradiated mouse lead to well-developed ossicles containing hematopoietic tissue (Haley and Brecher, '75). In contrast to such ectopic marrow-containing ossicles, boneless hematopoietic foci are unlikely to remain when the host recovers from radiation injury. (Table 1)

TABLE 1

Bone formation and resorption and hematopoietic activity

System	Irradiation: Removal or implant site	Irradiation: Body	Bone	Hematopoiesis
Marrow removal	-	-	+	+
Marrow removal	-	+	+	+
Marrow removal	+	-	-	-
Marrow implant	-	-	+	+
Marrow implant	-	+	+	+
Marrow implant	+	+	+/-	+/+ transient
Spleen implant	-	-	-	-
Spleen implant	+	+	-	+ transient

LOCAL REGULATORY FACTORS

Regulation of the widely disseminated medullary tissue involves the interplay of local and systemic mechanisms. It is known that the life history of pluripotential hematopoietic stem cells is determined by the nature of their surroundings, that hematopoietic activity is in many respects essentially a local option. Hematopoiesis is influenced by the character of the local blood flow (McClugage et al., '71) and proliferative activity even in the presumably normal steady state is geared to the cellularity of a circumscribed area of marrow (Patt and Maloney, '72c). Local control probably is epitomized best by the concept of specific hematopoietic inductive microenvironments (Curry and Trentin, '67). And in this respect cells of the stroma are generally regarded as the principal determinant of the quality of hematopoiesis. But how this is brought about remains a matter of conjecture as indicated by the following open questions: 1) What is the operational structure of one microenvironment relative to another? 2) Does a given microenvironment represent a transient state? 3) Does the stromal cell *per se* provide hematopoietic stem cell proliferative and/or commitment control? 4) Is there more than one type of regulatory stromal cell or a single type with specificity determined by ambient factors? Answers to these questions are germane to the central question of why medullary cavities of bone became the preferred site for normal hematopoiesis and to the related question of why new hematopoietic foci arise in intimate association with new bone in the absence of an obvious compensatory requirement.

An osseous environment would seem to contribute to normal hematopoiesis if only for the reason that it can influence the nature of the stromal cell distribution and microvasculature. It is not happenstance that the most active marrow is located in the interstices of cancellous bone and at the periphery of the medullary cavity of cortical bone where reticular cells are apparently concentrated on endosteal surfaces. Reference has already been made to the bone vicinity effect on CFU turnover (Lord et al., '75) and it is of interest that an osteogenic feeder layer promotes CFU cycling in bone marrow cultures (Chertkov et al., '74). It is also noteworthy that cells associated with the bone shaft provide a particularly rich source of granulocyte colony-stimulating activity (Chan and Metcalf, '72). Much of the blood that passes through marrow first circulates through bone and it has been suggested from time to time that agents derived from bone contribute to hematopoietic regulation (Röhlich, '41) but substantive evidence is lacking. As might be expected, there is a close relationship between marrow blood flow and marrow activity. More importantly, however, there appear to be regional differences in the character of the blood flow, erythropoietic foci being associated with higher rates of flow than granulopoietic (McClugage et al., '71; McCuskey et al., '72). Because of circulation through bone, blood returning to the periphery of the medullary cavity may have a relatively low oxygen tension and it has been proposed that this could favor granulopoiesis.

There is no unequivocal proof of stromal cell specificity in the regulation of stem cell commitment and we wonder whether the microcirculation is an important factor in the preponderance of erythropoietic foci relative to granulopoietic in spleen compared to bone marrow. We wonder too whether the splenic machinery for processing and disposal of injured and aged erythrocytes might also contribute to the formation of highly selective local environments for erythropoiesis as seen, for example, in rat spleen (Rauchwerger et al., 73). Although primarily granuloid colonies were evident within marrow implants in spleen parenchyma, it will be recalled that the marrow stroma was irradiated prior to implantation and that the recipient spleen was also irradiated (Wolf and Trentin, '68). Under these circumstances, a reasonably equivalent vascularity of the marrow implant and its neighboring splenic tissue is unlikely and for this reason alone, it may be difficult to ascribe differences in the quality of blood formation in the adjoining sites to differences in stromal cell specificity *per se*. Studies with Sl/Sl^d mice do provide definitive evidence of a stromal influence in respect to proliferation of committed erythroid stem cells (Wolf, '74). Although the hematopoietic defect in the Sl/Sl^d, manifest primarily as anemia, is suggestive of decreased production of a short-range specific proliferative stimulus by stromal cells, the mutant stroma could also produce a short-range inhibitory factor (Chui and Loyer, '75). As noted earlier, erythropoiesis can occur for a time when marrow fragments are placed in a foreign setting in an irradiated animal, presumably because surviving reticular cells provide a necessary growth factor. On the other hand, it is fairly easy to observe granulopoietic, but not erythropoietic, foci under similar conditions with marrow cell suspensions where there are apparently few viable reticular cells. However, these results do not necessarily point to a specific stroma-derived influence on erythroid progenitors, since there are probably alternate pathways for activation of granuloid precursors particularly after irradiation.

Although the evidence for stromal specificity is incomplete, there is no doubt that cells of the stroma play at least a general role in the management of marrow activity. A regenerated or newly formed marrow reflects the functional potential of its stromal elements; depopulation or heterotopic implantation of red or yellow marrow results in a similar type of marrow (Tavassoli et al., '74; Tavassoli and Crosby, '70). Although the precise mechanism underlying the conversion of red to yellow marrow is not understood, diminished blood flow is thought to be an important factor in bringing about what may be an epigenetic change in stromal cells (Van Dyke and Harris, '69). The evolving hematopoietic tissue consequent to evacuation or implantation of yellow marrow passes through a transient active phase, possibly because of a greatly increased blood flow. Significantly, hematopoietic activity can be sustained even in stroma derived from a yellow marrow implant if there is an externally-imposed continuing demand for increased blood cell production (Tavassoli et al., '74). In a hypertransfused animal the emerging marrow in a depopulated medullary cavity is devoid of recognizable erythroid cells (Maloney et al., unpublished data). Thus, despite the basically local nature of the regenerative program, the character of the new tissue is also dependent upon extrinsic controls.

Stromal cells constitute a long-lived slowly turning-over population, which no doubt accounts for their apparent radioresistance (Patt and Maloney, '75). Such cells are capable of intense proliferative activity in response to injury, but they have restricted migratory capability. Hence, it is usually the matrix that limits recovery of normal function. When a swath of marrow is disrupted without its removal, surviving reticular cells not only proliferate but also express their osteogenic potential and lay down new cancellous bone that serves as a temporary framework and source of cells for reconstruction of the disrupted sinusoidal matrix (Amsel et al., '69). An essentially similar picture is seen in an evacuated medullary cavity and in a heterotopic marrow implant. It will be recalled that mesenchymal cells of many tissues can be induced to form bone and while bone formation in such cases is not an intermediary process in regeneration of an injured or missing marrow, it does generate an environment for a new marrow. Several types of evidence suggest that bone formation is an obligatory step in the reconstitution of marrow as an organized tissue (Patt and Maloney, '72a, '72b; Tavassoli et al., '70) and this inference receives strong support from the fact that fragments of hematopoietic murine spleen do not have osteogenic potential and do not generate the requisite environment for sustained hematopoiesis upon heterotopic implantation. When a new marrow forms in an osseous setting, hematopoietic foci emerge first at sites of active bone resorption, but what is meant by requisite environment cannot yet be formulated in biophysical and biochemical terms.

In conclusion, present evidence is consistent with the thesis that stromal cells activate hematopoietic stem cells and regulate their proliferative behavior. It is not yet possible to say whether cells of the stroma also determine hematopoietic specificity or whether specific inductive microenvironments evolve in relation to the microvasculature and other factors as well. The ambience of bone surfaces apparently provides a source of cells that govern stem cell activity. The intertwining of the circulation of bone and marrow may also contribute a special quality to the environment. Whatever the reason(s), a normal stem cell arriving in the vicinity of endosteal surfaces has a much greater probability of initiating blood formation than a stem cell arriving in any non-osseous tissue in the absence of hematopoietic stress. Bone formation and resorption seem to be an essential feature of the regenerative program when the structure of marrow is disrupted or removed. It is suggested that the intermediation of bone plays a central role in restoration of the characteristic stromal environment.

ACKNOWLEDGMENTS

Work performed under the auspices of the U.S. Energy Research and Development Administration.

LITERATURE CITED

Amsel, S. and E. S. Dell. 1971a. The radiosensitivity of the bone-forming process of heterotopically grafted rat bone-marrow. *Int. J. Radiat. Biol. 20:* 119-127.

Amsel, S. and E. S. Dell. 1971b. Bone marrow repopulation of subcutaneously grafted mouse femurs. *Proc. Soc. Exp. Biol. Med. 138:* 550-552.

Amsel, S., A. Maniatis, M. Tavassoli and W. H. Crosby. 1969. The significance of intramedullary cancellous bone formation in the repair of bone marrow tissue. *Anat. Rec. 164:* 101-111.

Bloom, W., M. A. Bloom and F. C. McLean. 1941. Calcification and ossification. Medullary bone changes in the reproductive cycle of female pigeons. *Anat. Rec. 81:* 443-475.

Branemark, P. I., U. Breine, B. Johansson, P. J. Roylance, H. Rockert and J. M. Yoffey. 1964. Regeneration of bone marrow: A clinical and experimental study following removal of bone marrow by curettage. *Acta Anat. (Basel) 59:* 1-46.

Chan, S. H. and D. Metcalf. 1972. Local production of colony-stimulating factor within the bone marrow: Role of nonhematopoietic cells. *Blood 40:* 646-653.

Chertkov, J. L., N. L. Samoylina, N. A. Rudneva, A. M. Rakcheev and N. F. Kondratenko. 1974. Hemopoietic stem cells (CFU) kinetics in the culture. *Cell Tissue Kinet. 7:* 259-270.

Chui, D.H.K. and B. V. Loyer. 1975. Erythropoiesis in steel mutant mice: Effect of erythropoietin *in vitro*. *Blood 45:* 427-433.

Curry, J. L. and J. J. Trentin. 1967. Hemopoietic spleen colony studies. I. Growth and differentiation. *Develop. Bio. 15:* 395-413.

Drinker, C. K., K. R. Drinker and C. C. Lund. 1922. The circulation in the mammalian bone-marrow. *Amer. J. Physiol. 62:* 1-92.

Ford, C. E., H. S. Micklem, E. P. Evans, J. G. Gray and D. A. Ogden. 1966. The inflow of bone marrow cells to the thymus: Studies with part-body irradiated mice injected with chromosome-marked bone marrow and subjected to antigenic stimulation. *Ann. N.Y. Acad. Sci. 129:* 283-296.

Fried, W., S. Husseini, W. H. Knospe and F. E. Trobaugh, Jr. 1973. Studies on the source of hematopoietic tissue in the marrow of subcutaneously implanted femurs. *Exp. Hemat. 1:* 29-35.

Friedenstein, A. and A. I. Kuralesova. 1971. Osteogenic precursor cells of bone marrow in radiation chimeras. *Transplantation 12:* 99-108.

Friedenstein, A. J., K. V. Petrakova, A. I. Kurolesova and G. P. Frolova. 1968. Heterotopic transplants of bone marrow: Analysis of precursor cells for osteogenic and hematopoietic tissues. *Transplantation 6:* 230-247.

Haley, J. E. and G. Brecher. 1975. Personal communication.

Haley, J. E., J. H. Tjio, W. W. Smith and G. Brecher. 1975. Hematopoietic differentiative properties of murine spleen implanted in the omenta of irradiated and nonirradiated hosts. *Exp. Hemat. 3:* 187-196.

Kay, H.E.M. 1965. How many cell-generations? *Lancet II,* 418-419.

Knospe, W. H., J. Blom and W. H. Crosby. 1966. Regeneration of locally irradiated bone marrow. I. Dose-dependent, long term changes in the rat, with particular emphasis upon vascular and stromal reaction. *Blood 28:* 398-415.

Knospe, W. H., J. Blom and W. H. Crosby. 1968. Regeneration of locally irradiated bone marrow. II. Induction of regeneration in permanently aplastic medullary cavities. *Blood 31:* 400-405.

Knospe, W. H., S. A. Gregory, S. G. Husseini, W. Fried and F. E. Trobaugh, Jr. 1972. Origin and recovery of colony-forming units in locally curetted bone marrow of mice. *Blood 39:* 331-340.

Lord, B. I. and J. H. Hendry. 1972. The distribution of hemopoietic colony-forming units in the mouse femur and its modification by X-rays. *Brit. J. Radiol. 45:* 110-115.

Lord, B. I., N. G. Testa and J. H. Hendry. 1975. The relative spatial distribution of CFU_s and CFU_c in the normal mouse femur. *Blood 46:* 65-72.

Maloney, M. A., D. F. Bainton, P. L. Fong and H. M. Patt. Unpublished data.

Maloney, M. A. and H. M. Patt. 1975. On the origin of hematopoietic stem cells after local marrow extirpation. *Proc. Soc. exp. Biol. (N.Y.) 149:* 94-97.

McClugage, S. G., Jr., R. S. McCuskey and H. A. Meineke. 1971. Microscopy of living bone marrow *in situ.* II. Influence of microenvironment on hemopoiesis. *Blood 38:* 96-107.

McCulloch, E. A., L. Siminovitch, J. E. Till, E. S. Russell and S. E. Bernstein. 1965. The cellular basis of the genetically determined hemopoietic defect in anemic mice of genotype Sl/Sl^d. *Blood 26:* 399-410.

McCuskey, R. S., H. A. Meineke and S. F. Townsend. 1972. Studies of the hemopoietic microenvironment. I. Changes in the microvascular system and stroma during erythropoietic regeneration and suppression in the spleens of CF_I mice. *Blood 39:* 697-712.

McLean, F. C. and M. R. Urist. 1968. Bone: Fundamentals of the Physiology of Skeletal Tissue. Third edition. University of Chicago Press, Chicago, p. 12.

Meck, R. A., J. E. Haley and G. Brecher. 1973. Hematopoiesis versus osteogenesis in ectopic bone marrow transplants. *Blood 42:* 661-669.

Metcalf, D. and M.A.S. Moore. 1971. Haemopoietic Cells. North-Holland, Amsterdam, pp. 172-271, 448-465.

Meyer-Hamme, K., R.J. Haas and T. M. Fleidner. 1971. Cytokinetics of bone marrow stroma cells after stimulation by partial depletion of the medullary cavity. *Acta haemat. (Basel) 46:* 349-361.

Micklem, H. S., C. E. Ford, E. P. Evans and D. A. Ogden. 1975. Compartments and cell flow within the mouse haemopoietic system. I. Restricted interchange between haemopoietic sites. *Cell Tissue Kinet. 8:* 219-232.

Patt, H. M. and M. A. Maloney. 1970. Reconstitution of bone marrow in a depleted medullary cavity. In: *Hemopoietic Cellular Proliferation*. F. Stohlman, Jr., ed. Grune and Stratton, New York, pp. 56-66.

Patt, H. M. and M. A. Maloney. 1972a. Evolution of marrow regeneration as revealed by transplantation studies. *Exp. Cell Res. 71:* 307-312.

Patt, H. M. and M. A. Maloney. 1972b. Bone formation and resorption as a requirement for marrow development. *Proc. Soc. Exp. Biol. Med. 140:* 205-207.

Patt, H. M. and M. A. Maloney. 1972c. Relationship of bone marrow cellularity and proliferative activity: A local regulatory mechanism. *Cell Tissue Kinet. 5:* 303-309.

Patt, H. M. and M. A. Maloney. 1975. Bone marrow regeneration after local injury: A review. *Exp. Hemat. 3:* 135-148.

Rasmussen, H. and P. Bordier. 1974. Chapter II, Bone Cells - Morphology and Physiology. In: *The Physiology and Cellular Basis of Metabolic Bone Disease*. Williams and Wilkins Co., Baltimore, pp. 8-70.

Rauschwerger, J. M., M. T. Gallagher and J. J. Trentin. 1973. Role of the hemopoietic inductive microenvironment (H.I. M.) in xenogeneic bone marrow transplantation. *Transplantation 15:* 610-613.

Reddi, A. H. and C. B. Huggins. 1975. Formation of bone marrow in fibroblast-transformation ossicles. *Proc. Nat. Acad. Sci. 72:* 2212-2216.

Rohlich, K. 1941. On the relationship between the bone substance and hematopoiesis in the bone marrow. *Z. mikr.-anat. Forsch 49:* 425-464.

Seller, M. J. 1968. Transplantation of anaemic mice of the W- series with haemopoietic tissue bearing marker chromosomes. *Nature 220:* 300-301.

Siegel, C. D. 1970. Possible hematopoietic mechanisms in nonmammalian vertebrates. In: *Regulation of Hematopoiesis*. A. S. Gordon, ed. Appleton-Century-Crofts, New York, vol. 1, pp. 67-76.

Tavassoli, M. and W. H. Crosby. 1968. Transplantation of marrow to extramedullary sites. *Science 161:* 54-56.

Tavassoli, M. and W. H. Crosby. 1970. Bone marrow histogenesis: A comparison of fatty and red marrow. *Science 169:* 291-293.

Tavassoli, M., A. Maniatis and W. H. Crosby. 1970. Studies on marrow histogenesis. I. The site of choice for extramedullary implants. *Proc. Soc. Exp. Biol. Med. 133:* 878-881.

Tavassoli, M., A. Maniatis and W. H. Crosby. 1974. Induction of sustained hemapoiesis in fatty marrow. *Blood 43:* 33-38.

Tavassoli, M., R. J. Ratzan and W. H. Crosby. 1973. Studies on regeneration of heterotopic splenic autotransplants. *Blood 41:* 701-709.

Trueta, J. 1968. Studies of the Development and Decay of the Human Frame. Heinemann, London, p. 7.

Urist, M. R., P. H. Hay, F. Dubuc and K. Buring. 1969. Osteogenic competence. *Clin. Orthop. Rel. Res. 64:* 194-220.

Van Dyke, D. and N. Harris. 1969. Bone marrow reactions to trauma: Stimulation of erythropoietic marrow by mechanical disruption, fracture or endosteal curettage. *Blood 34:* 257-275.

Walker, D. G. 1975. Control of bone resorption by hematopoietic tissue: The induction and reversal of congenital osteopetrosis in mice through use of bone marrow and splenic transplants. *J. Exp. Med. 142:* 651-663.

Wolf, N. S. 1974. Dissecting the hematopoietic microenvironment. I. Stem cell lodgement and commitment, and the proliferation and differentiation of erythropoietic descendants in the Sl/Sl^d mouse. *Cell Tissue Kinet. 7:* 89-98.

Wolf, N. S. and J. J. Trentin. 1968. Hemopoietic colony studies: V. Effect of hemopoietic organ stroma on differentiation of pluripotent stem cells. *J. Exp. Med. 127:* 205-214.

Hemopoietic Inductive Microenvironments

JOHN J. TRENTIN

Division of Experimental Biology, Baylor College of Medicine, Houston, Texas 77025

The mouse has been extensively used in the study of hemopoiesis, because of the availability of useful genetic markers, inbred strains, and mutants for hemopoietic disorders. These studies have revealed that the bone marrow stem cell is pluripotent for the several lines of hemopoiesis, as well as for both T and B lymphocytes. Whereas at first it was believed that the hormone erythropoietin "induced" differentiation of the pluripotent stem cell into the erythroid line, it later became apparent that it acted instead on an erythropoietin-sensitive stem cell, one step removed from the pluripotent stem cell (Bruce and McCulloch, '64; Fried *et al.*, '66; Schooley, '66; Till *et al.*, '66; Marks *et al.*, '74). Granulopoietic stimulating factor also appears to act on already committed stem cells. Because no factors were known that directly controlled the important process of decision on the part of the pluripotent stem cell or its progeny, as to which line of differentiation to become committed to, the decision process was thought to be random (stochastic) (Till *et al.*, '64).

More recently, however, it became apparent, from detailed study of the early stages of differentiation of both spleen colonies and bone marrow colonies, in both normocythemic and polychythemic hosts, both before and after retransplantation, that "decision" was micro-geographically limited in both the spleen and the marrow (Trentin, '70). The results suggested that the reticuloendothelial stroma of the bone marrow and of the spleen somehow controlled the process of decision. Both spleen stroma and marrow stroma were found to be subdivided into four types of microenvironments, each dictating a different type of hemopoiesis. However, the ratio of the four types of microenvironments is different in marrow than in spleen. Transplantation of marrow stroma into the spleen, or of spleen stroma subcutaneously, showed that each type of organ stroma retained or regenerated its distinctive ability to support hemopoiesis and control decision (Wolf and Trentin, '68).

The remainder of this manuscript summarizes the evidence on which these conclusions are based.

Most of the discrete hemopoietic colonies that form in the spleen and bone marrow of heavily irradiated mice (1000R+) "thinly seeded" with exogenous bone marrow stem cells, are clonal in origin (Becker *et al.*, '63; Wolf and Trentin, '75) and derive from a stem cell that is pluripotent for at least four kinds of hemopoiesis, and for B and T lymphocytes (Nowell *et al.*, '70; Trentin, '71). Yet, for the first 8 to 10 days most colonies (clones) remain committed to a single line of hemopoietic differentiation (erythroid, neutrophilic granuloid, eosinophilic granuloid, or megakaryocytic), even though some may consist of a million or more cells (Curry and Trentin, '67; Curry *et al.*, '67; Trentin, '70; Trentin, '71). Of such early spleen colonies, approximately 60% are erythroid, 20% are neutrophilic granuloid, 10-15% are megakaryocytic, and 1% are eosinophilic granuloid colonies (Curry and Trentin, '67; Jenkins *et al.*, '72). The erythroid are the fastest growing and largest spleen colonies and become grossly visible by six days, when they can be dissected out for re-transplantation or other manipulation. The neutrophilic granuloid colonies are next largest and become grossly visible a day or two later. The eosinophilic granuloid colonies can rarely be seen grossly, and the megakaryocytic colonies never. After 8 to 10 days, an increasing percentage of the colonies develop a second line of hemopoietic differentiation (Trentin, '70; Trenton, '71). It has been established that the second line of differentiation arises endoclonally, rather than by immigration (Trentin, '71). The typical earliest mixed colony is a very large erythroid colony with granulopoiesis or megakaryocytopoiesis in a peripheral segment, suggesting that a fast growing and pluripotent erythroid colony has outgrown an erythroid microenvironment and interacted with a granuloid or a megakaryocytic microenvironment. The relative size of 10 day old spleen colonies, in terms of thousands of cells, was found to be 298 (100-750) for neutrophilic granuloid colonies, 873 (570-1,430) for erythroid colonies, and 2,120 (1,640-2,650) for mixed colonies (Curry and Trentin, '67).

The granuloid and erythroid microenvironments have distinctive and different distributions in the spleen, as indicated by the point of origin of these two colony types. The granuloid colonies originate adjacent to the splenic capsule or trabeculae, or in the radiation-depleted lymphoid follicles of the white pulp. The erythroid colonies may appear anywhere in the red pulp of the spleen, but not in the depleted lymphoid follicles, which they grow around rather than into (Curry and Trentin, '67).

Retransplanation of cells of early spleen colonies of either erythroid or of granuloid type into irradiated secondary hosts, yields spleen colonies of one or another of each of the four types, with only a slight increase in the proportion of secondary colonies of the same type as the primary colony (Curry *et al.*, '67; Lewis and Trobaugh, '64). This indicates that the stem cell of origin of the primary colony was pluripotent, and that while some of its progeny differentiated, others retained their pluripotency. It indicates further that most, but probably not all, of the secondary colonies were formed by pluripotent stem cells. It should be noted however, that unlike bone marrow cells, normal spleen cells transfused into a radiated primary host give rise to a much higher proportion of erythroid colonies (Wolf *et al.*, '72), many of which appear to be formed by a committed stem cell and to be unable to give rise to secondary repopulation.

If the primary irradiated recipient of bone marrow cells is hypertransfused to render it polychythemic and thus to eliminate endogenous erythropoietin production, the presumptive erythroid colonies grow very slowly and fail to differentiate (Curry *et al.*, '67). They remain very small and undifferentiated even beyond ten days. It is most instructive that they do not undergo differentiation into granuloid or other cell types at a time when they normally would, even though adjacent colonies of the other types are developing normally. Thus the limitation to one line of differentiation is geographic rather than temporal. If erythropoietin is administered at any time, they immeidately begin to proliferate rapidly and undergo erythroid differentiation. If erythropoietin is again withdrawn, they again become undifferentiated, but now consist of a larger number of undifferentiated cells only. These findings further indicate that under the influence of an erythroid microenvironment the progeny of the pluripotent cell of origin of the colony can only undergo erythroid differentiation, and this only under the additional influence of erythropoietin. However, residence in the erythroid microenvironment has induced some of the progeny of the pluripotent cell of origin into a state of erythropoietin sensitivity, a property not shared by pluripotent stem cells. Because of this observed inductive influence of the erythroid microenvironments, and the presumed inductive influence of the other types of microenvironments, the term hemopoietic inductive microenvironments (HIM) has been used (Curry *et al.*, '67). It should be remembered, however, that only for the erythroid HIM is there as yet formal proof of the inductive mechanism.

O'Grady *et al.* ('73) ingeniously attempted to "flood" the spleens of irradiated-hypertransfused mice (in which erythroid colonies do not develop) with granulopoiesis, by using a large inoculum of marrow cells. They found that they could not flood the spleens with granulopoiesis, and concluded that some areas of the mouse spleen are reserved for erythropoiesis only.

Of bone marrow derived spleen colonies, erythroid colonies outnumber granuloid colonies by a factor of about three (E:G colony ratio ~3) (Curry and Trentin, '67; Curry *et al.*, '67). However, of the hemopoietic colonies developing in the marrow cavities of such mice, erythroid colonies are only half as numerous as granuloid colonies (E:G colony ratio ~0.5) (Wolf and Trentin, '68). Is this the result of selective homing of different kinds of committed cells to spleen versus marrow, or does the ratio of stromal microenvironments of different types differ in spleen than in the marrow cavity? Irradiated mice were injected with large numbers of bone marrow cells, and 24 hours later those cells that had "homed" to the marrow cavities were harvested and retransplanted to irradiated secondary hosts. In the spleens of the secondary hosts they gave an E:G colony ratio of 2.8, and in the marrow cavity of the secondary hosts they gave an E:G colony ratio of 0.6. Thus there appears to be no selective homing of granuloid committed colony-forming cells to the marrow cavity of the primary host. Instead, the different E:G ratios reflect a preponderance of erythroid HIM in spleen and of granuloid HIM in the marrow cavity, acting on pluripotent stem cells. As noted below, the rat spleen does not appear to have granuloid or megakaryocytic HIM, but only an erythroid environment.

Whole mouse spleens transplanted subcutaneously to isogenic mice supported hemopoiesis, and gave an E:G colony ratio of 3.5, i.e. the same as the spleen *in situ* (Wolf and Trentin, '68).

Pieces of bone marrow stroma were trocar-implanted into the spleens of isogenic mice. They gave rise to hemopoietic colonies with an E:G colony ratio similar to that of bone marrow *in situ*, i.e. granuloid colonies predominated. In the surrounding spleen, erythroid colonies predominated. Unexpectedly, but most instructively, those individual "junctional" colonies (clones) that grew in both spleen stroma and marrow stroma invariably showed an abrupt change of hemopoietic type at the stromal junction, with erythropoiesis in the spleen stroma and granulopoiesis in the marrow stroma (Wolf and Trentin, '68).

The Steel anemic mutant mouse (Sl/Sl^d) which has normal stem cells and erythropoietin (Bernstein *et al.*, '68), and is not helped by stem cells from non-anemic littermates, was found to be curable by transplantation of spleen stroma from non-anemic littermates (Bernstein, '70; Trentin, '71). Spleen colony studies in anemic and non-anemic Steel littermates revealed that the spleens of the anemic littermates were almost devoid of erythroid microenvironments (Trentin, '71).

W/W^v genetic anemic mice have defective stem cells. They are cured by marrow cells from non-anemic littermates, or from congenic Sl/Sl^d anemic mice (Bernstein, '70; Bernstein *et al.*, '68). Parabiosis of Sl/Sl^d anemic mice with W/W^v anemic mice cures both partners. If separated, even after many months, W/W^v anemic partners remain cured but Sl/Sl^d partners quickly revert to anemic (Bernstein, personal communication). From this, one can conclude that whereas stem cells circulate, the erythroid HIM do not. The lack of mobility of the stromal cells that constitute the HIM can be deduced also from another set of observations. After initial hemopoietic regeneration of those segments of the femur or spleen of shielded rats or mice given very high doses of localized irradiation, there is a gradual and permanent late aplasia correlated with late damage to the reticuloendothelial stroma (Knospe *et al.*, '66; Jenkins *et al.*, '70). The line of demarcation between the hemopoietically active shielded portion of the femur or spleen, and the aplastic irradiated portion is generally quite abrupt, indicating little or no replacement of the radiation damaged stroma by migration of stromal elements from the unirradiated portion of the same organ.

It is important however, that the stromal cells that function as HIM can and do regenerate over a period of about 2 or 3 weeks in both bone marrow and spleen, following ischemic death due to transplanation, or to ligation of the splenic artery *in situ*, or following marrow currettage (Knospe *et al.*, '68; Knospe *el al.*, '72; Tavassoli and Crosby, '68; Wolf, '73; Wolf and Lagunoff, '74). The sequential stages of regeneration of the characteristic reticuloendothelial stromal structure of the organ have been described, and precede the return of hemopoiesis or the detectability of hemopoietic stem cells (CFUs).

Immunologically mediated injury to the spleen or marrow

stroma, by the graft-versus-host reaction, also impairs its ability to support hemopoiesis (Kitamura *et al.*, '73; Kitamura *et al.*, '70; Knospe *et al.*, '71).

Transfused rat bone marrow stem cells, forming spleen colonies in the spleens of irradiated rats, undergo only erythroid differentiation (E:G colony ratio = 490:0). To further test the determining influence of the organ stroma on differentiation, rat marrow cells were transfused into irradiated mice. Spleen colonies formed by rat cells in mouse spleens were found to be either erythropoietic or granulopoietic. The ratio of erythroid to granuloid colonies was approximately 3, i.e. the same as if formed by mouse cells, but the karyotype was that of the rat, further confirming the existence of functional HIM, and of their influence on stem cell differentiation (Rauchwerger *et al.*, '73). This indicated further that the mouse erythroid and granuloid HIM are capable of acting on rat stem cells.

Intraperitoneally implanted cellulose acetate membranes become coated with macrophages and fibroblasts, and support hemopoietic colony formation following irradiation and intraperitoneal injection of bone marrow cells. Granuloid colonies outnumber erythroid colonies (E:G colony ratio = 1:20) (Seki, '73). If the colonies so formed are pluripotent, the cells coating the membrane may include cells comparable to those of the granuloid (and erythroid?) HIM. However, if these colonies are formed by committed rather than pluripotent stem cells, the cells coating the membrane could represent cells producing the colony stimulating factor which is known to stimulate the growth of committed granuloid colony forming cells *in vitro*.

Hemopoiesis induced by injection of dispersed marrow cells into the omentum of irradiated mice (550 rads) is largely granulopoietic (Meck *et al.*, '73). The same is true of dispersed spleen cells (Haley *et al.*, '75). However, fragments of spleen implanted into the omentum of irradiated mice were predominantly erythropoietic at 10 days, with a shift to granulopoietic differentiation at 21 days (Haley *et al.*, '75).

What cells of the stroma function as HIM? One might suppose them to be the relatively radioresistant reticuloendothelial cells, i.e. macrophage-like cells. Friedenstein *et al.* ('74) have reported that they have been able to propagate *in vitro* the stromal cells of marrow and spleen responsible for transferring the microenvironment of the hemopoietic tissues, following retransplantation *in vivo*. After several passages *in vitro* the cultures lost all differentiated elements and appeared as cultures of fibroblasts; fibroblasts that retained the memory and ability to regenerate a structure and function characteristic of the organ they came from!

Much of the evidence for the existence of HIM has thus come from detailed study of the phenomenon of "homing" of transfused marrow cells to "reseed" the marrow cavity and spleen of lethally irradiated rodents (Trentin, '70). However, perhaps the first awareness on the part of the author that injury to hemopoietic organ stroma might impair its ability to support hemopoiesis, related to a clinical observation. A child with drug induced chronic pancytopenia was given marrow transfusions from an identical twin. In spite of genetic identity of the marrow donor, the transfused cells did not reseed and cure (Fernbach and Trentin, '62)! The thought occurred that perhaps the drug-induced injury was to the "home" necessary to sustain hemopoiesis.

Since that time an animal model of early recovery but late marrow failure after drug administration, resulting in chronic pancytopenia, has been reported (Morley and Blake, '74). By analogy with the above mentioned late local failure of marrow after high dose local irradiation, this model of late drug induced pancytopenia is undoubtedly the result of failure of the stromal elements to regenerate themselves after drug injury. As such, these pancytopenic mice should be cured by transplanation of whole isogenic spleens, but not by stem cell suspensions from either spleen or marrow.

Late persistent marrow hypocellularity after local X-irradiation of the rabbit femur can occur after as low as 1000 rads (Maloney and Patt, '72).

LITERATURE CITED

Becker, A. J., E. A. McCulloch and J. E. Till. 1963. Cytological demonstration of the clonal nature of spleen colonies derived from transplanted mouse marrow cells. *Nature 197:* 452-454.

Bernstein, S. E. 1970. Tissue transplantation as an analytical and therapeutic tool in hereditary anemias. *Am. J. Surgery 119:* 448-451.

Bernstein, S. E. (personal communication).

Bernstein, S. E., E. S. Russell and G. Keighley. 1968. Two hereditary mouse anemias (Sl/Sl^d and W/W^v) deficient in response to erythropoietin. *Ann. N.Y. Acad. Sci. 149:* 475-485.

Bruce, W. R. and E. A. McCulloch. 1964. The effect of erythropoietin stimulation on the hemopoietic colony-forming cells of mice. *Blood 23:* 216-232.

Curry, J. L. and J. J. Trentin. 1967. Hemopoietic spleen colony studies: I: Growth and differentiation. *Devel. Biol. 15:* 395-413.

Curry, J. L., J. J. Trentin and N. Wolf. 1967. Hemopoietic spleen colony studies: II. Erythropoiesis. *J. Exp. Med. 125:* 703-720.

Fernbach, D. J. and J. J. Trentin. 1962. Isologous bone marrow transplantation in an identical twin with aplastic anemia. Proceedings of the VIIIth International Congress of Hematology. *Pan Pacific Press, Tokyo, Vol. 1,* pp. 150-155.

Fried, W., D. Martinson, M. Weisman and C. W. Gurney. 1966. Effect of hypoxia on colony-forming units. *Exp. Hemat. 10:* 22.

Friedenstein, A. J., R. K. Chailakhyan, N. V. Latsinik, A. F. Panasyuk and I. V. Keiliss-Borok. 1974. Stromal cells responsible for transferring the microenvironment of the hemopoietic tissues. Cloning *in vitro* and retransplantation *in vivo*. *Transplantation 17:* 331-340.

Haley, J. E., J. H. Tjio, W. W. Smith and G. Brecher. 1975. Hematopoietic differentiative properties of murine spleen implanted in the omenta of irradiated and non-irradiated hosts. *Exp. Hemat. 3:* 187-196.

Jenkins, V. K., J. J. Trentin and N. S. Wolf. 1970. Radioresistance of the splenic hemopoietic inductive microenvironments (HIM). *Rad. Res. 43:* 212.

Jenkins, V. K., J. J. Trentin, R. S. Speirs and M. P. McGarry. 1972. Hemopoietic colony studies. VI. Increased eosinophil-containing colonies obtained by antigen pretreatment of irradiated mice reconstituted with bone marrow cells. *J. Cell. Physiol. 79:* 413-422.

Kitamura, Y., T. Kawata, A. Kanamaru and M. Seki. 1973. Suppression of erythropoiesis by simultaneous proliferation of alloantigen-sensitive units. *Exp. Hemat. 1:* 350-361.

Kitamura, Y., T. Kawata, O. Suda and K. Ezumi. 1970. Changed differentiation pattern of parental colony-forming cells in F_1 hybrid mice suffering from graft-versus-host disease. *Transplantation 10:* 455-462.

Knospe, W. H., J. Blom and W. H. Crosby. 1966. Regeneration of locally irradiated bone marrow. I. Dose dependent, long-term changes in the rat, with particular emphasis upon vascular and stromal reactions. *Blood 28:* 398-415.

Knospe, W. H., J. Blom and W. H. Crosby. 1968. Regeneration of locally irradiated bone marrow. II. Induction of regeneration in permanently aplastic medullary cavities.

Blood 31: 400-405.

Knospe, W. H., J. Blom, H. B. Goldstein and W. H. Crosby. 1971. Delayed bone marrow aplasia in rats protected against lethal irradiation by allogenic marrow transplantation. *Rad. Res. 47:* 199-212.

Knospe, W. H., S. A. Gregory, S. G. Husseini, W. Fried and F. E. Trobaugh, Jr. 1972. Origin and recovery of colony-forming units in locally curetted bone marrow of mice. *Blood 39:* 331-340.

Lewis, J. P. and F. E. Trobaugh, Jr. 1964. Hematopoietic stem cells. *Nature 204:* 589-590.

Maloney, M. A. and H. M. Patt. 1972. Persistent marrow hypocellularity after local X-irradiation of the rabbit femur with 1000 rad. *Rad. Res. 50:* 284-292.

Marks, P. A., L. Cantor, M. Cooper, D. Singer, G. Maniatis, A. Bank and R. A. Rifkind. 1974. Erythroid cell proliferation and differentiation: Action of erythropoietin. In: *Control of Proliferation in Animal Cells.* B. Clarkson and R. Baserga, eds. Cold Spring Harbor Laboratory, New York, Vol. 1, pp. 853-861.

Meck, R. A., J. E. Haley and G. Brecher. 1973. Hematopoiesis versus osteogenesis in ectopic bone marrow transplants. *Blood 42:* 661-669.

Morley, A. and J. Blake. 1974. An animal model of chronic aplastic marrow failure. I: Late marrow failure after Bulsulfan. *Blood 44:* 49-56.

Nowell, P. C., B. E. Hersch, D. H. Fox and D. B. Wilson. 1970. Evidence for the existence of multipotential lympho-hematopoietic stem cells in the adult rat. *J. Cell. Physiol. 75:* 151-158.

O'Grady, L. F., S. A. Knadle and J. P. Lewis. 1973. Evidence for limitation of granulopoiesis within the spleen. *Exp. Hemat. 1:* 251.

Rauchwerger, J. M., M. T. Gallagher and J. J. Trentin. 1973. Role of the hemopoietic inductive microenvironments (HIM) in xenogeneic bone marrow transplantation. *Transplantation 15:* 610-613.

Schooley, J. C. 1966. The effect of erythropoietin on the growth and development of spleen colony-forming cells. *J. Cell. Physiol. 68:* 249-262.

Seki, M. 1973. Hematopoietic colony formation in a macrophage layer provided by intraperitoneal insertion of cellulose acetate membrane. *Transplantation 16:* 544-549.

Tavassoli, M. and W. H. Crosby. 1968. Transplantation of marrow to extramedullary sites. *Science 161:* 54-56.

Till, J. E., E. A. McCulloch and L. Siminovitch. 1964. A stochastic model of stem cell proliferation, based on the growth of spleen colony-forming cells. *Proc. Nat. Acad. Sci. USA 51:* 29-36.

Till, J. E., L. Siminovitch and E. A. McCulloch. 1966. Growth and differentiation of marrow cells transplanted in anemic and plethoric mice of genotype W/W^v. *Exp. Hemat. 9:* 59.

Trentin, J. J. 1970. Influence of hematopoietic organ stroma (Hematopoietic Inductive Microenvironments) on stem cell differentiation. In: *Regulation of Hematopoiesis.* A. S. Gordon, ed. Appleton-Century-Crofts, New York, Vol. 1, pp. 161-186.

Trentin, J. J. 1971. Determination of bone marrow stem cell differentiation by stromal hemopoietic inductive microenvironments (HIM). *Amer. J. Path. 65:* 621-628.

Wolf, N. S. 1973. The hemopoietic microenvironment - its localization and capability for regeneration in Sl/Sl^d - congeneic +/+ spleen chimeras. *Exp. Hemat. 1:* 57-58.

Wolf, N. S., V. K. Jenkins and J. J. Trentin. 1972. Increase in proportion of erythropoietic spleen colonies in mice injected with spleen cells as compared with bone marrow cells. *Exp. Hemat. 22:* 37-38.

Wolf, N. S. and D. Lagunoff. 1974. Autoregeneration of splenic structure and function. *Abstracts, Third Annual Meeting, International Society for Experimental Hematology.*

Wolf, N. S. and J. J. Trentin. 1968. Hemopoietic colony studies. V. Effect of hemopoietic organ stroma on differentiation of pluripotent stem cells. *J. Exp. Med. 127:* 205-214.

Wolf, N. S. and J. J. Trentin. 1975. Linearity of counts of endogenous and exogenous colonies in mouse bone marrow. *Exp. Hemat. 3:* 54-56.

Discussion, Session IV - Stem Cells in Hemopoietic and Lymphoid Tissue II

Dr. Osmond's presentation

Dr. Moore commented that by using radioiodinated Fab fragments of anti-IgM he experienced a plateau labeling of bone marrow cells with somewhat lower labeling index, as compared to the linear increase in labeling found with increasing concentrations of unfragmented anti-IgM. Dr. Osmond responded by stating that the linearity in labeling with the technique in question could not be dismissed as a technical artifact. For example, a weakly labeled component indicative of low surface IgM density also emerged at a single antiglobulin concentration, when the grain count distribution of small lymphocytes in bone marrow was compared with those in peripheral lymphoid tissues. Furthermore, using the same technique, a plateau labeling was seen in bone marrow small lymphocytes during post-irradiation regeneration (at 3 wk after 150 R irradiation).

Dr. Moore raised the question whether a proportion of marrow lymphocytes could represent monocyte precursors, and thus account partly for the Fc-receptor bearing cells. He mentioned that in an Abelson-virus induced lymphoma line, cells had Fc receptors and were Ig +ve in culture; they produced lysozymes and could be induced to differentiate into macrophages. Dr. Osmond replied that his reference to C3 or Fc positive cells related only to small lymphocytes clearly identifiable on morphological grounds in cytocentrifuge preparations; rosettes with central monocytoid cells were also seen in the marrow but not included here. Dr. Moore mentioned his work with Kincade and that of Phillips and Miller which showed that destruction of marrow environment by Strontium 89 did not reduce either the number or functional capacity of B cells appearing after subsequent whole body X-irradiation and reconstitution with either fetal liver or bone marrow cells. Dr. Osmond felt that this finding was in no way contradictory to his results; while bone marrow, under normal conditions, was a site of large scale production of small lymphocytes that mature into functional B cells with time, the marrow environment may not be obligatory for such maturation. Probably, in normal animals the maturation is completed after migration to peripheral lymphoid tissues, while in animals with no functional marrow environment it may occur wholly in peripheral organs, notably the spleen.

Dr. Micklem raised the issue of the extent of lymphocyte migration from the marrow to other lymphoid organs. Dr. Osmond responded that his previous studies with local marrow labeling with ^{3}HTdR could not provide a quantitative estimate. By relating the number of circulating granulocytes which are marrow derived to the labeled lymphocytes, one could only say that there was a large scale lymphocyte migration out of the marrow. Some homed to peripheral lymphoid tissues, and such homing was somewhat increased after local antigenic stimulation. Dr. Everett mentioned that he found a substantial homing of the marrow derived small lymphocytes into the spleen, in particular the red pulp. Dr. Osmond added that in relative terms the extent of homing was found to be high in the spleen, moderate in the lymph nodes and very low in the thymus. Dr. Micklem commented that using parabiosis or part body irradiation, the input from the marrow to the lymph nodes (directly as well as indirectly via thymus) could be estimated as 4.5 - 5.0% per day; thus, many of the 10^8 cells produced per day in the marrow of 4 week old mice in Dr. Osmond's study would be spare on that basis. Dr. Osmond agreed and emphasized that one must distinguish between two different streams of cell traffic: migration of mature small lymphocytes directly, and the migration of stem cells, whose progeny are released from the thymus at a later interval.

Dr. Miller's presentation

In reply to several technical questions (from Drs. Moens and Brugal), Dr. Miller listed the following attributes of the cell sorting system: cells were confined to columns of about 10 micrometers across, seen by a laser beam focussed onto a spot of about 70 micrometers across; there was no cell damage, and approximately 10^4 cells could be processed per second without having any coincidence problem. Dr. Osmond wanted to know whether the cell cycle properties of Dr. Miller's "Pre-B" cell were known; for example, from the use of the thymidine suicide technique. Dr. Miller said that he had no such direct information, although the shape of the sedimentation profile suggested that it was largely out of cycle.

Dr. Everett's presentation

Dr. Lala wanted to know whether the proportion of stem cells amongst Dr. Everett's "monocytoid" cells seen in the exudate or other tissues, was known. Dr. Everett indicated that a precise estimate was not available and referred to Lord's study on buffy coat in irradiated animals which was rich in these monocytoid forms, and also showed a 10 fold repopulating potential compared to normal bone marrow.

Dr. Leblond expressed his concern about the fact that the nuclear chromatin in Dr. Everett's "candiate stem cells" was dense in contrast to diffuse chromatin pattern of identifiable stem cells in most tissues excepting the satellite cell of the muscle in the adult animal. While the animal is young, and satellite cells are actively dividing and contributing to muscle growth, the chromatin remains diffuse. He suggested that one should possibly examine young animals to see whether the chromatin is diffuse or dense in hemopoietic stem cells. Dr. Chui (Hamilton) mentioned that he did not find any cell having similar morphology (comparable to those shown by Dr. Everett) in 13 day fetal liver that had a high CFU content.

Dr. Patt's presentation

Dr. Lamerton pointed out that the spleen does not need an osseous environment to produce blood cells. Dr. Patt responded that there is rather little hemopoiesis in most spleens. He emphasized that the osseous environment is not an "obligatory" but a "preferred" one in normal mammals, and ectopic hemopoiesis without bone formation is seen mostly in diseased conditions such as leukemias and myeloproliferative disorders. Although important insights into the hemopoietic regulators have been obtained with studies on the spleen, he suggested that in some respects the spleen might be an artifact, and it might not be an ideal model for extracting general principles in hemopoiesis.

Dr. Leblond enquired whether the osteoclasts in the endosteum may have a regulatory role by releasing free calcium. Dr. Patt emphasized that calcium might play an important role in proliferation of marrow elements as suggested by Rohlich in 1941, and also shown by Whitfield, Rixon and their associates (from Chalk River and Ottawa) in a number of studies. He also mentioned his frustrating literature search on various bone diseases to examine whether there is any association of hemopoietic status of the bone to demineralizing conditions; unfortunately no information on hemopoiesis was available.

Dr. Leblond mentioned that Bloom's hypothesis about transformation of connective tissue cells into osteoblasts and vice-versa in birds during the course of the cycle was not substantiated in recent studies by Dr. LeDuarin in France; by reciprocal transplantation of bones between developing chicken and quail (whose cells can be identified by distinctive nuclear morphology), the osteoblasts were always found to be donor-type and not the host type. Dr. Patt listed several experiments indicating that there must be some cellular elements which retain the potential for bone formation. For example, fibroblasts derived from bone marrow after several generations of *in vitro* culture, when implanted under the skin or renal capsule could lead to bone formation; transplantation of bladder epithelium in some sites (as shown by Huggins many years ago) could also do the same thing. Similarly, there could be elements in marrow that had the osteogenic potential. Dr. Leblond wondered whether the osteoblasts could be derived from circulation in all these experiments. Dr. Patt alluded to the ongoing work of Walker at Johns Hopkins on the osteopetrotic mouse having no bone cavity; the osteopetrosis can be cured either by injection of hemopoietic tissue or by parabiosis. In this case, as postulated by some workers, monocytes may have the ability to transform into osteoblasts.

Dr. Trentin's presentation

Dr. Till stated that Dr. Trentin's evidence did not, to his satisfaction, resolve the issue whether the microenvironment was indeed "inductive". Does the very early division of CFU-S produce committed progenitors such as CFU-C or CFU-E in a random fashion or in a specific manner as determined by the environment? Since, at such early interval, the splenic colony size would be too small both for morphological study and recovery of enough cells for *in vitro* assay, he felt that the issue may never be resolved. Dr. Trentin emphasized that his experiments definitely showed a microenvironmental limitation of colony growth *in vivo* into one type or the other. If this limitation was different from induction, one had to invoke highly differential concentrations of materials such as CSA or erythropoietin. Since these were known to be circulating molecules, in Dr. Trentin's view, this would seem very unlikely. He felt that his experiments showing the "geographic" rather than "temporal" limitation of development of apparently undifferentiated cells into erythroid colonies suggested that the induction for erythropoiesis was proven to be determined geographically. Dr. Till mentioned that Connie Gregory in his laboratory looked at some early spleen colonies that were morphologically identifiable as purely erythroid but still produced measurable numbers of CFU-C *in vitro;* similarly measurable numbers of CFU-E were produced by cells from colonies morphologically identified as purely granulocytic. These numbers were significant and could not be explained away as results of contamination with migrants into the colonies. Thus, Dr. Till felt that the "HIM" (Hemopoietic Inductive Microenvironment) and "HER" (Hemopoiesis is Engendered at Random) controversy was not over!

General Discussions

Dr. E. Daniels (Dept. of Anatomy, McGill Univ.) commented on current isolation techniques for marrow stromal elements *in vitro* and presented some of his findings in guinea pigs. He illustrated two morphologically distinct types of monolayers of adherent or migrant cells developing on coverglasses placed underneath semisolid agar in which either bone marrow cells or marrow depleted bone fragments were suspended (Fig. 1A and B).

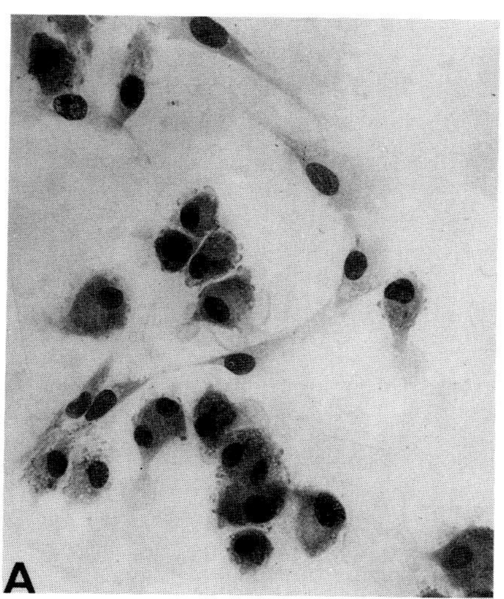

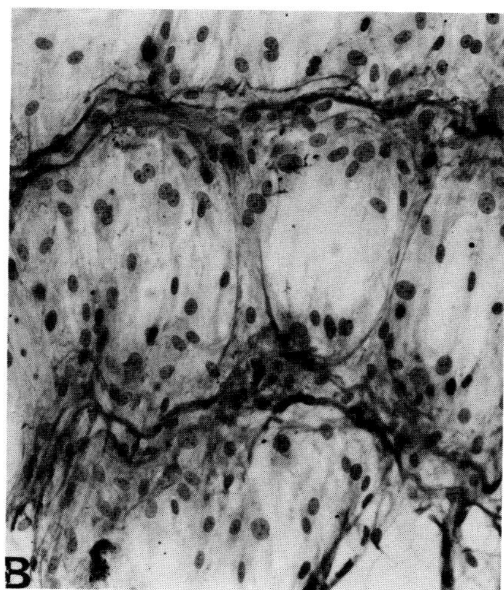

Fig. 1A and B. Monolayers derived from guinea pig marrow cell suspensions (A) and from diaphyseal bone fragments physically depleted of marrow (B).

When bone marrow cell suspensions were used, monolayers of adherent cells developed on the coverglasses concurrent with the development of predominantly macrophage colonies in agar. Cells in these monolayers soon became confluent, and some were binucleate or multinucleate. They showed phagocytic properties as indicated by the presence of agar inclusions in the cytoplasm. When marrow depleted bone fragments were cultivated, monolayers on the coverglasses developed as direct outgrowths from these fragments. These showed extensive networks of extracellular material in contrast to the purely cellular monolayers derived from marrow cell suspensions. Both these morphologically distinct cellular monolayers exhibited colony stimulating activity for CFU-C growth.

Session V
Stem Cells in the Testis

Spermatogonial stem cells and their behaviour in the seminiferous epithelium of rats and monkeys

Y. CLERMONT AND L. HERMO

Department of Anatomy, McGill University, Montreal, Quebec, Canada

In seminiferous tubules of adult mammals, spermatogonia continuously produce spermatocytes that yield spermatids which in turn metamorphose into spermatozoa. The spermatogonial population must, therefore, be maintained by renewal. Regaud ('01) in his classical studies on rat seminiferous tubules clearly identified the problem when he tried to recognize amongst the various types of spermatogonia those which served as *"stem cells"* from those that were *"differentiating cells"*, i.e., those committed to produce spermatocytes. From his studies, he concluded that the "dust-like" spermatogonia, now called type A, served as stem cells undergoing renewal, while the "crust-like" spermatogonia, now identified as type B, were committed to give rise to spermatocytes. This conclusion is still generally valid today. However, despite a number of extensive investigations (reviewed in Clermont '72), the exact behaviour of spermatogonial stem cells remains a matter of controversy (Huckins, '72). It is the purpose of this article to present quantitative data on the spermatogonial population of rats and monkeys that will permit a description of the behaviour of the various types of spermatogonia and in particular demonstrate the existence of two categories of stem cells referred to as the "reserve" and "renewing" stem cells (Clermont and Bustos-Obregón, '68). The terms *"stem cells"* and *"differentiating cells"* used in the present article have been defined as follows by Leblond *et al.* ('67): *"stem cells* are unspecialized cells which have the ability to divide in such a way as to give rise to new stem cells and to differentiating cells, while *differentiating cells* include all the cells that are at various steps toward specialization; these may divide to give rise to other differentiating cells but not to new stem cells."

METHOD OF STUDY

The histological characteristics of the seminiferous epithelium permit an analysis of the spermatogonial population in function of time. Indeed, spermatogonia, spermatocytes and spermatids at various steps of their respective development form cellular associations of fixed composition that correspond to stages of a cycle of the seminiferous epithelium; the cycle being defined as "a complete series of the successive cellular associations appearing in any one area of a seminiferous tubule" (Leblond and Clermont, '52). In both rats and monkeys, the criteria used to identify the cell associations or stages of the cycle are the steps of development of the spermatids as seen in periodic acid-Schiff-hematoxylin stained sections. This method stains vividly the developing acrosomic system of the spermatids and, therefore, permits a rapid and accurate identification of both the steps of spermiogenesis and the cell associations to which the spermatids belong. Using this approach the cycle of the seminiferous epithelium was subdivided into fourteen stages in the rat and twelve stages in the monkey (Leblond and Clermont, '52; Clermont and Leblond, '59).

In our studies both transverse sections of testes stained with PA-Schiff-hematoxylin and dissected tubules, fixed, stained with hematoxylin and mounted *in toto* were utilized. In the latter type of material, the stages of the cycle were more difficult to identify, since the PA-Schiff technique could not be used due to the thickness of the preparation. For this reason, other criteria were utilized, such as the type of spermatogonia, the step of spermatocyte development and the position of late spermatids within the epithelium, thus allowing the recognition of the stages of the cycle (Clermont and Bustos-Obregón, '68; Clermont, '69; Clermont and Hermo, '75).

In the present study, enumeration of resting and dividing spermatogonia was performed in whole mounts of both rat and monkey *(Cercopithecus aethiops)* seminiferous tubules, and the counts were expressed as the average number of spermatogonia per unit surface area of tubular wall (or frame). In addition, the labeling indices of the various types of spermatogonia, in particular of the type A cells, were calculated following scoring of labeled and unlabeled cells in radioautographed sections from rats and monkeys *(Macaca mulatta)* sacrificed at various time intervals after one or more injections of ^{3}H-thymidine.

MORPHOLOGY OF SPERMATOGONIA

In both rats and monkeys, spermatogonia can be classified into one of two main categories, referred to as the type A and type B spermatogonia. Type A cells can be identified by their nucleus which contains a uniformly distributed finely granulated chromatin (euchromatin), while the type B spermatogonia can be recognized by the presence of flakes of heavily stained chromatin (heterochromatin) which is associated with the nuclear envelope and nucleolus (Fig. 1).

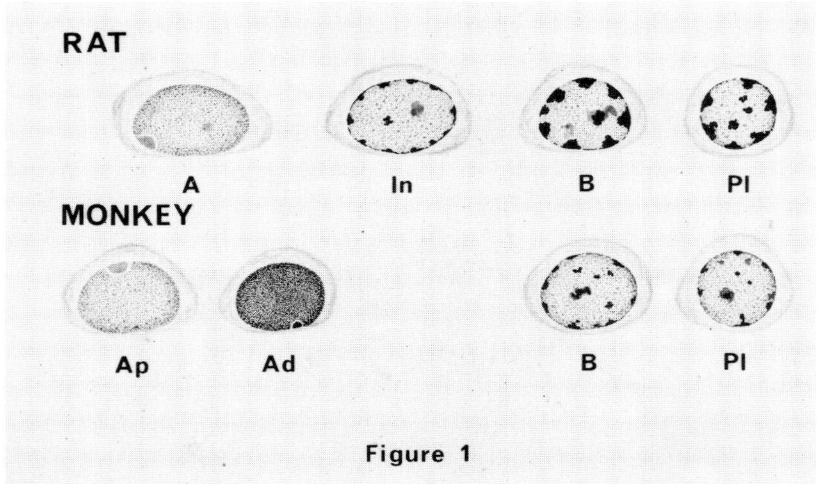

Fig. 1. Drawings showing the characteristic appearance of the nucleus of the various types of spermatogonia and of preleptotene spermatocytes (Pl) in rat and monkey.
A, In and B: type A, Intermediate and type B spermatogonia of the rat.
Ap, Ad and B: pale type A, dark type B and type B spermatogonia of the monkey.

While all type A spermatogonia have the same general appearance in the rat, the type A cells in the monkey can be readily subdivided into two distinct classes, respectively referred to as the pale type A (Ap) and the dark type A (Ad) spermatogonia. The former show a nucleus containing a granulated but pale stained chromatin, while the latter present a nucleus containing a fine densely packed and fairly deeply stained enchromatin (Figs. 1, 2 and 3).

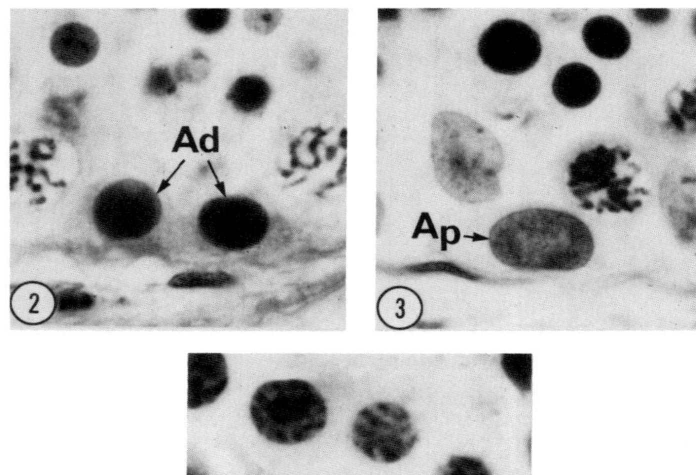

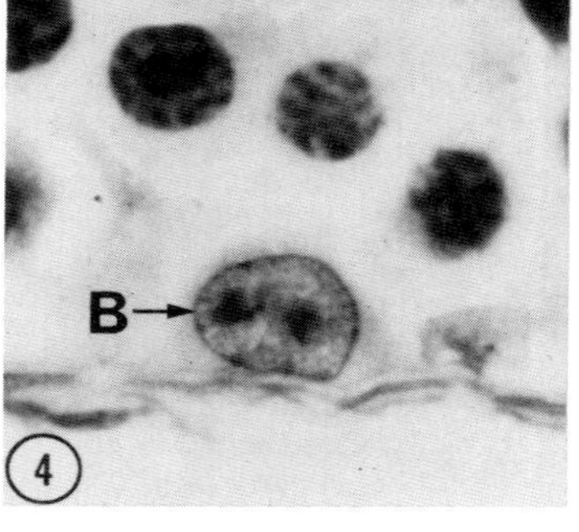

Figs. 2, 3, 4 show the various types of spermatogonia along the limiting membrane of monkey seminiferous tubules. The following cells are shown: dark type A spermatogonia with their nucleus containing a homogeneous deeply stained chromatin (Ad, Fig. 2); a pale type A spermatogonium with a nucleus containing a finely granulated pale stained chromatin (Ap, Fig. 3); a type B spermatogonium showing accumulations of deeply stained chromatin along the nuclear membrane and on the nucleolus (B, Fig. 4).

In both rats and monkeys there are, as indicated by the cell counts, several generations of type B-like spermatogonia. In the rat there are two generations, the youngest being referred to as Intermediate type spermatogonia and the oldest as type B spermatogonia (Clermont and Leblond, '53). In each case, both show heterochromatin clumps along the nuclear envelope, however, those of the Intermediate cells are smaller than those of the type B cells (Fig. 1). In the monkey there are four generations of type B cells (B_1-B_4). All have a nucleus presenting the same general appearance, although they show, from the younger (B_1) to the older generation (B_4), a progressive decrease in size of the nucleus accompanied by a

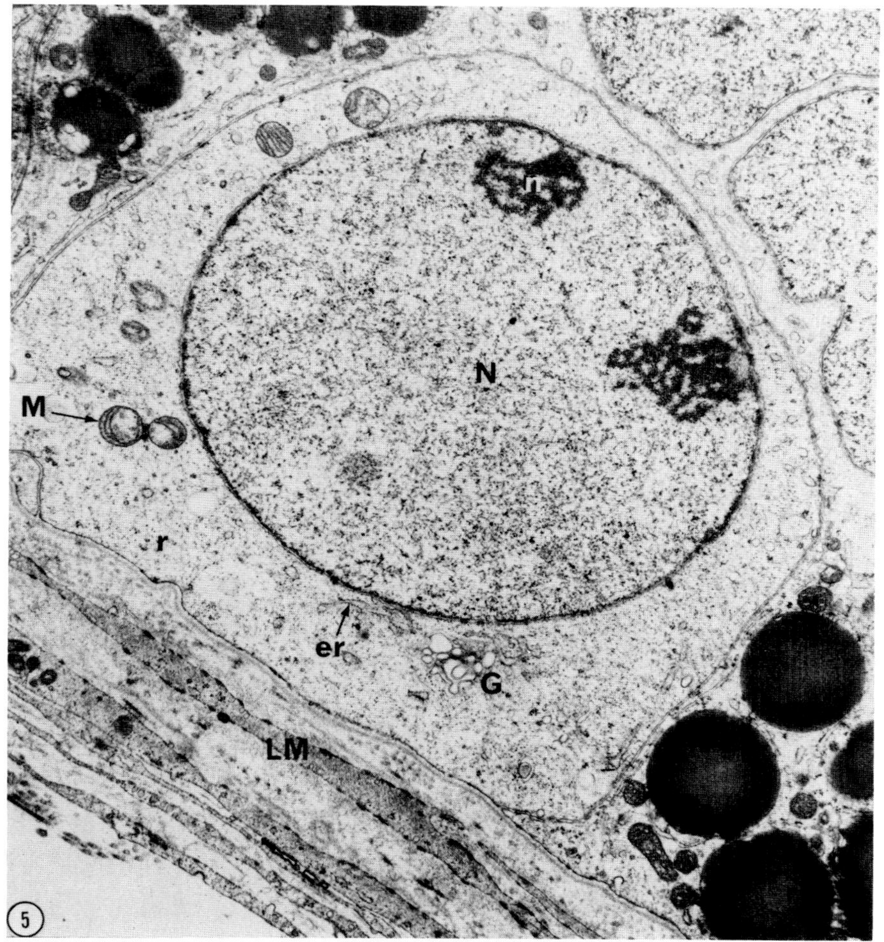

Fig. 5. Electron microphotograph showing a type A spermatogonium (Ap) along the limiting membrane (LM) of a monkey seminiferous tubule. The nucleus (N) shows a diffuse chromatin and two nucleoli (n). The cytoplasm contains a few organelles, i.e., mitochondria (M), a poorly developed Golgi apparatus (G), some cisternae of endoplasmic reticulum (er) and ribosomes (r).

---

progressive increase in the amount of heterochromatin (Figs. 1, 4).

While structural differences are noted in the nucleus of the various types of spermatogonia, the cytoplasm, as seen under the electron microscope, is found to show similar features in all spermatogonia of a given species. In addition to a poorly developed Golgi apparatus, accompanied by a pair of centrioles, next to the nucleus, the cytoplasm contains a few mitochondria and a scanty, poorly developed endoplasmic reticulum. There is, however, an abundance of free ribosomes and some polysomes (Fig. 5).

TOPOGRAPHICAL ARRANGEMENT OF TYPE A SPERMATOGONIA ALONG THE LIMITING MEMBRANE OF SEMINIFEROUS TUBULES

In whole mounts of rat seminiferous tubules, the type A spermatogonia showed the following characteristic topographical arrangement. At all stages of the cycle of the seminiferous epithelium some types of spermatogonia were isolated or paired, while others formed larger groups composed of an even number of cells (Fig. 6). From stage II to IX of the cycle, the cells of a given group of type A cells were compact and variable in size (groups 4, 8 and 16 were most frequent), while in stages X to XIV when type A cells divide, the groups of type A cells were larger and the cells of each

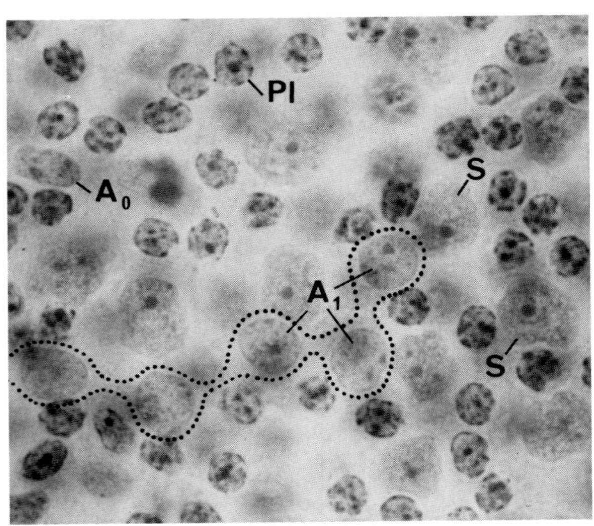

Fig. 6. Photograph of a tubular whole mount of the rat; the spermatogonial layer is in focus. In addition to the nuclei of Sertoli cells (S), recognized by their distinctive nucleolus and the nuclei of the numerous preleptotene spermatocytes (Pl), the type A spermatogonia are visible. An

group were more widespread along the tubular wall. The grouped arrangement of the latter cells was particularly obvious at the time of their division (e.g. in stage IX), since they usually underwent mitosis in a quasi-synchronous manner. The cells of a given group were most probably joined by open intercellular bridges (Dym and Fawcett, '71; Moens and Go, '72). The isolated and paired type A spermatogonia were referred to as type A_0 cells (for reasons to be given below), while the grouped type A cells were identified as type A_1-A_4, since there were four consecutive generations of such cells (Clermont and Bustos-Obregón, '68). In monkeys the type Ap and Ad spermatogonia, present throughout the cycle of the seminiferous epithelium, were segregated into homogeneous groups which were variable in size and composed of an even number of cells (Clermont, '69).

SPERMATOGONIAL PEAKS OF MITOSIS DURING THE CYCLE OF THE SEMINIFEROUS EPITHELIUM

When a large number of tubular cross sections are quantitatively surveyed or when whole mounts of tubules are analyzed, it soon becomes evident that spermatogonia do not divide at random throughout the cycle of the seminiferous epithelium, but that they tend, on the contrary, to divide at specific stages of the cycle. Thus in the rat, spermatogonia divide mainly in stages I (late), IV, VI, IX, XII and XIV of the cycle. In the two species of monkey presently considered (*Macaca mulatta, Cercopithecus aethiops)*, spermatogonia divide in stages II, IV, VI, IX-X, and XII of the cycle. It has been noted that, in the rat, the Intermediate and type B spermatogonia divide in stages IV and VI respectively, while type A cells divide at stages I (late), IX, XII and XIV of the cycle (Clermont and Leblond, '53). In the species of monkeys mentioned above, the type A (Ap) spermatogonia divide in stages IX-X of the cycle, while type B spermatogonia are the dividing cells in stages II, IV, VI and XII of the cycle (Clermont, '69). The quantitative data to be considered below support these conclusions.

---

isolated type A_0 spermatogonium is present in the field (A_0) as well as a linear group of type A_1 spermatogonia (A_1). This group of A_1 cells, indicated by a dotted line, continued on the left of the present field, but was out of focus.

COUNTS OF SPERMATOGONIA AT VARIOUS STAGES OF THE CYCLE

For the purpose of the present discussion, the data used will be those collected from whole mounts of seminiferous tubules, and the cell counts will be expressed as the number of cells per frame (7,225 μm^2) of tubular limiting membrane {Details of these data for the rat are given in the paper of Clermont and Hermo ('75), while that for the monkey *(Cercopithecus aethiops)* are presented in the work of Clermont ('69)}. In both rats and monkeys, counts were performed at those stages of the cycle where mitotic activity was minimal, i.e., at stages I (early), II-III, V, VII-VIII, X-XI and XIII for the rat and stages I, III, V, VII-VIII and XI for the monkey. It should be remembered that in the rat a distinction between the isolated or paired type A cells (A_o) and the grouped type A cells (A_1-A_4) can only be performed on tubular whole mounts, since the identification of either type is based on their topographical arrangement along the tubular limiting membrane; hence their identification in tubular cross sections cannot be ascertained. In contrast, the type Ad and Ap spermatogonia of the monkey can be readily identified in both sections and whole mounts of seminiferous tubules.

In the rat, the average number of isolated type A_o spermatogonia per frame remained constant throughout the cycle of the seminiferous epithelium; the small variation between the cell numbers was not statistically significant (Table I).

TABLE I

Number of Spermatogonia per Frame in the Rat

Stages of the Cycle	*Type* A_o isolated	*Type* A_o paired	*Type* A_1-A_4	*Type In, B* or Spermatocytes
II-III	0.50	0.40	1.9 (A_1)	13.4 (In)
V	0.47	0.40	2.9 (A_1)	26.0 (B)
VII-VIII	0.42	0.40	2.8 (A_1)	50.7 (Sptc)
X-XI	0.37	0.17	5.9 (A_2)	
XIII	0.34	0.17	5.7 (A_3)	
I	0.38	0.40	7.8 (A_4)	

In the case of the paired type A_o cells, except for the low values in stages X-XI and XIII, the average number was the same at each of the remaining stages of the cycle and, concurrently, of the same order of magnitude when compared to the values obtained for the isolated type A_o cells. It was also observed that the overall mitotic index of isolated and/

or paired type A_0 cells was very low in normal adult rats, i.e., 0.13%. As for the grouped type A spermatogonia (Table I), the type A_1 cells appeared following the peak of mitosis taking place in late stage I of the cycle, and their number remained constant (2.8) over the duration of stages V and VII-VIII of the cycle. Then following the peak of mitosis in stage IX, the number of type A cells doubled (5.9). This clearly indicated that all type A_1 cells divided to yield type A_2 cells. In stage XIII of the cycle, the number of type A cells counted (A_3) remained unchanged despite the mitotic peak observed in stage XII of the cycle. It was noted, however, that in stage XII of the cycle a large number of type A cells degenerated, usually in clusters, as they entered mitosis as already noted by Clermont ('62), Clermont and Bustos-Obregon ('68) and Huckins ('71a). This, therefore, explained the poor yield of type A cells at stage XIII of the cycle. Following the stage XIV peak of mitosis, the number of type A cells (A_4) at stage I reached a value of 7.8, a figure which was not twice that of the type A_3 cells counted in stage XIII. Once again, degeneration of type A cells was noted during this peak of mitosis, thus accounting for the relatively low yield of type A cells at stage I. Following the peak of mitosis observed late in stage I of the cycle, Intermediate and type A_1 cells were found along the tubular limiting membrane. The total number of Intermediate and type A_1 cells in stage II-III of the cycle was approximately twice the number of type A_4 spermatogonia, thus indicating that, during the stage I peak of mitosis, all type A_4 cells divided. The intermediate spermatogonia divided in stage IV to give rise to type B spermatogonia, the latter in turn divided in stage VI to yield preleptotene spermatocytes. This is indicated by the doubling in the number of cells at each one of these last two peaks of mitosis. It is clear, therefore, and this is generally accepted, that the Intermediate and type B spermatogonia are committed to produce spermatocytes and should, therefore, be considered as "differentiating" spermatogonia. The overall mitotic index of type A_1-A_4 cells was found to be 0.63% (Clermont and Hermo, '75).

Considering on the one hand the constancy of the average number of type A_0 spermatogonia per unit area of tubular wall at the various stages of the cycle (this is particularly true for the isolated type A_0 cells; the low number of paired type A_0 cells in stages X-XI and XIII was considered to be due to the difficulty of identifying these cells in presence of the numerous widespread type A_2 and A_3 cells along the tubular wall), and on the other hand the low mitotic index of the type A_0 cells, it was concluded that this class of type A cells, although they may divide sporadically, could not account for the production of the large number of type A_1 cells found along the tubular wall at each cycle of the seminiferous epithelium. The type A_0 cells were thus considered to be generally quiescent and to constitute a population of "reserve stem cells" (Clermont and Bustos-Obregón, '68; Clermont and Hermo, '75). The reserve function of the type A_0 cells was evident during the repair phase of the spermatogonial population following X-irradiation. Indeed, in such a material the radioresistant type A_0 spermatogonia became mitotically active and contributed to the restoration of the spermatogonial population. Once this was achieved, some type A cells returned to their resting reserve stem cell condition (Dym and Clermont, '70).

The proliferative activity of the type A_1-A_4 spermatogonia was apparent both from morphological observations and quantitative data, both of which indicated that this population of cells was self-renewing at each cycle of the seminiferous epithelium. New type A_1 spermatogonia made their appearance along the wall of the tubule with Intermediate type spermatogonia immediately after the peak of mitosis of type A_4 cells late in stage I of the cycle. Cell numbers indicated that Intermediate and type A_1 spermatogonia resulted from the mitoses of type A_4 cells (i.e., the number of Intermediate plus A_1 cells equalled twice the number of A_4 cells). Furthermore, the labeling index of type A cells calculated from tubular sections in stage IV of the cycle following multiple injections of ^{3}H-thymidine reached a maximal value of 68%, a percentage equivalent to the proportion of type A_1 cells calculated from the total number of type A_0 + A_1 cells (obtained from whole mounts) at this same stage of the cycle (Clermont and Hermo, '75). This clearly indicated that all type A_1 cells collected their label during the peak of mitosis of type A_4 cells at stage I of the cycle. Thus the type A_1-A_4 cells are elements, which at each cycle of the seminiferous epithelium, undergo a series of divisions which lead to the formation of new type A_1 cells and to Intermediate type spermatogonia.

The type A_1-A_4 spermatogonia should, therefore, be considered as "renewing stem cells". This conclusion is not accepted by Huckins ('71a, b, c) working on the rat, nor Oakberg working on the mouse ('71); these investigators stated that the isolated type A spermatogonia divided throughout the cycle of the seminiferous epithelium to give rise to new isolated type A as well as to other type A cells, which through a series of consecutive mitoses eventually resulted in the formation of Intermediate type spermatogonia.

In the monkey *(Macaca mulatta* and/or *Cereopithecus aethiops)* the cell counts (Table 2) indicated that the average number of type Ad and Ap cells was on the one hand, constant throughout the cycle of the seminiferous epithelium and, on the other hand, of the same order of magnitude for both classes of cells (Ap:Ad = 1:1). Following the stage IX-X peak of mitosis, type B_1 cells appeared and their average number per frame was equal to that of the type Ap (or Ad) cells. During the following peaks of mitosis, taking place consecutively in stages XII, II and IV of the cycle, the number of type B cells doubled each time and finally, following the last mitosis of type B cells in stage VI of the cycle, the number of spermatocytes enumerated was exactly twice the number of progenitor type B_4 spermatogonia. It is obvious from these data that the type B_1-B_4 spermatogonia are differentiating cells leading to the production of spermatocytes.

TABLE 2

Number of Spermatogonia per Frame in the Monkey

Stages of the cycle	Type Ad	Type Ap	Type B or Spermatocytes
XI	3.5	3.3	3.3 B_1
I	3.2	3.2	7.5 B_2
III	3.4	2.8	15.0 B_3
V	3.5	3.1	28.4 B_4
VII-VIII	3.5	3.5	53.7 Sptc

The spermatogonia dividing in stage IX-X of the cycle were the pale type A spermatogonia and this was evident from the study of this class of cells in segments of tubular whole mounts at stages VII, VIII and IX of the cycle. During these stages, while the type Ad spermatogonia did not show signs of nuclear changes, the nucleus of type Ap spermatogonia progressively increased in size and eventually showed an early transformation of the chromatin characteristic of prophase. Correspondingly, in radioautographed sections of testes from animals sacrificed 3 to 5 hours after injection with ^3H-thymidine, the labeling index of the type Ap spermatogonia was found to be as high as 36% with most of the cells labeled in stages VIII and IX of the cycle, while the labeling index of the dark type A spermatogonia (Ad) was negligible and less than 0.1%. Therefore, it was obvious that while the type Ad spermatogonia were non-proliferative cells or cells that were dividing very rarely, the type Ap cells were in fact dividing at stages IX-X of each cycle of the seminiferous epithelium to produce an equal number of new type Ap cells and type B_1 spermatogonia.

CONCLUSION

Morphological and quantitative studies on the spermatogonial population in seminiferous tubules of rats and monkeys at various stages of the cycle of the seminiferous epithelium led to the following conclusions. There are in these two animal species and possibly in all mammals, two main categories of spermatogonia; the type A with a nucleus showing a uniform finely granulated euchromatin, and the type B with a nucleus containing dense coarse granules of heterochromatin. Whereas the type A spermatogonia are considered to be undifferentiated cells and fall within the category of stem cells, there appears to be two distinct classes, namely the reserve and the renewing type A. The reserve type A cell (type A_o in the rat, type Ad in the monkey), as judged by their constant number per unit surface area of tubular wall at the various stages of the cycle, their low mitotic activity and low labeling index at various intervals after ^{3}H-thymidine injection, do not appear, in normal adult animals, to be actively engaged in the production of spermatocytes. These cells may divide sporadically to maintain the pool of stem cells that could be partially depleted due to cell degeneration and in this way contribute to maintaining the population of spermatogonial stem cells in steady state. This process could be accelerated in instances of massive depletion of the type A spermatogonial population such as is taking place after X-irradiation. On the other hand, the renewing type A spermatogonia (type A_1-A_4 in the rat, type Ap in the monkey) are cells which periodically, during each cycle of the seminiferous epithelium, undergo one or more mitoses that eventually yield new type A renewing stem cells and more differentiated cells committed to the production of spermatocytes.

LITERATURE CITED

Clermont, Y. 1962. Quantitative analysis of spermatogenesis in the rat. A revised model for the renewal of spermatogonia. *Am. J. Anat. 111:* 111-129.

Clermont, Y. 1969. Two classes of spermatogonial stem cells in the monkey (*Cercopithecus aethiops*). *Am. J. Anat. 126:* 57-72.

Clermont, Y. 1972. Kinetics of spermatogenesis in mammals: seminiferous epithelium cycle and spermatogonial renewal. *Physiol. Rev. 52:* 198-236.

Clermont, Y. and E. Bustos-Obregon. 1968. Re-examination of spermatogonial renewal in the rat by means of seminiferous tubules mounted *in toto*. *Am. J. Anat. 123:* 237-248.

Clermont, Y. and L. Hermo. 1975. Spermatogonial stem cells in the albino rat. *Am. J. Anat. 142:* 159-175.

Clermont, Y. and C. P. Leblond. 1953. Renewal of spermatogonia in the rat. *Am. J. Anat. 93:* 475-502.

Clermont, Y. and C. P. Leblond. 1959. Differentiation and renewal of spermatogonia in the monkey, *Macacus rhesus*. *Am. J. Anat. 104:* 237-272.

Dym, M. and Y. Clermont. 1970. Role of spermatogonia in the repair of the seminiferous epithelium following X-irradiation of the rat testis. *Am. J. Anat. 128:* 265-282.

Dym, M. and D. W. Fawcett. 1971. Further observations on the numbers of spermatogonia, spermatocytes, and spermatids connected by intercellular bridges in the mammalian testis. *Biol. Reprod. 4:* 195-215.

Huckins, C. 1971a. The spermatogonial stem cell population in adult rats. I. Their morphology, proliferation and maturation. *Anat. Rec. 169:* 533-558.

Huckins, C. 1971b. The spermatogonial stem cell population in adult rats. II. A radioautographic analysis of their cell cycle properties. *Cell Tissue Kinet. 4:* 313-334.

Huckins, C. 1971c. The spermatogonial stem cell population in adult rats. III. Evidence for a long-cycling population. *Cell Tissue Kinet. 4:* 335-349.

Huckins, C. 1972. *Spermatogonial Stem Cells in Rodents*. In: Biology of Reproduction - Basic and Clinical Studies. J.T. Velardo and B.A. Kasprow (eds.). III Pan American Congress of Anatomy, New Orleans. pp. 395-421.

Leblond, C. P. and Y. Clermont. 1952. Definition of the stages of the cycle of the seminiferous epithelium in the rat. *Ann. N.Y. Acad. Sci. 55:* 548-573.

Leblond, C. P., Y. Clermont and N. J. Nadler. 1967. The pattern of stem cell renewal in three epithelia (Esophagus, Intestine and Testis). *Proc. Can. Cancer Res. Conf. 7:* 3-30.

Moens, P. B. and V.L.W. Go. 1972. Intercellular bridges and division patterns of rat spermatogonia. *Z. Zellforsch. 127:* 201-208.

Oakberg, E. F. 1971. Spermatogonial stem cell renewal in the mouse. *Anat. Rec. 169:* 515-532.

Regaud, C. 1901. Etudes sur la structure des tubes seminiferes et sur la spermatogenese chez les mammiferes. *Arch. Anat. Micro. Morph. Exp. 4:* 101-156, 231-280.

Spermatogonial Stem Cell Renewal in the Mouse as Revealed by ^{3}H-Thymidine Labeling and Irradiation[1]

E. F. OAKBERG AND CLAIRE HUCKINS

Biology Division, Oak Ridge National Laboratory,
Oak Ridge, Tennessee 37830
and
Department of Cell Biology,
Baylor College of Medicine,
Houston, Texas 77025

The mechanism of spermatogonial stem cell renewal presented here has unfolded over the past few years as a result of long-term ^{3}H-thymidine (^{3}H-TdR) labeling (Oakberg, '71a; Huckins, '71b,c); cell mapping (Huckins, '71a); cell counts (Huckins, '71a; de Rooij, '73); ^{3}H-TdR labeling in tubule whole mounts (Huckins, '71b,c); cell survival and repopulation studies after irradiation and chemicals (Oakberg, '68; van Keulen and de Rooij, '74); and ^{3}H-TdR labeling in conjunction with irradiation (Oakberg, '71b). These experiments have indicated that the single (A_s) spermatogonium functions as the stem cell. Many questions of stem cell behavior remain unanswered, such as the relationship between the rapid- and slow-cycling components of the population, the kinetics of the long-cycling cells, control of stem cell numbers, and the point of commitment to differentiation. Though the earlier steps of stem cell behavior are not yet clear, the most reasonable interpretation of the data is the model proposed by Huckins ('71a). The A_s spermatogonia divide to form either new A_s cells or a pair of cells with a cytoplasmic bridge. Subsequent divisions of the A_{pr} generate

---

[1] Research sponsored by the U. S. Energy Research and Development Administration under contract with the Union Carbide Corporation, and by NIH Grant No. HD-07655.

chains of A_{al} spermatogonia which differentiate into the A_1. The successive generations of A_2, A_3, A_4, In, and B spermatogonia undergo differentiation and division to form primary spermatocytes.

The above model evolved concurrently with radiation studies designed to characterize and elucidate the behavior of the radiation resistant A spermatogonia. Study of spermatogonial repopulation after 100 R had suggested that some surviving A spermatogonia must have already been "programmed" at the time of irradiation. Furthermore, it was realized that the low mitotic index of A spermatogonia in the irradiated testis could arise through selective survival of cells with an inherently low mitotic rate. Long-term labeling experiments to test these hypotheses already were under way when Clermont and Bustos-Obregon ('68) published their two-stem-cell model of spermatogonial renewal in which there were an acyclic A_0 and a renewing, cycling population of A_1-A_4 spermatogonia. The A_4 were given the role of pivotal cells, forming either more differentiated In or less differentiated A_1 spermatogonia. We repeatedly had demonstrated the high radiation sensitivity of A_1-A_4 spermatogonia (Oakberg, '59) while high radiation resistance would be expected for the acyclic A_0 cells. It was immediately apparent that labeling with ^{3}H-TdR before irradiation would be a critical test of the two-stem-cell model. For if the model were correct, only A_1-A_4 spermatogonia should label, and should subsequently be killed by radiation. The surviving A_0 cells should not be labeled. However, data demonstrating labeled A spermatogonia 5 days after radiation doses as high as 1000 R had been available for some time (Oakberg, '64). Preliminary results from long-term labeling experiments in control mice revealed heavily labeled cells as long as 10 days (more than one cycle of the seminiferous epithelium) after ^{3}H-TdR injection. On the basis of these data, we suggested (Oakberg, '68) that the A_0 spermatogonia were the active stem cells of the seminiferous epithelium. Also, Hilscher ('64) had observed that labeled A spermatogonia persisted for as long as 13 days in the normal adult rat. Morphological analysis and cell-cycle studies of tubule whole mounts have revealed the stem cell role of the A_s spermatogonia of the rat (Huckins, '71a) and the existence of a long-cycling compartment of A_s cells (Huckins, '71c). Also, data demonstrating the survival of labeled spermatogonia in the irradiated rat testis are now accumulating (Huckins, unpublished data).

In our previous studies we relied on sectioned material. For this reason, the present experiments included both sections and whole mounts in order to compare results from the two techniques. Both methods demonstrate the survival of A spermatogonia labeled with ^{3}H-TdR prior to irradiation, and the survivors are almost exclusively A_s spermatogonia. The data are consistent with the hypothesis that the stem cells are in continuous cycle, and that the total cycle time of some A_s spermatogonia is long. There is no evidence for a noncycling reserve stem cell (A_0) population.

MATERIALS AND METHODS

All mice used in these experiments were 12-week-old F_1 hybrid (101 X C3H) males. Both sections and whole mounts were made from the same animal: one testis was fixed for sectioning, the contralateral testis was used for whole mounts. For sections, testes were fixed for seven hours in Zenker-formol, washed for 12—15 hours in running tap water, embedded in paraffin, sectioned at 5 μm, stained by the PAS technique, dipped in Kodak NTB_2 emulsion, exposed for four weeks at 4°C, developed six minutes in D-170, and counterstained in Ehrlich's hematoxylin. $HgCl_2$ was removed by the procedure of Kopriwa and Huckins ('72). All procedures were rigidly controlled, for proper identification of spermatogonia cannot be made with confidence unless the slides are of excellent quality. For tubule whole mounts, segments of tubules were fixed for two hours in Bouin's fluid, and autoradiographs were prepared by the technique of Huckins and Kopriwa ('69).

Specific activity of the ^{3}H-TdR varied from 2 to 5 Ci/mmol. Dilutions were made in physiological saline to give the desired dose of 17.5 μCi/0.25 ml. All injections were made intraperitoneally.

Radiation exposures of 150 and 300 R were made with a Norelco 300 kV X-ray machine operated at 200 kV, 10 mA, with inherent filtration of 0.2 mm Cu, ∿90 R/min, and a target-testis distance of 53.5 cm. All experiments were run concurrently in order to minimize variability.

Stages of spermatogenesis were classified according to the cohorts of the differentiating spermatogonia. For several reasons, this classification is more meaningful than the 12 stages based on acrosome morphology of spermatids (Oakberg, '56a) which we have used previously: spermatogonial generations span more than one stage of acrosome development, data of sections and whole mounts are more readily compared, and analysis of spermatogonia with a low frequency per cross section is facilitated. These six stages in the cycle of the seminiferous epithelium are defined in table 1.

TABLE 1

Stages in the cycle of the seminiferous epithelium of the mouse as identified by the differentiating spermatogonia.

Stage	Spermatogonial type	Comparable stages based on acrosome development	Approximate duration in hours*
1	A_1	Late VI - Late IX	62
2	A_2	Late IX - Late XI	29
3	A_3	Late XI - Early I	29
4	A_4	Early I - Late II	29
5	In	Late II - Late IV	29
6	B	Late IV - Late VI	29

* *Calculated from the data of Monesi (J. Cell Biol., 14: 1-18, 1962) and Oakberg (Am. J. Anat., 99: 507-516, 1956).*

RESULTS

In sections, the A_s spermatogonia can be identified on the basis of an oval, darkly-staining granular nucleus with poorly defined nucleoli (Figs 1 and 2). In whole mounts, they appear as isolated cells with ovoid nuclei and dust-like chromatin (Clermont and Bustos-Obregon, '68). The morphology of the A_s spermatogonia in controls is identical to that of the cells surviving 150 R, both in sections (Figs 1 and 2) and in whole mounts (Figs 3 and 4). Also, during regeneration, the A_{a1} cells reappear at the same time at which they normally are generated (Figs 5 and 6). It is especially important to note that these cells appear in stage 3, in conjunction with A_3 spermatogonia *before* the appearance of the A_4. That this is not an artefact of the sections is demonstrated by the whole mount in figure 7. The morphology of the labeled spermatogonia 8.5 days after ^{3}H-TdR injection in the mouse is the same in both control and irradiated mice (Figs 8 and 9), and rats (Figs 10 and 11).

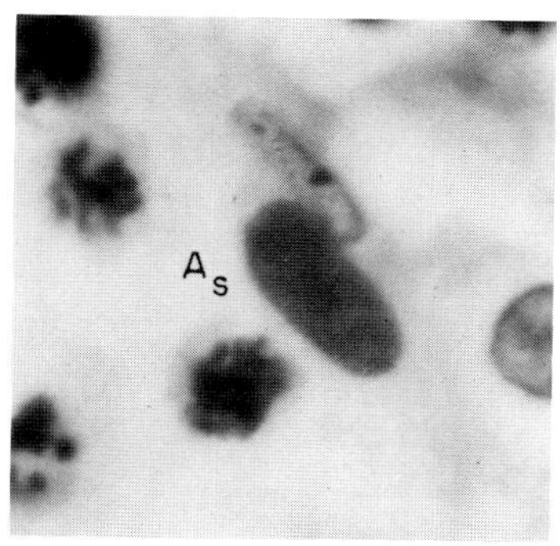

Fig. 1. A_s *spermatogonium of the mouse, control, stage 3.*

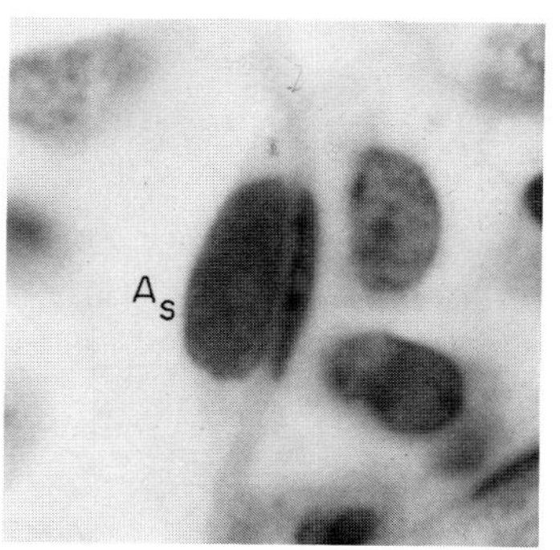

Fig. 2. A_s *spermatogonium of the mouse 207 hours after 500 R, stage 3.*

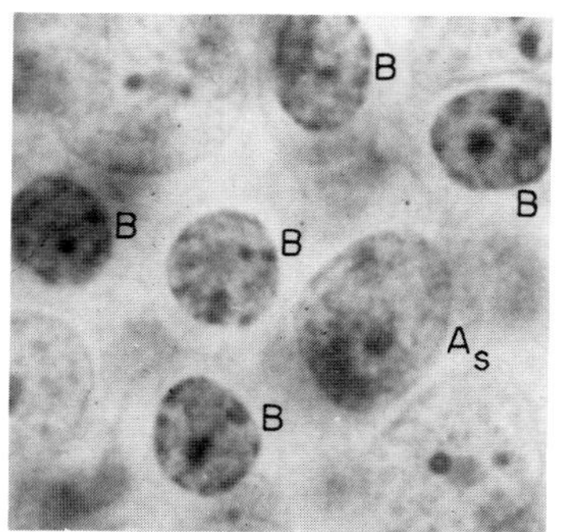

Fig. 3. A_s *spermatogonium of the mouse in tubule whole mount, control, stage 6.*

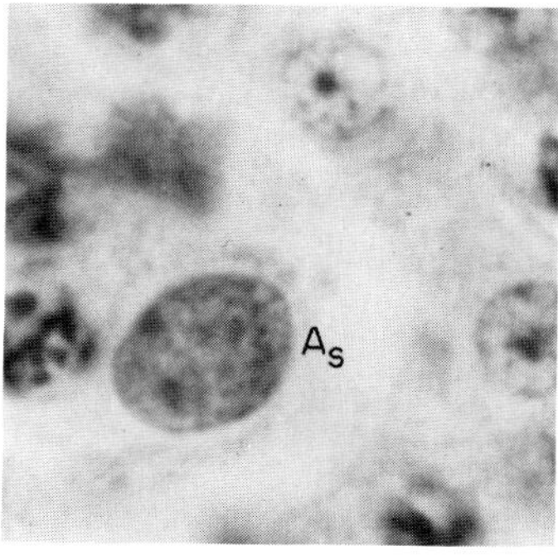

Fig. 4. A_s *spermatogonium of the mouse in whole mount 48 hours after 150 R, stage 6. Note absence of B spermatogonia.*

In the tubule whole mounts, extensive spermatogonial degeneration is seen in the first day after exposure to 150 R. By two days the spermatogonia have almost disappeared from stages 4 and 5. Of those cells which remain, most are undifferentiated A_s and A_{pr}, although a few short chains persist.

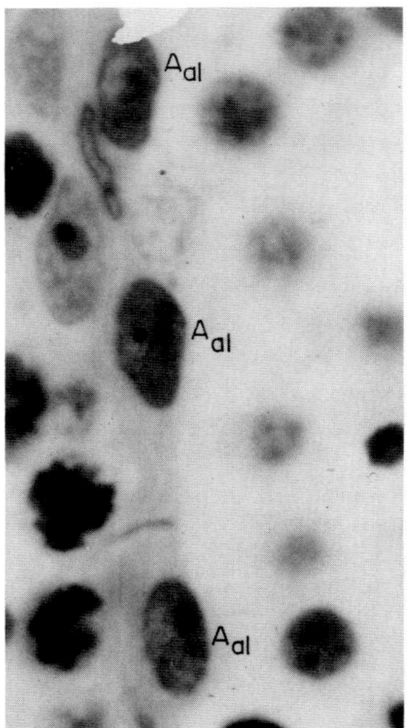

Fig. 5. A_{al} spermatogonia of the mouse in early prophase, stage 3, control.

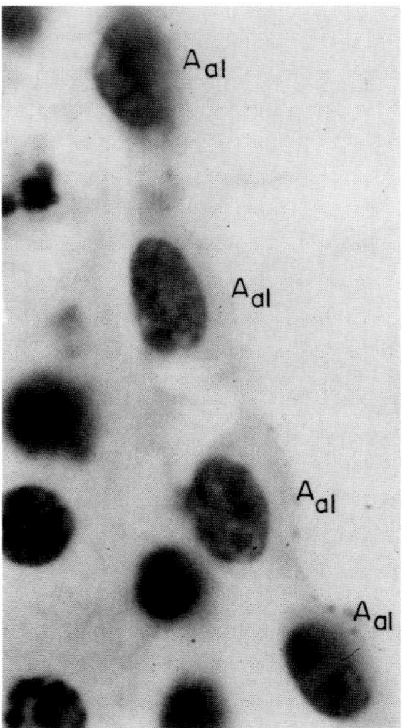

Fig. 6. Early prophase of A_{al} spermatogonia of the mouse in stage 3, 207 hours after 300 R.

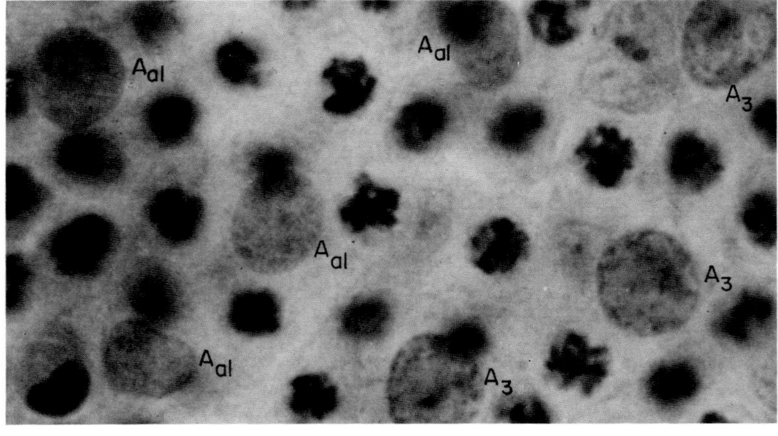

Fig. 7. Whole mount showing A_{al} spermatogonia in association with A_3 spermatogonia.

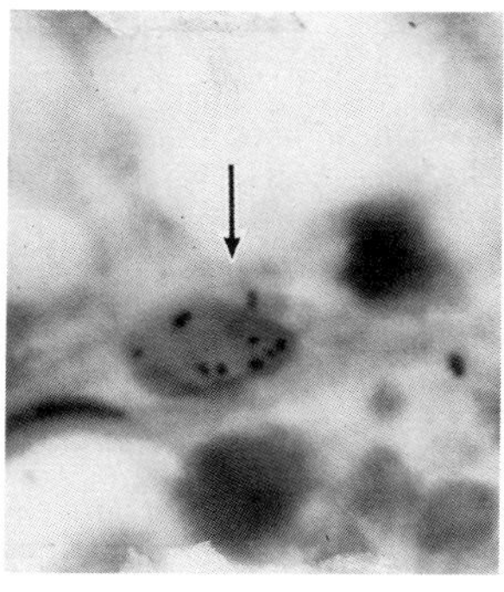

Fig. 8. Labeled A_s spermatogonium of the mouse, 207 hours after ^{3}H-TdR, stage 3.

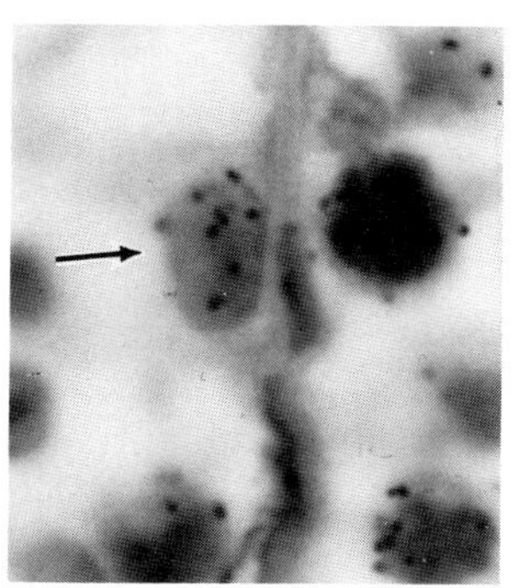

Fig. 9. Labeled A_s spermatogonium of the mouse, 207 hours after ^{3}H-TdR, 48 hours after 150 R, stage 4.

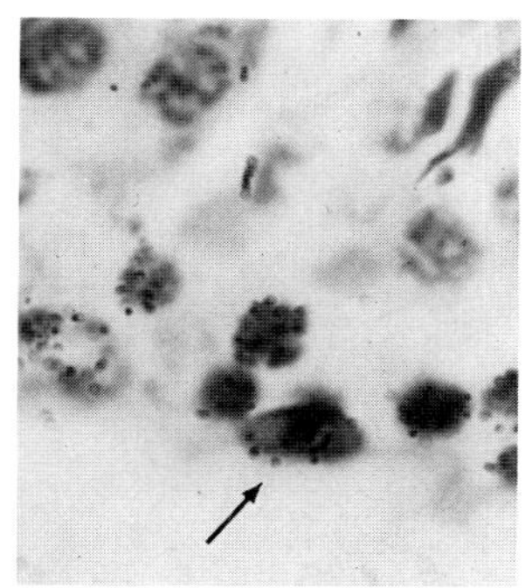

Fig. 10. Labeled A_s spermatogonium of the rat, 11 days after ^{3}H-TdR.

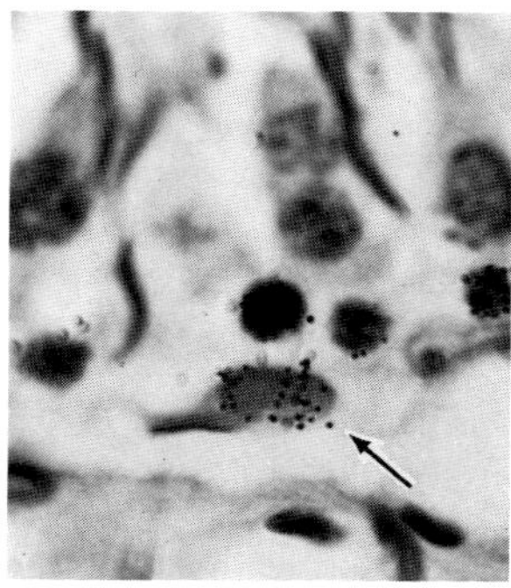

Fig. 11. Labeled A_s spermatogonium of the rat, 11 days after ^{3}H-TdR, 7 days after 330 R ^{60}Co γ rays.

Some of these show overt signs of degeneration. Incorporation of ^{3}H-TdR by some A_s spermatogonia is observed as early as two days after irradiation, and rarely, mitotic figures appear at this interval. (It is important to remember that almost the full complement of spermatogonia associated with preleptotene spermatocytes (stage 1) is still present at two days, for cell death among A_{al} and A_l cells occurs primarily during late interphase or early prophase in the first postirradiation division at the end of stage 1.] Minimal numbers of spermatogonia are reached at three days in stages 4 and 5; they are almost exclusively the undifferentiated A_s cells. There are some pairs, but no chains of cells. A number of these spermatogonia have begun to divide; the mitotic index is 1.98. Interestingly, the total number of 0.64 spermatogonia per frame is comparable to the value of 0.47 A_s spermatogonia observed for control mice (Table 2).

TABLE 2

Number of undifferentiated A spermatogonia in stages 5 and 6 in control and irradiated mice. Data based on analysis of tubule whole mounts.

Group				
dose	Interval (days)	Frames	ΣA	$\overline{NA}$*
Control†	—	325	1272	3.50 ± 0.17
150 R-irradiated	2	385	852	1.26 ± 0.14
	3	308	215	0.64 ± 0.05
	5	793	631	0.81 ± 0.08
	8.5	287	696	2.54 ± 0.21
300 R-irradiated	2	155	96	0.60 ± 0.05
	3	355	161	0.51 ± 0.06
	5	424	162	0.37 ± 0.06

* *Corrected to 20 Sertoli cells per counting frame (7225 μ^2).*
† *Counted from stage 6 (among B spermatogonia) only; of the average of 3.50 A spermatogonia per frame, 0.47 are A_s.*

Exposure to 300 R gives an effect similar to 150 R, but degeneration is more profound. By two days, only A_s spermatogonia can be seen in whole-mount tubules , and it remains so until five days, when these cells begin to divide. It is of special note that the total number of A_s cells is comparable to the number of A_s spermatogonia in control animals (table 2).

The data of tables 3 and 4 extend the information in table 2 by demonstrating that significant numbers of the surviving A spermatogonia are labeled with ^{3}H-TdR. Stages 1 and 6 were omitted for irradiated mice in table 3 and the 48-hour interval of table 4 because these stages contain many lethally damaged spermatogonia which have not yet divided. Even with these precautions, cell counts of irradiated testes always will give an overestimate of survival because of the mixture of cells which already have begun to divide with those which have not yet degenerated.

TABLE 3

*Frequency of labeled cells 48 hours after 150 R; ^{3}H-TdR given 159 hours before irradiation**.

Stage	6 X 17.5 μCi ^{3}H-TdR†		17.5 μCi ^{3}H-TdR	
	Number of cells	% labeled	Number of cells	% labeled
Control				
1	103	30.1	111	9.0
2	116	36.2	120	8.3
3	107	41.1	89	14.6
4	104	49.0	117	19.7
5	117	45.3	109	10.1
6	54	37.0	57	1.8
150 R-irradiated				
2	98	27.6	80	16.3
3	42	50.0	44	20.5
4	45	62.2	72	27.8
5	66	51.5	67	7.4

* *Total time from ^{3}H-TdR injection to killing was 207 hours, or one cycle of the seminiferous epithelium.*

† *Injections were given intraperitoneally every 12 hours; interval to irradiation was calculated from time of last injection.*

Only rare survivors of A_2-A_4 spermatogonia were seen after 150 R, and they are easily identified. The data of table 3 are almost exclusively A_s spermatogonia. The matching whole-mount data confirm this observation. With a single ^{3}H-TdR injection, frequency of labeled cells was highest at stages 3 and 4, which are the same stages at which A_3 and A_4 spermatogonia are dividing. This confirms previous observations (Huckins, '71b; Oakberg, '71a) that this is a time of active mitotic activity among the undifferentiated spermatogonia: it also points to the time of origin of the long-cycling spermatogonia. A high frequency of labeling was also observed at stage 5 after multiple injections. The primary difference between single and multiple injections is the higher proportion of labeled cells in the latter. The frequency of labeled cells is usually higher in irradiated mice than in controls because all the A_{a1} cells have been killed.

TABLE 4

Frequency of labeled cells 48 and 80 hours after 300 R; ^{3}H-TdR given 24 hours before irradiation.

	Time after 300 R			
	48 hours*		60 hours†	
Stage	Number of cells	% labeled	Number of cells	% labeled
Control				
1	120	10.0	97	21.6
2	124	1.6	67	1.5
3	109	21.1	68	7.4
4	121	14.9	73	21.9
5	70	14.3	58	32.8
6	56	32.1	37	35.1
300 R-irradiated				
2	59	6.8	56	10.7
3	36	0.0	35	5.7
4	32	6.3	31	0.0
5	34	55.9	33	33.3
6	—	—	47	40.4

* *72 hours after ^{3}H-TdR injection.*
† *84 hours after ^{3}H-TdR injection.*

Only A_s spermatogonia survived 300 R (Tables 2 and 4). Labeling was very strongly dependent upon the stage of the cycle of the seminiferous epithelium. Cells labeled during stages 3 and 4 gave rise to the 56% labeled spermatogonia in stage 5 at 48 hours, and the 40% labeling at stages 5 and 6 at 60 hours. Just as after 150 R (Table 3), this represents the time of peak mitotic activity of the undifferentiated spermatogonia. With 300 R, none of the cells labeling during stages 1 and 2 in the cycle survived, for no labeled cells were observed either at stage 3 at 48 hours, or in stage 4 at 60 hours (Table 4).

DISCUSSION

The single type A spermatogonia have been identified in both whole mounts and sections as the primary cell type surviving 150 R and they are the only surviving spermatogonial type after 300 R (Tables 2-4). Likewise, Dym and Clermont ('70) identified the A_0 as the radiation-resistant cell, though their counts were made at a tubule stage and time interval that included surviving cells as well as lethally damaged A_{a1} and A_1 spermatogonia which had not yet degenerated. They speculated that the A_0 spermatogonia were stimulated to "break" their dormancy in order to replace the depleted "cycling" spermatogonial population. This is at variance with the results of Withers et al. ('74), which suggested that the kinetics of surviving stem cells was normal as long as 14 days after irradiation. Also, if ^{3}H-TdR is given before irradiation, a significant proportion of the surviving cells are labeled (Tables 3 and 4, Oakberg '64, '71b) and the whole-mount technique clearly reveals these to be single cells (Fig. 4). Therefore, the A_s spermatogonia must have been in cycle *prior* to the radiation insult. Furthermore, many of the surviving cells are labeled even after high radiation doses, where cell survival is very low (Oakberg '64, '71b). Similar data are now being evolved for the rat (Huckins, unpublished data), where labeled A_s spermatogonia survive 330 R of ^{60}Co gamma rays. Furthermore, the surviving cells appear to be long-cycling. This agreement of results in mouse and rat was predictable on the basis of the similarity of stem cell renewal in the two species.

The above data constitute further proof that some of the A_s spermatogonia in both mouse and rat are long-cycling (Huckins, '71c; Oakberg, '71a). They incorporate ^{3}H-TdR at the same time in the cycle of the seminiferous epithelium, labeled cells persist for the same period of time, and the shape of labeling curves is the same in irradiated and control animals. Therefore, one would not expect 100% labeling of these cells in experiments performed to date, for ^{3}H-TdR has not been given for the total cell cycle. Also, the long-cycling population is small, and one would expect to see very few divisions. This is demonstrated by high radiation doses, where the survivors come primarily from the long-cycling population and divisions are infrequent. In the model of stem cell renewal proposed by Huckins ('71a) and Oakberg ('71b) the stem cells need divide only once during each cycle of the seminiferous epithelium, and therefore we would predict the low mitotic index for A_s spermatogonia reported by Clermont and Hermo ('75).

Since the cycle in the mouse is of 207 hours duration (Oakberg, '56b), scoring at that time permits the observation of cells in the same stage of the cycle of the seminiferous epithelium in which treatment occurred. The data (Tables 3 and 4) indicate a high frequency of labeled cells in the A_s population in both control and irradiated mice after 150 R. This radiation dose is essentially an LD_{100} for A_{pr}, A_{al}, and all generations of differentiating spermatogonia, and morphology of the surviving cells in both sections and whole mounts shows that they are either stem cells or cells recently derived from stem cells. That both controls and irradiated mice show similar labeling indicates comparable stem cell behavior (Oakberg, 71b). The 207-hour interval is especially significant, for, according to the dual-stem-cell concept of Clermont and Bustos-Obregon ('68), the cycling (A_4) spermatogonia should have divided four times in this interval, and the label would have been diluted by 2^4. The fact that labeling is as high as 41-49% in controls, and 50-62% in irradiated mice after multiple injections, and 10-20% and 10-28% after a single injection of ^{3}H-TdR (Table 3) demonstrates that the A_4 spermatogonium cannot have a stem cell role. These data agree with the observation of Huckins ('71a) in the rat and van Keulen and de Rooij ('74) in the mouse that the A_1 spermatogonia are derived from A_{al} cells, some of which divide at the same time as the A_4 division. On the other hand, some A_{al} divisions continue until stage 5 of the cycle. These were observed earlier by Hilscher *et al.* ('69), and were considered to be A_5 spermatogonia. The data of Huckins ('71a) clearly revealed these divisions as belonging to the A_{al}. In agreement with Monesi ('62), there is no evidence for formation of anything but In spermatogonia by the division of the A_4 cells.

In the two-stem-cell model proposed by Clermont and Bustos-Obregon ('68), the spermatogonia were divided into two populations, the noncycling A_0 (reserve stem cell) and the "renewing stem cells" - which included the A_1-A_4 spermatogonia. The A_4 cells were given the pivotal role, being able to form either the differentiated In or less differentiated A_1 types. Several difficulties with this model have not yet been resolved. Principally these are: (i) that formation of the A_1 from A_4 spermatogonia requires dedifferentiation of a cell well advanced in a developmental se-

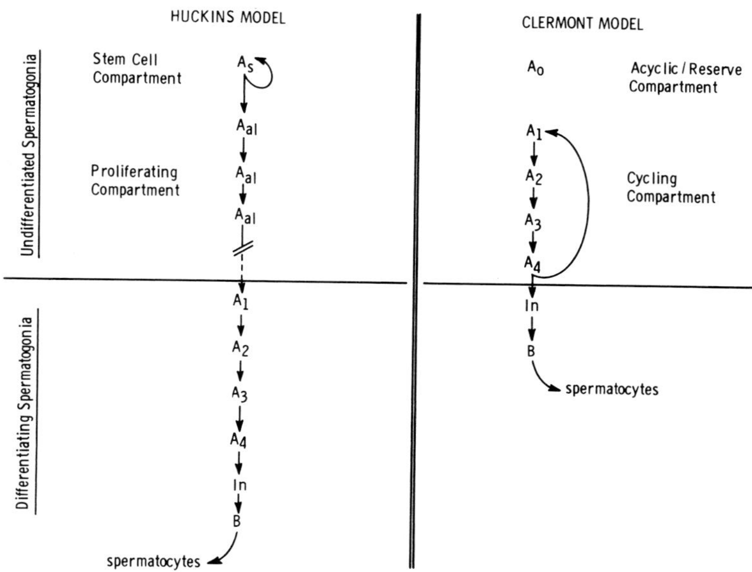

Fig. 12. Comparison of the model of spermatogonial stem cell renewal proposed by Huckins ('71a) with that proposed by Clermont and Bustos-Obregon ('68).

quence [in fact, the A_4 spermatogonia were designated as In spermatogonia by Clermont ('62) prior to 1968], (ii) the mechanism by which chains of cells 2^n in length are separated from the syncytium of A_4 spermatogonia has not been explained, (iii) the role of the A_{al} spermatogonia which are intermingled with A_3 and A_4 has not been discerned, and (iv) cell cycle kinetics of undifferentiated spermatogonia is different from that of the differentiated types (Huckins, '71b). It is clear from the studies of Huckins ('71a), however, that the A_{al}, and not the A_4, are the progenitors of the A_1 spermatogonia (Fig. 12). In this model, the A_s, A_{pr}, and A_{al} are indicated as undifferentiated cells on the basis of nuclear morphology, though differentiation probably occurs at the stem cell level. A_1-A_4 spermatogonia are considered to be differentiated cells in contrast to the Clermont-Bustos Obregon ('68) scheme (Fig. 12). Many questions of basic stem cell behavior remain, but the model proposed by Huckins ('71a) provides a logical interpretation of spermatogonial stem cell renewal which is compatible with our understanding of spermatogenesis in the rest of the animal kingdom [see Hannah-Alava ('65) for a review]. Finally, there is reason to expect that all stem cell renewal systems should be basically similar, and the model proposed here also meets this requirement. It implies, however, that at least one cycle of the seminiferous epithelium is required for formation of A_1 from A_s spermatogonia, and that the full duration of spermatogenesis in the adult requires more than four cycles of the seminiferous epithelium.

LITERATURE CITED

Clermont, Y. 1962. Quantitative analysis of spermatogenesis of the rat: A revised model for the renewal of spermatogonia. *Am. J. Anat. 111:* 111-129.

Clermont, Y. and E. Bustos-Obregon. 1968. Re-examination of spermatogonial renewal in the rat by means of seminiferous tubules mounted *in toto*. *Am. J. Anat. 122:* 237-248.

Clermont, Y. and L. Hermo. 1975. Spermatogonial stem cells in the albino rat. *Am. J. Anat. 142:* 159-176.

Dym, M. and Y. Clermont. 1970. Role of spermatogonia in the repair of the seminiferous epithelium following X-irradiation of the rat testis. *Am. J. Anat. 128:* 265-282.

Hannah-Alava, A. 1965. The premeiotic stages of spermatogenesis. *Advan. Genet. 13:* 157-226.

Hilscher, W. 1964. Beitrage zur Orthologie und Pathologie der "Spermatogoniogenese" der Ratte. *Beitr. Pathol. Anat. Allg. Pathol. 130:* 69-132.

Hilscher, B., W. Hilscher, and W. Maurer. 1969. Autoradiographische Untersuchungen uber den Modus der Proliferation und Regeneration des Wistarratte. *Z. Zellforsch. 94:* 593-604.

Huckins, C. 1971a. The spermatogonial stem cell population in adult rats. I. Their morphology, proliferation and maturation. *Anat. Rec. 169:* 533-558.

Huckins, C. 1971b. The spermatogonial stem cell population in adult rats. II. A radioautographic analysis of their cell cycle properties. *Cell Tissue Kinet. 4:* 313-334.

Huckins, C. 1971c. The spermatogonial stem cell population in adult rats. III. Evidence for a long-cycling population. *Cell Tissue Kinet. 4:* 335-349.

Huckins, C. and B. M. Kopriwa. 1969. A technique for the radioautography of germ cells in whole mounts of seminiferous tubules. *J. Histochem. Cytochem. 17:* 848-851.

Keulen, C.J.G. van, and D. G. de Rooij. 1974. The recovery from various gradations of cell loss in the mouse seminiferous epithelium and its implications for the spermatogonial stem cell renewal theory. *Cell Tissue Kinet. 7:* 549-558.

Kopriwa, B. M. and C. Huckins. 1972. A method for the use of Zenker-formol fixation and the periodic acid Schiff staining technique in light microscope radioautography. *Histochemie 32:* 231-244.

Monesi, V. 1962. Autoradiographic study of DNA synthesis and the cell cycle in spermatogonia and spermatocytes of mouse testis using tritiated thymidine. *J. Cell Biol. 14:* 1-18.

Oakberg, E. F. 1956a. A description of spermiogenesis in the mouse and its use in analysis of the cycle of the seminiferous epithelium and germ cell renewal. *Am. J. Anat. 99:* 391-414.

Oakberg, E. F. 1956b. Duration of spermatogenesis in the mouse and timing of stages of the cycle of the seminiferous epithelium. *Am. J. Anat. 99:* 507-516.

Oakberg, E. F. 1959. Initial depletion and subsequent recovery of spermatogonia of the mouse after 20 R of gamma rays and 100, 300, and 600 R of X-rays. *Radiat. Res. 11:* 700-719.

Oakberg, E. F. 1964. The effects of dose, dose rate and quality of radiation on the dynamics and survival of the spermatogonial population of the mouse. *Jap. J. Genet. 40:* 119-127.

Oakberg, E. F. 1968. Radiation response of the testis. *Excerpta Medica Internatl. Congr.* Ser. No. 184. Progress in Endocrinology, 1070-1076.

Oakberg, E. F. 1971a. Spermatogonial stem-cell renewal in the mouse. *Anat. Rec. 169:* 515-532.

Oakberg, E. F. 1971b. A new concept of spermatogonial stem-cell renewal in the mouse and its relationship to genetic effects. *Mutat. Res. 11:* 1-7.

Rooij, D. G. de. 1973. Spermatogonial stem cell renewal in the mouse. I. Normal situation. *Cell Tissue Kinet. 6:* 281-287.

Withers, H. R., N. Hunter, H. T. Barkley, Jr., and B. O. Reid. 1974. Radiation survival and regeneration characteristics of spermatogenic stem cells of mouse testis. *Radiat. Res. 57:* 88-103.

A New Approach to Stem Cell Research in Spermatogenesis

P. B. MOENS AND A. D. HUGENHOLTZ

Department of Biology, York University, Downsview, Ontario, Canada

As early as 1899 J. H. McGregor, working on spermatogenesis in the urodele *Amphiuma*, noted the cytoplasmic bridges between the cells and postulated that 'the "bridges" are to be recognized as remnants of central spindles and when the cell contains two or more bridges each represents a past mitotic division'. Recent electron microscope studies on reproductive cells of a wide variety of animals, representing most phyla, have verified McGregor's assumption (Dym and Fawcett, '71; King and Akai, '71; MacKinnon and Basrur, '70; Mahowald, '71). The persistent bridges between germ cells which have entered the developmental pathway leading to oocyte or spermatocyte formation provide a record of the mitotic events in the history of a given clone. It follows that if clones and cytoplasmic bridges can be recorded for tissues in successive stages of development the manner of renewal and proliferation of germ cells can be visualized.

The feasibility of germ cell kinetic studies from three-dimensional models based on electron microscopy of serially sectioned tissue was first demonstrated by Koch, Smith and King ('67). They showed that the 16 cystocytes of *Drosophila melanogaster* ovaries are connected to one another through cytoplasmic bridges so that the two central cells have 4 bridges each, 2 cells of the next order have 3 bridges each, the next 4 cells have 2 bridges each, and each of the remaining 8 cells has a single bridge (Fig. 2, 16 cells). The pattern of bridges suggests that the syncytium is produced by four successive cell divisions.

We have applied this type of analysis to spermatogenesis in the Locust, *Locusta migratoria*. Spermatogonia are generated from stem cells at the distal end of elongate seminiferous sacs. Going proximally, one encounters groups of 2,4,8,

16 and 32 interconnected spermatogonia and groups of 64 spermatocytes. Each group is enclosed by a membrane and is referred to as a "cyst" (Fig. 1). The cells within a cyst develop synchronously (Moens, '70).

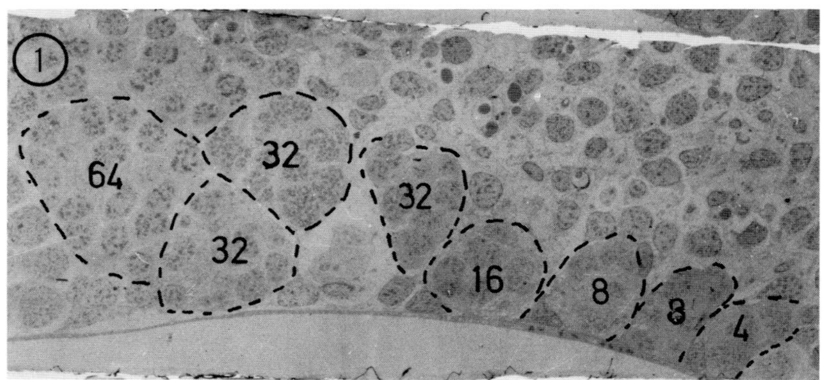

Fig. 1. Seminiferous tubule of Locusta migratoria. One section of a series used to determine the number of cells in each cyst and the connections between the cells of a cyst.

The 3-dimensional analysis of *Drosophila* ovaries with polytrophic oogenesis and *Locusta* testes with cystic spermatogenesis shows that the distribution of bridges is regulated so that all bridges of previous divisions congregate on one side of each cell prior to the next division. A symmetrical pattern results (Fig. 2).

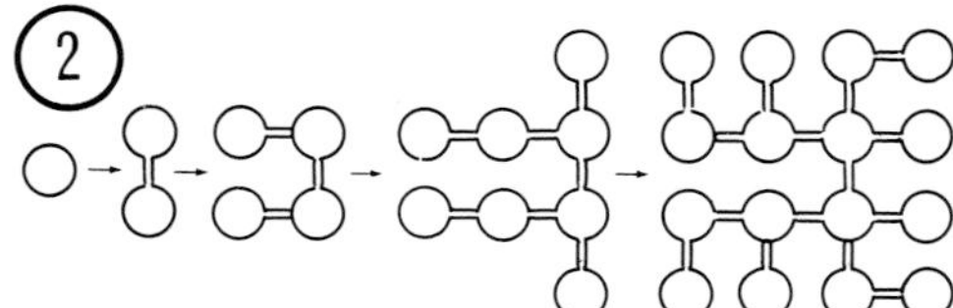

Fig. 2. Schematic representation of syncytial organization in Drosophila oogenesis and Locusta spermatogenesis.

The symmetry breaks down when the cyst goes from 32 spermatogonia to the 64 spermatocytes. These observations indicate that:

a) Spermatogonia which remain connected by bridges and form a cyst are committed to the spermatogenic developmental pathway. They do not function any longer as stem cells.

b) In the 20 cysts analysed by us, cell numbers are powers of 2 and all cells are part of a single syncytium. It follows that no cell losses had occurred in those cysts.

With this background we attended a spermatogenesis workshop at McGill in the fall of 1970 where Dr. Leblond reviewed mammalian spermatogenesis. It occurred to us that some of the uncertainties in the study of the process might be resolved if the committed spermatogonia of the rat also remain interconnected by cytoplasmic bridges. We found out at that meeting that the occurrence of bridges had in fact been reported in mammals by Fawcett ('61) who had diagrammed two connected spermatocytes giving rise to 8 connected spermatids. Encouraged by this information we obtained some rat seminiferous tubule embedded by Dr. V.L.W. Go from Dr. I. B. Fritz's laboratory. While on sabbatical leave at the Hubrecht Laboratory in Utrecht, Netherlands, one of us (P.B.M.) cut series of 400 to 600 tangential consecutive sections and mapped all the bridges between the spermatogonia and spermatocytes. These results have been published (Moens and Go, '71) and they show that as many as 22 spermatogonia and 74 spermatocytes can be interconnected. The results prompted a more extensive study (A. Hugenholtz, Master's thesis) of seminiferous epithelium in different stages of development. In addition information was obtained on the patterns of cell division of syncytia passing through mitotic and meiotic telophase, and on the ultrastructural morphology of cells and bridges of singles, pairs, small and large syncytia. Some of the numerical data are presented below.

Unlike *Locusta* the syncytia of committed spermatogonia in the rat are not surrounded by a membrane and the distribution of bridges to "old" and "new" cells is not regulated in the rat. As a result the rat syncytia occur as chains or irregularly branching structures (Fig. 3).

In addition to the spermatogonial groups of 2, 4 and 8 cells (Fig. 3A, h, m and r) groups of 3, 5, 6 and 7 spermatogonia were regularly observed. These groups probably correspond to the aligned spermatogonia discussed by Huckins ('71b) in the rat and by Oakberg ('71) in the mouse. But since bridges are not visible in most light microscope preparations their contention that aligned spermatogonia occur in groups of 4, 8 and 16 cells only must represent an expectation rather than an observation.

The origin of, for example, a 7-celled syncytium needs clarification. In the absence of observed degenerating cells or broken bridges we are inclined to believe that at a given division cycle not all cells in the syncytium divide. We assume that a given cell can temporarily arrest in G_1 but that because of its open connections with the rest of the syncytium which does undergo a normal division cycle it will progress its developmental stage with the rest of the syncytium. Thus a B-spermatogonium may become a premeiotic spermatocyte without itself undergoing a division. We have observed occasional non-dividing cells in a syncytium consisting of cells in telophase of mitosis. In labelling experiments, the G_1 arrested nuclei should not take up H^3-thymidine while the rest of the syncytium does. We are in the process of checking this prediction. All cells in S-phase will take up label and will all proceed to mitosis, but the total number of cells will less than double. This is in agreement with the reports by Huckins ('71a) and Clermont ('72).

Although cell losses may locally reduce cell numbers, systematic degeneration is rendered unlikely by the preponderance of large syncytia (Fig. 3B, syncytium C and D). Regular cell losses would result in continuous fragmentation of the syncytium into small pieces. Losses of terminal cells on the other hand would not cause fragmentation, but so far no evidence for selective terminal degradation has been reported.

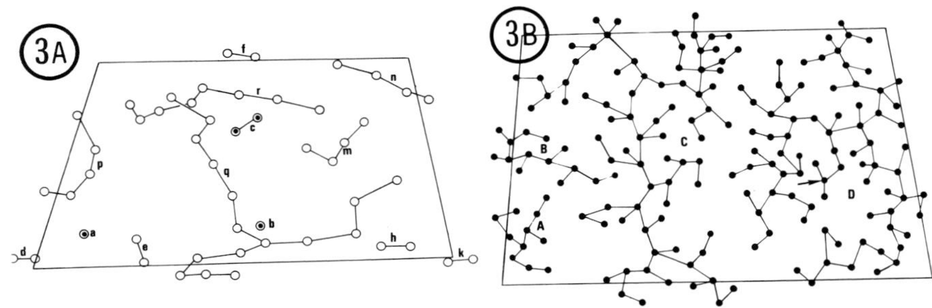

Fig. 3. Example of spermatogonial and spermatocyte syncytia in .05 mm^2 seminiferous epithelium of stage I of rat cytospermatogenesis. A. Spermatogonia (circles) and their cytoplasmic bridges (lines). Cells outside the boundary lines may have additional bridges. Cells a, b and c are dense spermatogonia. B. The spermatocyte syncytia A, B, C and D overlie the spermatogonia in 3A. The arrow marks an incomplete terminal branch (from Moens and Hugenholtz, 1975).

In the studies on germ cell kinetics based on these ultrastructure observations, the syncytium is considered as the logistic unit of spermatogenesis. We assume that there is a 1:1 ratio for the number of spermatogonial syncytia generated and the number of spermatocyte syncytia produced. Areas such as shown in Fig. 3 produce at most 2 spermatocyte syncytia. At stage I the area could be expected to contain 2 spermatogonial syncytia (which will eventually replace the 2 spermatocyte syncytia) and some stem cells to produce new spermatogonial syncytia (Fig. 3A). There appears to be an excess of spermatogonial syncytia. If this region had reached stage VII after 3 mitotic cycles, syncytium q would have become the predominant spermatocyte syncytium. Syncytium m could at that time reach a maximum of 32 cells, which is too small to qualify it as a spermatocyte syncytium. It could do so at a subsequent stage VII with the proviso that it does not participate at all division cycles. As such syncytium m resembles the undifferentiated spermatogonia of Huckins ('71b) which divide out of phase with the main mitotic cycles. In a stage VI area we in fact observed two nondividing spermatogonial syncytia while all others were dividing (Moens and Hugenholtz, '75). We predict that the development syncytium r will be retarded on the grounds that we have not observed overlapping spermatocyte syncytia. The general conclusion is that a given area of seminiferous epithelium may have a greater supply of spermatogonial syncytia than are immediately required to replace maturing spermatocyte syncytia.

In the composition of stem cells and spermatogonial syncytia we observed unexpected differences among areas of seminiferous epithelium which were identical in terms of spermatid and spermatocyte criteria: such results and their interpretation are summarized in Fig. 4B.

Although Ba and Bb are in about the same developmental stage Bb has a large number of single and twin spermatogonia. Ba, on the other hand, has few singles but does have more advanced spermatogonial syncytia. Diagram 4B accommodates these observations in the general framework of spermatogenesis (4A). The model implies that for a given area there is a periodic buildup of spermatogonia which then supply several quanta of spermatocytes, and when the spermatogonia are depleted a new buildup takes place.

Research supported financially by the National Research Council of Canada.

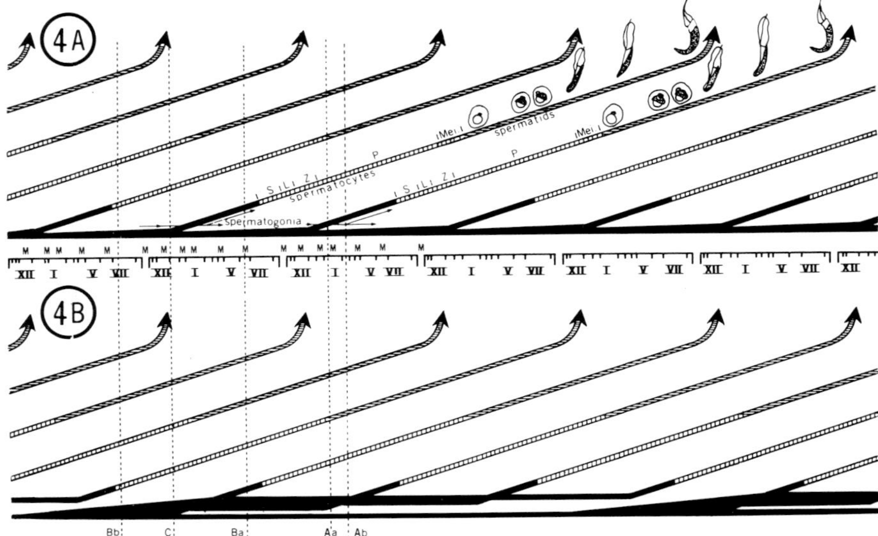

Fig. 4. Diagrammatic representation of sperm production. *A, a base population of spermatogonia continues while periodically some cells of the population enter the spermatogenic developmental pathway. Such cells pass through successively more advanced spermatogonial stages, then become spermatocytes, which pass through S-phase, S, leptotene, L, zygotene, Z, pachytene, P, and the meiotic divisions, Mei. Each spermatocyte produces 4 spermatids which undergo pronounced morphological changes, some of which have been sketched in. The developmental stages are indicated in Roman numerals and the mitotic peaks indicated by M (Clermont, '72; Huckins, '71b). Vertical lines connect the cell types found at the various stages.*

B, whereas the model of 4A predicts that given stages have comparable spermatogonial populations, electron microscope observations show quantitative and qualitative differences in the spermatogonia populations of stages which are identical in terms of spermatid and spermatocyte morphology. The diagram attempts to accommodate such differences in spermatogonial composition while leaving the general model of spermatogenesis unaltered. The thickness of a single bar represents about 40 syncytia per mm^2 *of epithelium. There are, for example, at (Ba) 40 single and twin spermatogonia (black bar at the bottom of the figure). Then there are 120 syncytia of more advanced spermatogonia which can supply 3 successive quanta of spermatocytes (broad black band). Next are*

LITERATURE CITED

Clermont, Y. 1972. Kinetics of spermatogenesis in mammals: Seminiferous epithelium cycle and spermatogonial renewal. *Physiol. Rev. 52:* 198-236.

Dym, M. and D. W. Fawcett. 1971. Further observations on the number of spermatogonia, spermatocytes, and spermatids connected by intercellular bridges in the mammalian testis. *Biol. Reprod. 4:* 195-215.

Fawcett, D. W. 1961. Intercellular bridges. *Exp. Cell Res. Suppl. 8:* 174-187.

Huckins, C. 1971a. Cell cycle properties of differentiating spermatogonia in adult Sprague-Dawley rats. *Cell Tissue Kinet. 4:* 139-154.

Huckins, C. 1971b. The spermatogonial stem cell population in adult rats. I. Their morphology, proliferation and maturation. *Anat. Rec. 169:* 533-558.

King, R. C. and H. Akai. 1971. Spermatogenesis in *Bombyx mori* I. The canal system joining sister spermatocytes. *J. Morph. 134:* 47-56.

Koch, E. A., P. A. Smith and R. C. King. 1967. The division and differentiation of *Drosophila* cystocytes. *J. Morph. 121:* 55-70.

MacKinnon, E. A. and P. K. Basrur. 1970. Cytokinesis in the gonocysts of the drone honey bee (*Apis mellifera* L.). *Can. J. Zool. 48:* 1163-1166.

Mahowald, A. P. 1971. The formation of ring canals by cell furrows in *Drosophila*. *Z. Zellforsch. mikrosk. Anat. 118:* 162-167.

McGregor, J. H. 1899. The spermatogenesis of *Amphiuma*. *J. Morph. 15:* 57-104.

Moens, P. B. 1970. Premeiotic DNA synthesis and the time of chromosome pairing in *Locusta migratoria*. *Proc. Nat. Acad. Sci. 66:* 94-98.

Moens, P. B. and V.L.W. Go. 1971. Intercellular bridges and division patterns of rat spermatogonia. *Z. Zellforsch. mikrosk. Anat. 127:* 201-208.

Moens, P. B. and A. D. Hugenholtz. 1975. The arrangement of germ cells in the rat seminiferous tubule: an electron microscope study. *J. Cell Sci. 19:* in press.

the B spermatogonia which are just dividing to produce spermatocytes. Further towards the interior of the tubule are pachytene spermatocytes and next 2 layers of spermatids, one early, one further developed (Moens and Hugenholtz, 1975).

Oakberg, E. F. 1971. Spermatogonial stem-cell renewal in the mouse. *Anat. Rec. 169:* 515-532.

Studies of the Genetic Implications of Abnormal Spermatozoa

W. R. BRUCE

Ontario Cancer Institute,
Toronto, Canada

My interest in spermatogenesis was kindled by the beautiful histological and radioautographic studies of Leblond and Clermont ('53, '55). These authors clearly described the morphological and temporal steps which the spermatogonia take through the meiotic process to make the mature sperm. The description seemed to beg the question: What controls the maturation of these seemingly identical, complex, beautiful and energetic creatures?

The control of spermatogenesis is undoubtedly complex. There is, of course, the physiological environment of the testis: the unique temperature, the close association with the Sertoli cells, the blood-testis barrier, the cell associations and the proximity to the Leydig cells. Undoubtedly each of these factors is important in the control of the spermatogenesis. But of more fundamental importance are the underlying genetic controls that had evolved at the time of the emergence of the earliest eukaryotes and have persisted in a remarkably conserved pattern to the present. Can these genetic controls be manipulated in such a way as to give us an insight into the differentiation process by which sperm are created? Several lines of evidence, including data to be presented here, indicate that this may be possible. For instance, it is known that structure of mouse sperm is under strict genetic control; the shape of the sperm of a given strain of mouse is highly specific and characteristic of that strain (Beatty, '70), and studies with mosaic mice have shown that the shape of the sperm is determined by the genotype of the spermatogonia and not that of cells that surround the spermatogenic cells (Burgoyne, '75). Further, we know that several mutations affecting shape and maturation have been defined (Bennet, Gall et al, '71) and factors on the Y chromosome have been shown to be responsible for morphological characteristics of sperm (Krzanowska, '69). In this publication we will review data which indicate that it is possible to induce mutations affecting sperm shape, or more correctly, it is possible to induce mutations affecting the efficiency with which normally shaped sperm are produced.

Amongst the sperm in the ejaculate, the vas deferentia, or the cauda epididymis, there is always a fraction of cells with misshapen heads. These cells can be recognized at a glance as being abnormal and the fraction of these forms is found to differ markedly from strain to strain. For instance, we found 3% of inbred DBA mice had abnormal forms while 15% of the sperm from C58 mice were abnormal (Bruce, Furrer and Wyrobek '74). The number of abnormalities in hybrids was generally lower than the number seen in hybrids; for instance, the hybrid of C57 X C3H mice had only about 1% abnormal sperm.

Irradiation of the testis with X or gamma rays resulted in elevations in the fraction of abnormal forms (Bruce, Furrer and Wyrobek, '74). The elevation was dependent on both the time following radiation and the dose of radiation used. The abnormalities appeared in 2-3 weeks following irradiation, reached a maximum at 4-8 weeks, and then decreased to an above-background level for the remainder of the animal's life. This result suggested that radiation was producing no significant effects on mature sperm, or indeed on late spermatids. But effects were seen when primary spermatocytes were irradiated and definite, though less pronounced effects were seen following the irradiation of spermatogonia. An increase in abnormalities was readily evident following a dose of 30 rads and the dose-effect curves for sperm abnormalities corresponded to dose-effect curves for genetic effects of irradiation. We thus suggested that sperm abnormalities were a consequence of genetic effect on primary spermatocytes and spermatogonia. That is, that radiation-induced mutations increased the background of errors in the normal maturation of spermatogenic cells and thus the fraction of dysmorphic sperm produced. We have now looked at this suggestion from several points of view. While the studies are not complete, present data indicates that the induced elevations of sperm abnormalities are indeed a consequence of mutations in spermatocytes and spermatogonia.

We at first argued that if irradiation produced its effect on sperm abnormality by mutation of germ cells, then we might well expect that other mutagens besides radiation might do the same. To determine whether this was the case we took groups of mice and exposed these animals to graded doses of 25 different agents with known mutagenic and non-mutagenic activity (Wyrobek and Bruce, '75). As expected, methyl methane sulfonate, ethylmethane sulfonate, and many other mutagenic agents produced elevations in sperm abnormalities that persisted for 5 to 10 weeks following treatment. These studies have now been extended to include nearly 60 agents and in almost every case known mutagens have been found to produce elevations in sperm abnormalities, while non-mutagens produced none (Heddle and Bruce, in preparation).

We have also compared the effects of the 60 mutagens and non-mutagens for their production of micronuclei (Heddle and Bruce, in preparation). We did this because, when we started this work, we felt that the most likely class of mutations affecting sperm shape would be chromosomal aberrations. Micronuclei arise from chromosomal fragments which are not included in the daughter nuclei at mitosis. Thus if sperm abnormalities result from chromosome aberrations we might expect a general agreement in the dose-effect curves for the two phenomena. Although this was generally the case, several agents gave no increase in micronuclei but did give rise to marked elevations in the fraction of sperm abnormalities. Such agents included lead acetate and methyl cholanthrene. The results suggest that the mechanism which leads to sperm abnormalities and micronuclei are related but are perhaps not exactly the same. However, as micronuclei were scored for cells of the bone marrow while abnormalities were scored for cells in the testis it was possible that a difference in the disposition of the mutagens might explain the difference between the two results.

Evidence that sperm abnormalities were not a consequence of chromosomal aberrations *per se*, however, was derived from examinations of mice with known Robertsonian and reciprocal translocations (Wyrobek, Heddle and Bruce). The fraction of abnormal sperm in mice, homozygous or heterozygous for translocations, were nearly always within the range for normal mice. The results thus showed that sperm can be aneuploid or otherwise grossly unbalanced in chromosomal complement, and could still have a normal sperm shape. This suggests that if the fraction of abnormal sperm is elevated for genetic reasons, it is unlikely that this is a consequence of aneuploidy or translocations of the genetic material in the sperm but is more likely a consequence of point mutations or deletions.

Our most recent studies of the transmission of induced sperm abnormalities support this genetic interpretation (Hugenholtz, Heddle and Bruce in preparation). In these studies we have used male mice irradiated to a dose of 300 rads. In the presterile period these mice were mated with unirradiated females and sham-irradiated control male mice were similarly mated. The sperm of the male progeny of both crosses were examined when they had reached maturity. As anticipated, the frequency of sperm abnormalities in the two groups was significantly different - the sperm of mice born from fathers that had been irradiated contained more abnormalities. A few progeny from irradiated fathers had over 50% abnormalities and nearly all the remainder had a small but significant increase in sperm abnormalities. Subsequent mating of the population of F1 males with their sisters produced two groups of F2 mice which persisted in showing the effect of irradiation. We have now examined in more detail 3 individual induced abnormalities in male mice. One such mutation results in mice with 100% of sperm of abnormal form. The other two have 40 to 50% abnormalities. At the present time (at 3 generations) it appears that these mutations are segregated and transmitted as Mendelian traits. The results thus indicate that induced sperm abnormalities result from mutation though they do not yet suggest a mechanism.

But to return to the theme of the conference - stem cells and their maturation. It is our hope that induced mutations for elevated sperm abnormalities may give us a technique for manipulating the genetic factors which normally control the spermatogenesis process. We are a long way from understanding this genetic control but we hope that by the orderly collection of many mutations affecting spermatogenesis we will have the tools for a genetic approach to the study of this most interesting differentiation system.

LITERATURE CITED

Beatty, R. A. 1970. The genetics of the mammalian gamete. *Biol. Rev. 45:* 73-119.

Bennet, W. I., A. M. Gall, J. L. Southard and R. L. Sidman. 1971. Abnormal spermiogenesis in quaking, a myelin-deficient mutant mouse. *Biol. Reproduction 5:* 30-58.

Bruce, W. R., R. Furrer and A. J. Wyrobek. 1974. Abnormalities in the shape of murine sperm after acute testicular X irradiation. *Mutation Res. 23:* 381-386.

Burgoyne, P. S. 1975. Sperm phenotype and its relationship to somatic and germ line genotype: A study using mouse aggregation chimeras. *Develop. Biol. 44:* 63-76.

Clermont, Y. and C. P. Leblond. 1953. Renewal of spermatogonia in the rat. *Am. J. Anat. 93:* 475.

Clermont, Y. and C. P. Leblond. 1955. Spermiogenesis of man, monkey, ram and other mammals as shown by the "periodic acid-Schiff" technique. *Am. J. Anat. 96:* 229.

Heddle, J. A. and W. R. Bruce. In preparation.

Hugenholtz, A., J. A. Heddle and W. R. Bruce. In preparation.

Krzanowska, H. 1969. Factor responsible for spermatozoan abnormality located on the Y chromosome in mice. *Genet. Res. 13:* 17-42.

Wyrobek, A. J. and W. R. Bruce. 1975. Chemical induction of sperm abnormalities in mice. *Proc. Nat. Acad. Sci. 72 (11):* 4425-4429.

Wyrobek, A. J., J. A. Heddle and W. R. Bruce. Chromosomal abnormalities and the morphology of mouse sperm. *Cdn. J. Genetics Cytol.* In press.

Discussion, Session V -

Stem Cells in the Testis

Dr. Clermont's presentation

In response to questioning, Dr. Clermont indicated that the dark and pale cells, which incidentally are found also in man, correspond to the A_0 and A_1-A_4 series he has described in the rat. The 1:1 ratio of pale and dark cells does not, however, extend beyond the primates. A discussion followed in which several questioners sought clarification of the relationship between the cycle of the seminiferous epithelium and the cell cycle of the component cells. Dr. Clermont pointed out that the seminiferous cycle is a histological, not a cytological, process in which a particular region of a tubule progresses through time through a series of developmental cell associations.

Dr. Leblond asked Dr. Clermont to substantiate again his claim that there are in the testis both renewing and reserve stem cells, as he had shown in the rat. Dr. Clermont felt that the data he had presented for the monkey did indeed indicate two populations.

Dr. Oakberg's presentation

In reply to a question from Dr. Bruce about possible reutilisation of thymidine, Dr. Oakberg said that after 300R there is essentially no uptake for 48-72 hours. The surviving stem cells take almost one complete seminiferous cycle to get back into the cell cycle - this time available for repair may account for their radioresistance.

Dr. Clermont asked whether the labelling data were obtained with whole mounts or sections. Dr. Oakberg replied that whole mounts had been used for morphology but at present the autoradiographic information came from sections; autoradiographs of whole mounts have not yet been completely analysed.

Dr. Withers spoke about the response of the testis to split doses of 600R and 600R separated by intervals up to 210 days. His conclusion was that stem cell regeneration after the first dose is much slower than the regeneration of the differentiating cell population (Withers *et al.*, 1974. Radiation Research 57: 88-103). He corrects the count of spermatogonia for tubule shrinkage by relating it to the count of Sertoli cells. On this basis there is no transient overshoot. Dr. Clermont commented that in the rat recovery of the A_0 cell population requires two cycles of the epithelium after 300R, longer after higher doses.

Dr. Moens' presentation

Dr. Lala queried the impact of syncytia held over from differentiating on the cytological appearance of subsequent stages. Dr. Moens pointed out that these cells would be a very minor component of the total cell population present. He was not able to explain for Dr. Fowler why syncytia did not consist always of 2^n cells. Perhaps one degenerates off the end, or one fails to enter S.

Dr. Leblond expressed strongly the view that this paper claimed too much from the technique, that the number of spermatogonia is precise at each time, and that cell degeneration does occur. Dr. Moens replied that he was confident that in his serial-sectioning techniques he would not miss connections.

Dr. Oakberg claimed that after 500 or 1000R you see only singles, no pairs, and that pairs reappear after 207 hours. He asked Dr. Clermont how he would suggest chains of A_1 cells of 2^n length could form from a syncytium of A_4 cells. Dr. Clermont replied that in our present state of ignorance we could not predict how these bridges behave.

Dr. Bruce's presentation

Dr. Bruce told Dr. Leblond that he had not yet begun to trace back to earlier stages the development of abnormal sperm. This would be done when they had mice producing 100% of one abnormality. However, the time of appearance of abnormal sperm after irradiation indicated that the primary spermatocyte, or perhaps the later spermatogonial stages, were the site of action. Long term effects implicate earlier spermatogonia.

In man sperm abnormalities are seen both spontaneously and following 100R; man is more sensitive than the mouse. Attempts to culture testis long-term have not succeeded.

Discussion between Drs. Oakberg and Bruce covered factors affecting sperm transport through the epididymis - where after radiation sperm may persist for a long period (Bruce has a paper coming out in Nature) - and selection pressure against some gametes at meiosis, transport in the female genital tract, and fertilisation.

Session VI

Stem Cells in Growth, Aging and Neoplasia

Hemopoietic Stem Cells During Embryonic Development and Growth

M.A.S. MOORE AND G.R. JOHNSON

Sloan-Kettering Institute for Cancer Research, New York, New York 10021, U.S.A., and the Walter and Eliza Hall Institute, Melbourne, Australia.

Development of hemopoietic stem cells in fetal liver and the initiation of lymphopoiesis and myelopoiesis in various organs have been explained by two polarised concepts. The yolk sac migration theory recognizes that hemopoietic stem cells develop *de novo* in the yolk sac blood islands and, by subsequent proliferation and migration, colonize the developing primary lymphoid organs, the thymus and bursa of Fabricius (or its mammalian equivalent), and the fetal liver, spleen and, finally, bone marrow (Moore and Owen, '65, '67). According to the alternate view, hemopoietic stem cells develop locally within hemopoietic organs by transformation of mesenchymal or epithelial cells (Rifkind et al., '69). Whereas evidence for the latter theory has been mainly based on interpretation of static morphology, the cell migration concept has been supported by the application of tracer techniques utilizing chromosome markers, tritiated thymidine labelling or differences in interphase nuclear morphology in a variety of parabiotic or grafting situations. Most of these studies have been performed in avian systems due to the ease with which vascular anastamosis can be established by parabiosis at varying stages of embryonic development. Such experimental manipulations are not readily applicable in mammalian hemopoietic development, however numerous naturally occurring examples of placental vascular anastamosis between twin embryos have been reported and in all cases the extensive hemopoietic chimerism observed in post natal life supports the concept of *in utero* stem cell migration (Metcalf and Moore, '71). Two major differences between avian and mammalian

This work was supported in part by grants NCI 08748-11A, CA-17085 and the Jean Shaland Foundation.

hemopoietic development are 1) the absence in mammals of a specific, localized site of B lymphocyte ontogeny equivalent to the bursa of Fabricius and 2) the important hemopoietic function of the mammalian fetal liver which is not a hemopoietic tissue in avian development, being replaced by the major continuing hemopoietic function of the avian yolk sac.

In an attempt to define the importance of the mammalian yolk sac, (despite its transitory hemopoietic function), in the establishment of the hemopoietic system we have analyzed the ontogeny of both multipotential stem cells (CFU-s) and committed granulocyte-monocyte progenitor cells (CFU-c) in mouse fetal development. Both stem cell populations were detected in yolk sac blood islands at 7-8 days gestation, in the circulation by 9-10 days and in the embryonic liver by 10.5 days (Metcalf and Moore, '71). This critical phase of transition from extraembryonic to intraembryonic hematopoiesis was reproduced by organ culture of intact presomite embryos maintained for 3 days *in vitro* with essentially normal development of embryonic organ systems and extraembryonic membranes (Moore and Metcalf, '70). *In vitro* hemopoietic development closely parallelled that in utero and the same transition from extraembryonic to intraembryonic hemopoiesis was observed. In cultures of presomite embryos deprived of associated yolk sac, normal organ development was observed but no evidence of hemopoiesis or hemopoietic stem cell development was obtained. In contrast, cultures of presomite yolk sac showed marked erythropoiesis with increased numbers of CFU-c and CFU-s (Moore, M.A.S., unpublished observation). These observations together with analysis of the stem cell population doubling time during the early phase of hepatic hemopoiesis strongly indicated that yolk sac migration of CFU-s and CFU-c was responsible for initiation of hemopoiesis in the fetal liver. These conclusions may be criticized due to the artificial situation provided by *in vitro* culture of intact embryos which may not reproduce the normal ontogenetic sequence seen *in vivo*, and by the fact that removal of extraembryonic membranes may disrupt normal hemopoietic development in the embryo via a mechanism unrelated to a requirement for yolk sac stem cell migration.

In vitro studies have demonstrated that hepatic hemopoietic tissue can develop if tissues are explanted after the 28 somite stage but not if explanted prior to this stage despite normal parenchymal development (Johnson and Jones, '73).

GRAFTING OF FETAL HEPATIC TISSUE TO ADULT SYNGENEIC RECIPIENTS

In order to overcome some of the viability problems associated with *in vitro* organ culture, fetal hepatic tissue was grafted under the kidney capsule of adult syngeneic mice. Intact liver obtained from fetuses younger than 11.5 days gestation (36 somites) was implanted under the kidney capsule. With older hepatic tissue, portions of liver were placed on a graft bed prepared by shaving off a thin slice of host renal cortex. In all grafts initially older than 28 somites, a marked morphological change was observed in graft hemopoiesis with the predominant pattern of erythropoiesis being replaced by granulopoiesis and megakaryocytopoiesis (Fig. 1a). Within 48 hours of grafting, 45% of 13-day fetal liver hemopoietic elements were granulocytic at all stages of differentiation in contrast to only 5.5% of granulopoietic cells in fetal liver of comparable gestation age *in utero* (Fig. 1b). A similar dramatic transition in hepatic morphology was seen in grafts to nude mice or lethally irradiated syngeneic recipients.

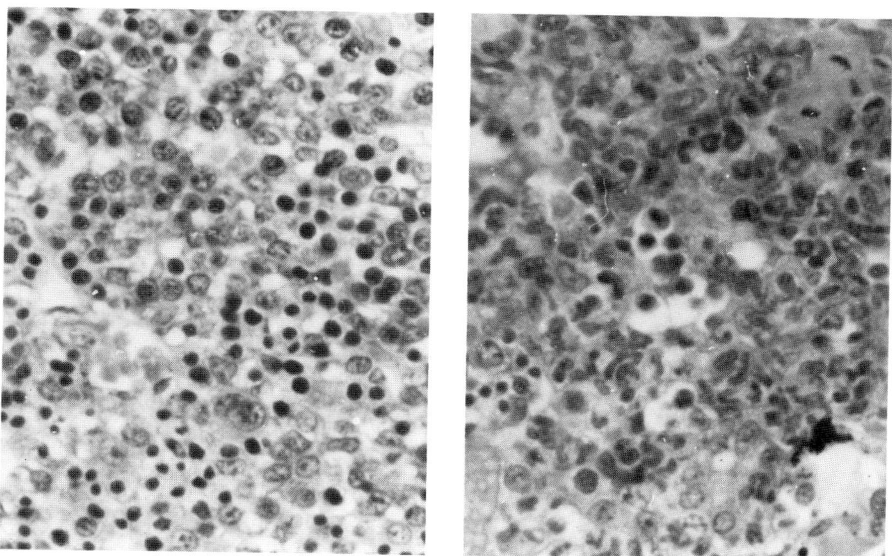

Fig. 1a. Section of 15-day fetal liver, erythroid differentiation predominantes. 1b. Section of 13-day fetal liver grafted onto a syngeneic kidney for 2 days. Granulocytes in various stages of differentiation and mitotic figures are apparent.

The latter observation confirmed that the granulopoiesis was of donor graft origin rather than host infiltration. Hemopoiesis in grafts older than 28 somites declined rapidly and by 10 days post grafting only connective tissue remained. This observation suggests that the fetal hemopoietic microenvironment of the liver is programmed to involute or, alternatively, the influence of the adult environment is incompatible with persistence of hepatic hemopoiesis.

With hepatic grafts younger than the 28 somite stage, hepatic differentiation continued even when sampled 10-12 days post grafting with parenchymal and duct system development but absence of hemopoietic tissue. Absence of hemopoiesis in such grafts may be attributed to either lack of requirement for additional hemopoietic environments (in normal or nude mice) or lack of available hemopoietic stem cells (in lethally irradiated recipients). Pre-28 somite liver was consequently grafted under the kidney capsule of syngeneic Sl/Sl^d mice which have a normal stem cell population but possess a defective hemopoietic microenvironment. Under such grafting situations grafts obtained from embryos as early as the 22 somite stage displayed areas of both erythropoiesis and granulopoiesis (Fig. 2) presumably due to seeding of host stem cells.

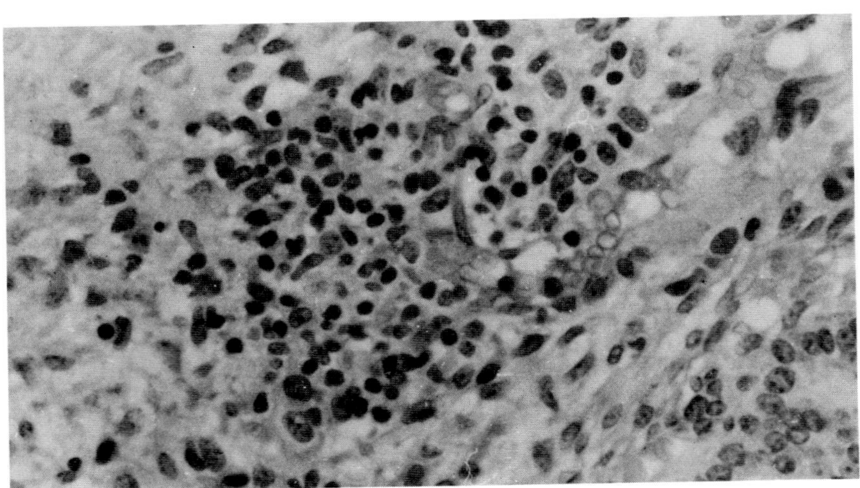

Fig. 2. Section of a 24-somite CBA fetal liver graft after 7 days under the kidney capsule of a Sl/Sl^d host. Note population of the graft by hemopoietic cells.

The ability of exogenous stem cells to seed fetal hepatic tissue was tested by grafting post-28 somite liver to syngeneic hosts subsequently irradiated (850R) and reconstituted with 10^7 syngeneic bone marrow cells. One week post irradiation and marrow injection, the hepatic graft displayed prominent erythroid, megakaryocytic and granulocytic colonies, the latter being most frequent. Control grafts placed in hosts subsequently irradiated but not grafted with bone marrow showed no evidence of endogenous colony formation.

These results indicate that fetal liver grafts either pre- or post-28 somite stage can sustain normal hematopoiesis provided that, in the case of pre-28 somite liver, a source of exogenous stem cells is available. Thus the *in vivo* grafting studies support the *in vitro* organ culture observations indicating that yolk sac derived stem cells colonize the hepatic rudiment of the mouse embryo at approximately the 28 somite stage of development.

FETAL REGULATION OF STEM CELL PROLIFERATION AND DIFFERENTIATION

The preceding observations present a paradox involving the influence of a hemopoietic microenvironment and the role of systemic humoral regulators in determining stem cell differentiation. In yolk sac and fetal liver, granulopoietic stem cells are present in numbers approaching or exceeding their incidence in adult marrow and yet granulopoiesis is absent in murine yolk sac, and fetal liver is almost exclusively an erythropoietic organ. Furthermore, the rapid transformation to extensive granulopoiesis in fetal liver grafts in adult syngeneic recipients and the predominance of granulocytic colonies over erythroid colonies following irradiation and stem cell seeding of such grafts would indicate a predominance of granulocytic microenvironments. In the context of the role of microenvironmental influences determining multipotential stem cell differentiation it is unlikely that this concept can account for the predominant if not exclusive pattern of erythropoiesis seen in fetal yolk sac and liver for the following reasons (a) the yolk sac and liver microenvironments are as effective as adult bone marrow in supporting differentiation of multipotential stem cells into the granulocyte-monocyte committed stem cell compartment, (b) following grafting into the adult environment fetal liver becomes a predominately granulopoietic tissue, (c) colony formation in the irradiated-reconstituted fetal liver graft is predominately granulopoietic rather than erythroid.

For these reasons it is more plausible to consider that the restriction in fetal granulopoiesis is due to a systemic inhibition of granulopoietic differentiation or absence of the appropriate differentiation stimulus provided in the adult environment.

Analysis of the proliferative status of CFU-c in fetal hemopoietic tissue by tritiated thymidine killing procedures *in vitro* has shown that fetal CFU-c proliferation in species of short gestation (rabbit, mouse, rat) approached or exceeded that observed in adult marrow (Fig. 3). In contrast, in species of long gestation (human, monkey, calf, lamb, guinea pig) a period of variable duration was observed when fetal liver CFU-c entered a non-cycling G_o or blocked G_1 phase (Moore and Williams, '73a).

In vitro cycle activation of fetal liver CFU-c has been obtained following brief, 1-3 hour exposure of the cells to colony stimulating factor (Moore and Williams, '73b), however the inability of the cells to proliferate and differentiate within the fetal liver cannot be attributed to absence of CSF since this factor is present in substantial quantities in fetal serum. These observations suggest that fetal granulopoiesis may be suppressed by a systemic inhibitory factor which is absent in the adult environment.

Buoyant density and velocity sedimentation separation of fetal hemopoietic tissue have shown that multipotential stem cells (CFU-s) and committed precursors (CFU-c) can be distinguished from their adult counterparts on the basis of larger mean volume and lighter buoyant density (Moore and Williams, '73a; Haskill and Moore, '70). A bias to macrophage differentiation has also been reported for fetal CFU-c (Moore and Williams, '73a) and early fetal CFU-s can sustain a more rapid population doubling and have an intrinsically greater self-renewal capacity than adult stem cells (Metcalf and Moore, '71; Micklem *et al.*, '72).

The existence of hemopoietic cell populations uniquely adapted to the requirements of particular stages of development has long been recognized in the case of erythropoiesis. It would appear that similar adaptations also characterize hemopoietic stem cells and that the sequence of migration and maturation of fetal stem cells into populations displaying adult characteristics can be experimentally produced by transferring fetal cells or organs into an adult environment.

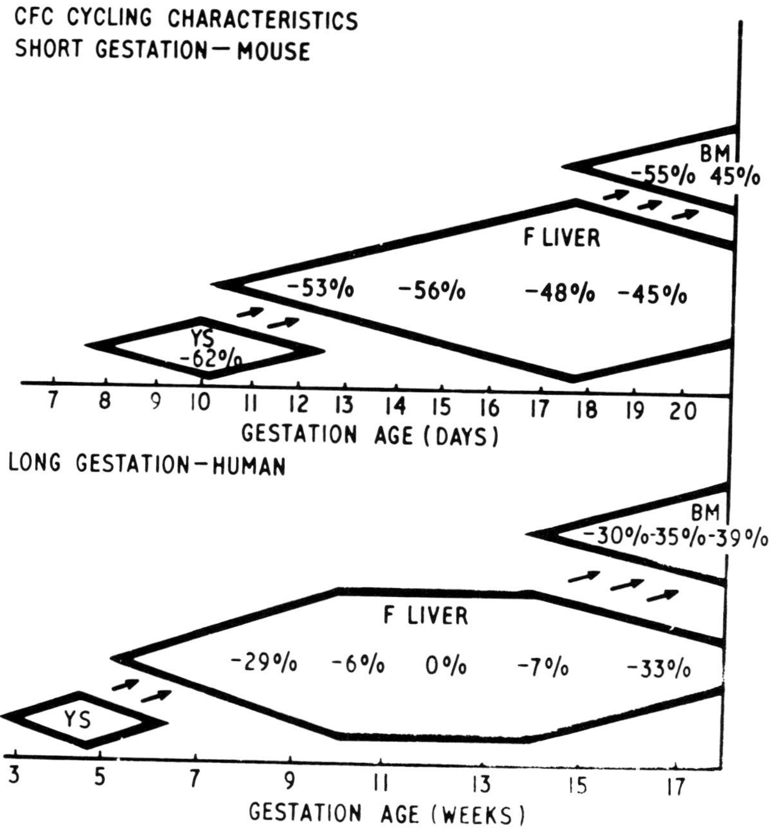

Fig. 3. The proliferative status of CFU-c in different hemopoietic tissues as a function of gestation age in a comparison of mouse and human. Percentages are the fraction of CFU-c in DNA synthesis as determined by the in vitro H^3TdR killing technique. Note the period from 10-15 weeks gestation when human fetal liver CFU-c enter a non-cycling state. The shape of the outlined areas approximates to the expansion, plateau, and decline of hemopoiesis in yolk sac, liver and marrow as a function of age. Arrows indicate direction and timing of stem cell migration streams.

LITERATURE CITED

Haskill, J. S. and M.A.S. Moore. 1970. Two dimensional cell separation: Comparison of embryonic and adult haemopoietic stem cells. *Nature 226:* 853.

Johnson, G. R. and R. O. Jones. 1973. Differentiation of the mammalian hepatic primordium *in vitro*. I. Morphogenesis and the onset of haematopoiesis. *J. Embryol. exp. Morph. 30:* 83.

Metcalf, D. and M.A.S. Moore. 1971. *Haemopoietic Cells*. North Holland Publishing Co.

Micklem, H. S., C. E. Ford, E. P. Evans, D. A. Ogden and D. S. Papworth. 1972. Competitive *in vivo* proliferation of foetal and adult haematopoietic cells in lethally irradiated mice. *J. Cell Physiol. 79:* 293.

Moore, M.A.S. and D. Metcalf. 1970. Ontogeny of the haemopoietic system: Yolk sac origin of *in vivo* and *in vitro* colony forming cells in the developing foetal liver. *Brit. J. Haematol. 18:* 279.

Moore, M.A.S. and J.J.T. Owen. 1965. Chromosome marker studies on the development of the haemopoietic system of the chick embryo. *Nature 208:* 956.

Moore, M.A.S. and J.J.T. Owen. 1967. Stem cell migration in the developing myeloid and lymphoid systems. *Lancet 1:* 658.

Moore, M.A.S. and N. Williams. 1973a. Analysis of proliferation and differentiation of foetal granulocyte-macrophage progenitor cells in haemopoietic tissue. *Cell Tissue Kinet. 6:* 461.

Moore, M.A.S. and N. Williams. 1973b. Functional, morphological and kinetic analysis of the granulocyte-macrophage progenitor cell. In: *Hemopoiesis in Culture*. W. Robinson, ed. Grune and Stratton, New York. p. 17.

Rifkind, R. A., D. Chui and H. Epler. 1969. An ultrastructural study of early morphogenetic events during the establishment of foetal hepatic erythropoiesis. *J. Cell Biol. 40:* 343.

Ageing of Haematopoietic Stem Cell Populations in the Mouse

H. S. MICKLEM AND D. A. OGDEN[1]

Immunobiology Unit, Department of Zoology, University of Edinburgh

Although many workers have attempted to study the behaviour of haematopoietic cell populations during ageing, no clear and consistent picture has yet emerged. For example, it is still not agreed whether the stem cell pool is subject to ageing processes, and, if so, whether this is normally a limiting factor determining mammalian lifespan.

In this paper we shall consider evidence from several experimental systems, and in particular the use of serial transfer as a tool to study haematopoietic ageing.

Haematopoietic stem cells can be directly enumerated by the method of McCulloch and Till ('62). Although the relationship is not 1:1, the number of spleen colony-forming units (CFU-S) found in this assay is proportional to the number of stem cells in the cell population under test. Stem cells and their descendants can also be followed, with less numerical precision but over a longer period, by their capacity to repopulate the depleted bone marrow cavities of lethally irradiated recipients. Such experiments require the use of a chromosome marker system to distinguish between donor and host cells; more complex experimental systems involving the separate identification of three cell populations can yield additional information.

CFU-S IN RELATION TO AGE

Studies of the numbers of CFU-S in ageing mice have generally not revealed very striking changes, any proportional decline in CFU-S being offset by an increase in the overall cellularity of the bone marrow (Chen, '71). Metcalf and Moore

[1]Present address: Department of Anatomy, St. Andrews University, St. Andrews, Fife, U.K.

('71) reported a 5-10-fold rise in the proportion of femoral CFU-S in 130-week old, compared to 8-week old CBA male mice. We have found an approximately 2-fold increase in 146-week old females of the same strain (unpublished data).

STEM CELL QUALITY IN RELATION TO AGE

There is substantial evidence that the quality of stem cells varies between embryonic and adult mice. Spleen colony cells, derived originally from stem cells from various sources, were serially transferred through irradiated hosts by Metcalf and Moore ('71). The numbers of transfers achieved, at 11-14 day intervals, before CFU renewal ceased, were as follows:

9-day yolk sac	7
10-day foetal liver	6
15-day foetal liver	4
neonatal liver	4
8-week adult marrow	3
130-week adult marrow	3

Schofield ('70) showed that the doubling time for foetal liver-derived CFU-S was shorter than for adult marrow-derived CFU-S. When foetal and adult stem cells were injected together into irradiated recipients, the 'foetal' population outgrew the 'adult' over a period of 10 weeks (Micklem *et al.*, '72). An experiment of similar design, comparing bone marrow from young and old donors showed no such difference (unpublished data). Functional tests of marrow populations derived from young and old donors also showed no differences (Micklem *et al.*, '73; Harrison, '75).

As Metcalf and Moore ('71) have pointed out, the seemingly paradoxical similarity between young and old bone marrow may plausibly be explained in terms of the mitotic history of the stem cells. To produce the young adult number of stem cells by division from the fertilized ovum may take upwards of 50 divisions. Thereafter, considerably less than that number may suffice to maintain haematopoiesis throughout adult life. If, therefore, a stem cell's proliferative capacity is influenced by its mitotic history, the greatest changes would occur during development.

'ARTIFICIAL AGEING': SERIAL TRANSFER EXPERIMENTS

Several attempts have been made to estimate the functional lifespan of the mouse haematopoietic system by serially transferring bone marrow cells through irradiated recipients.

What in principle happens is that a small number of stem cells are injected; these multiply and regenerate normal haematopoiesis; later their descendants can be injected into further recipients to repeat the process, and so on in series. The outcome can be interpreted either in terms of the number of months of life the haematopoietic system will sustain (Micklem and Loutit, '66; Harrison, '73) or in terms of the number of mitoses through which the stem cell population has to pass to effect regeneration at each transfer. The latter approach demands more data - for example, on bone marrow cellularity and CFU content - and some unproven assumptions. It is, however, worth attempting since calculations based simply on time almost inevitably underestimate the population's capacity to support life under steady-state conditions. The number of transfers has always proved to be a more constant limiting factor than the total time involved. Simple considerations of time do, however, show that a cell population can maintain haematopoiesis for at least 5 years - almost twice as long as the normal lifespan of the mouse itself (Micklem and Loutit, '66; Harrison, '73). But it is not immortal, since even when transferred at long intervals the donor population eventually loses its reproductive capacity (Micklem and Loutit, '66; Micklem *et al.*, '73). In these respects it resembles other normal tissues (Krohn, '62; Daniel *et al.*, '68).

The results of Lajtha and Schofield ('72) stand in some contrast. They showed that even after 6 transfers at 8-week intervals, stem cells were able to multiply as rapidly as ever. The authors attributed this to the relatively long interval between transfers compared with those used by some earlier workers. However, persistence of the original population could not be verified in these experiments, and one or more of the irradiated hosts may have contributed to the end-population. Barnes, Ford and Loutit (personal communication, '68) had in fact achieved 21 serial transfers of one line over an 8-year period, but the original donor identified by the T6 marker chromosome disappeared after the 6th transfer and was replaced by host cells many of which were recognizable clones carrying radiation-induced chromosome rearrangements. Survival was promoted by supplementing the bone marrow with fresh lymph node cells at each transfer (Barnes *et al.*, '62). Micklem *et al.* ('73) subsequently found host regeneration after 3-4 generations of transfer, although this was not sufficient to allow survival of 5th generation recipients.

We have recently extended these experiments (Ogden and Micklem, '76). CBA mice were lethally irradiated and injected with cells from two, separately identifiable, syngeneic sources (carrying 1 and 2 T6 marker chromosomes respectively). The bone marrow was serially transferred at 8-10 week intervals. The use of two markers meant that the fate of two donor populations could be directly compared within the same recipients. In the first experiment, of which a preliminary account was given earlier (Micklem *et al.*, '73), 'young' (3-month) cells gradually outgrew 'old' (24-month) over the course of 4 serial transfers. Subsequent experiments, however, have led us to conclude that this result was not related to the age of the original donors. In a second experiment, exactly the opposite happened. Further investigation suggested that, as serial transfer proceeded, a dwindling number of clones contributed to repopulation (Table 1). This occurred in spite of transferring large numbers of CFU-S, and was indicated (Micklem *et al.*, '75a) by the extreme variability of the proportions of the donor populations. In none of the experiments did the donor cells persist beyond the 6th transfer.

Host mitoses began to be detected in considerable numbers in the 3rd-4th transfer generations. This, too, indicated that few clones could be active in haematopoiesis. Only 0.01% of the mouse's stem cells (100 or less) would be expected to survive 900rad (McCulloch and Till, '62). We injected upwards of 900 CFU-S to each mouse - about 10^4 stem cells on average. Therefore, the fact that host cells can be competitive at all, especially since they must carry some burden of radiation-induced damage, argues that few of the donor stem cells have engendered haematopoietic clones of any size. It should be noted that even when normal bone marrow cells are injected the number of clones which are active at any one time is, for reasons which are not well understood, quite small (Wallis *et al.*, '75; Micklem *et al.*, '75a): with serially transferred marrow it evidently becomes still smaller.

This reduction in the number of effective clones suggests that the injected stem cells have, on average, decreasing powers of self-renewal. The clones which were active presumably reached an unusually large size, since in other experiments (Ogden and Micklem, '76) it was shown that after 4 serial transfers the marrow had still regenerated normal numbers of CFU-S in the bone marrow; moreover, no deficiency of erythropoiesis was noted before the 5th transfer generation. Only in some immunological functions could a progressive decline be observed at earlier stages; this immunological disfunction was predictable in the light of earlier results (Barnes *et al.*, '62).

TABLE 1

*Mitotic cells in the bone marrow of lethally irradiated CBA mice injected with normal or serially transferred syngeneic bone marrow.**

Transfer	Cells injected ($x10^{-6}$)	CFU injected	Time killed (weeks)	Mitoses Donor, % of total		Mitoses T6/+, % of total	Mitoses T6/+, % of donor
1	10.0	1350	1	100	100	53	55
			3	100	100	52	51
			5	100	100	47	60
			8+	100	100	63	66
			24	100	100	50	62
2	10.0	2700	1	98	97	46	51
			3	100	100	39	56
			8+	93	94	45	46
			17	89	94	49	50
			50	100	100	44	46
3	2.4	n.t.	1	91		31	
			8+	90	73	52	87
			18	98	95	45	62
4	7.5	4800	1	78	73	42	54
			3	78	56	22	14
			8+	59	72	12	29
			37	95	13	77	100
5	3.8	900	1	62	65	30	16
			2	54	57	4	9
			4	46	47	1	5
			9+	62	67	45	97
6**	5.6	280	1	75		93	

* The original donors were 3-month old CBA-T6 and F_1 CBA x CBA-T6 females, carrying 2 and 1 T6 marker chromosome respectively. All week-1 data are from spleen; otherwise from bone marrow.

+ These mice were used as donors for the next transfer.

** No recipients survived beyond 10 days after irradiation.

n.t. = Not tested.

THE MITOTIC BURDEN ACCUMULATED DURING SERIAL TRANSFER

With certain assumptions, it is possible to calculate the number of mitoses through which, on average, the stem cell pool would have to pass to effect full regeneration at each serial transfer. These assumptions are: 1) that each injected stem cell is equally likely to contribute to regeneration; and 2) that the probability (p) of a dividing stem cell producing two daughter stem cells, as opposed to differentiating cells, is 0.62 (Vogel et al., '68). Under such conditions, the number of stem cells (y) generated by each initial donor stem cell is equal to $(2p)^n$, where n = the number of mitoses undergone. Thus, $n = \log y/\log(2p)$. Since the approximate number of stem cells regenerated and the initial number injected are known, n can be calculated to be between 14 and 20 for each transfer. On that basis, the pool could pass during serial transfer through some 65 - 85 mitoses, in addition to the 50 - 60 required during development. This total would indicate that CBA mice have a more than ample reserve of stem cell capacity to last throughout life.

Unfortunately, the assumptions on which this calculation is based are undoubtedly over-simplified. The value of p may change as serial transfer proceeds. Even if it remains constant at 0.62, 63% of dividing stem cells will be expected to differentiate terminally at an early stage (Vogel et al., '68). Moreover, stem cells may not be equally sensitive to mitogenic stimuli, so that some may remain quiescent for undetermined periods, leaving the burden of regeneration to be carried by others; this could result in the sequential appearance and decline of large clones. Persuasive estimates of the proliferative capacity of stem cell populations cannot therefore be made at present. All that can be said with confidence is that, under conditions of serial transfer, the capacity is strictly limited.

LIMITATIONS OF THE SERIAL TRANSFER SYSTEM

The mortality of haematopoietic cell populations can only be asserted if it is assumed (1) that all bone marrow stem cells are transplantable and (2) that they can potentially function as effectively after transfer as before. However, there is evidence that CFU-S have a well-defined spatial distribution in the bone marrow, and that CFU-S which are in cycle (as judged by ^{3}H-thymidine suicide) are distributed differently from those which are not (Lord, Testa and Hendry, '75). It is not known whether this distribution is of fundamental importance in determining the behaviour of individual stem cells, or whether it is re-established after intravenous injection of stem cells to irradiated hosts. Cairns ('75) has suggested a mechanism which would reduce the likelihood of error-accumulation in the genome of the stem cell pool. This depends on the micro-anatomical siting of the most primitive stem cells. If such a system were operative in the bone marrow, and if the crucial stem cells failed to re-establish themselves in the correct microenvironment after injection, then serial transfer experiments might give too low an estimate of the potential lifespan of haematopoietic populations.

DESIRABLE QUALITIES OF A STEM CELL HIERARCHY

Any dividing cell population is liable to accumulate changes in the genome due to mutation or mis-copying, and some of these may result in a cumulative decrease in the fidelity with which new DNA, RNA and proteins are synthesized (Orgel, '73). Although repair mechanisms may be active, the inherent dangers would be further reduced if the genome of the most primitive stem cells could be safeguarded in some way. This could be done, as suggested by Cairns ('75), by arranging for all the original DNA to go to one of the two daughter cells at division, thus preserving one of the daughters free from any copying errors. Or it could be done by making the most primitive stem cells relatively difficult to trigger into mitosis. The daughters, grand-daughters, etc. would become increasingly likely to respond to mitogenic stimuli. In this way, once a clone had started to proliferate, it would become increasingly likely to continue proliferating; the most primitive stem cells would only be activated occasionally as established clones declined and failed to respond sufficiently to haematopoietic demands. In this way a clonal succession would be established, with haematopoiesis at any one time being largely dependent on a small number of clones. Evidence in favour of this last point has already been mentioned. The following are essential differences which would be expected between the most primitive ('A') and the most differentiated ('Z') stem cells.

	A -------	Z
Frequency of cycle	Low	High
Ratio, self-renewal/terminal differentiation	High	Low
Potential size of descendant clones	Large	Small
Susceptibility to error accumulation	Low	High

The transition between 'A' and 'Z' could either be a gradient or a series of steps of unknown number. For the present scheme, one intermediate step ('M') would be the minimum. 'M' stem cells would then be the main foundation of current haematopoiesis in the adult; 'A' would be the reserve, activated during development and regeneration, but largely quiescent under steady-state conditions; and 'Z' would be effete. Indeed, the 'Z' category can only be considered to be stem cells to the extent that they can engender spleen colonies, clones of some 10^6 terminally differentiating cells whose production involves about 20 doublings. Heterogeneity of bone marrow stem cells is not a new idea. Schofield ('70), for example, interpreted his data in terms of two CFU-S populations, distinguished by different turnover times. The suggested transitions could be based on micro-environment, length of mitotic history, or both.

THE FATE OF EFFETE ('Z') STEM CELLS

The 'simplest' fate for a 'Z' stem cell would be terminal differentiation. This may, however, be too simple. The existence of CFU-S in the peripheral blood is well known. We have recently shown that few, if any, of these re-enter the bone marrow and contribute to haematopoiesis (Micklem et al., '75b). They also have a low capacity for self-renewal when assayed as CFU-S or in terms of bone marrow recolonization (Micklem et al., '75c). A large proportion of them are subject to ^{3}H-thymidine suicide (Gidali et al., '74). In all these characters they differ from many of the stem cells in bone marrow, and the characters are ones which we should expect to find in effete stem cells. The circulating stem cells may reasonably be regarded as an overflow of an effete sub-population from the bone marrow. Several variables appear to influence the release of cells from the bone marrow (Lichtman and Weed, '72; Chamberlain et al., '75). They include the membrane properties of the cells themselves as well as the permeability of the sinusoidal walls; hormonal factors are also important (Khaitov et al., '75).

Little is known of the half-life of 'stem' cells in the bloodstream, so that calculation of the numbers produced daily is not yet practicable.

CONCLUSIONS

Under conditions of serial transfer, at 8-week intervals or longer, haematopoietic cell populations have strictly limited regenerative capacity, failing after six transfers or less. After 3-4 transfers, there is evidence that a decreasing number of clones are contributing to haematopoietic regeneration, although these maintain normal erythropoiesis. No consistent differences have been detected between the stem cells present in the bone marrow of young and old mice. We suggest that this may well be because the most primitive stem cells divide very seldom and are thus relatively immune from genetic error and other 'ageing' influences, while aged and incipiently effete stem cells are continuously expelled into the bloodstream.

Serial transfer systems are obviously artificial. The important question is: Does serial transfer merely accelerate a normal ageing process, by rapidly lengthening the mitotic history of the stem cell pool? Or does it introduce fundamental abnormalities, by causing the loss of the more primitive stem cells or by deranging the haematopoietic microenvironment? In the former case, serial transfer provides a useful model for the study of ageing processes; in the latter case, its use is more limited. The answer remains unknown.

LITERATURE CITED

Barnes, D.W.H., J. F. Loutit and H. S. Micklem. 1962. Secondary disease of radiation chimaeras: a syndrome due to lymphoid aplasia. *Ann. N.Y. Acad. Sci. 99:* 374-385.

Cairns, J. 1975. Mutation selection and the natural history of cancer. *Nature 255:* 197-200.

Chamberlain, J. K., L. Weiss and R. I. Weed. 1975. Bone marrow sinus cell packing: a determinant of cell release. *Blood 46:* 91-102.

Chen, M. G. 1971. Age-related changes in hematopoietic stem cell populations of a long-lived hybrid mouse. *J. Cell Physiol. 78:* 225-232.

Daniel, C. W., K. B. DeOme, J. T. Young, P. B. Blair and L. J. Faulkin. 1968. The *in vivo* lifespan of normal and preneoplastic mouse mammary glands: a serial transplantation study. *Proc. Nat. Acad. Sci. 61:* 53-60.

Gidali, J., I. Feher and S. Antal. 1974. Some properties of circulating hemopoietic stem cells. *Blood 43:* 573-580.

Harrison, D. E. 1973. Normal production of erythrocytes by mouse marrow continuous for 73 months. *Proc. Nat. Acad. Sci. 61:* 53-60.

Harrison, D. E. 1975. Normal function of transplanted marrow cell lines from aged mice. *J. Gerontol. 30:* 279-285.

Khaitov, R. M., R. V. Petrov, B. B. Moroz and G. I. Bezin. 1975. The factors controlling stem cell recirculation. I. Migration of hemopoietic stem cells in adrenalectomized mice. *Blood 46:* 73-77.

Krohn, P. L. 1962. Heterochronic transplantation in the study of ageing. *Proc. Roy. Soc. B157:* 128-147.

Lajtha, L. G. and R. Schofield. 1971. Regulation of stem cell renewal and differentiation: possible significance in ageing. *Adv. Gerontol. Res. 3:* 131-146.

Lichtman, M. A. and R. I. Weed. 1972. Alteration of the cell periphery during granulocyte maturation: relationship to cell function. *Blood 39:* 301-316.

Lord, B. I., N. G. Testa and J. H. Hendry. 1975. The relative spatial distribution of CFUs and CFUc in the normal mouse femur. *Blood 46:* 65-72.

McCulloch, E. A. and J. E. Till. 1962. The sensitivity of cells from normal mouse bone marrow to gamma radiation *in vitro* and *in vivo*. *Radiat. Res. 16:* 822-832.

Metcalf, D. and M.A.S. Moore. 1971. *Haemopoietic Cells.* North Holland, Amsterdam.

Micklem, H. S., N. Anderson and E. Ross. 1975c. Limited potential of circulating haemopoietic stem cells. *Nature 256:* 41-43.

Micklem, H. S., C. E. Ford, E. P. Evans and D. A. Ogden. 1975a. Compartments and cell flows within the mouse haemopoietic system. I. Restricted interchange between haemopoietic sites. *Cell Tissue Kinet. 8:* 219-232.

Micklem, H. S., C. E. Ford, E. P. Evans, D. A. Ogden and D. S. Papworth. 1972. Competitive *in vivo* proliferation of foetal and adult haematopoietic cells in lethally irradiated mice. *J. Cell. Physiol. 79:* 293-298.

Micklem, H. S. and J. F. Loutit. 1966. *Tissue Grafting and Radiation.* Academic Press, New York.

Micklem, H. S., D. A. Ogden, E. P. Evans, C. E. Ford and J. G. Gray. 1975b. Compartments and cell flows within the mouse haemopoietic system. II. Estimated rates of interchange. *Cell Tissue Kinet. 8:* 233-248.

Micklem, H. S., D. A. Ogden and A. C. Payne. 1973. Ageing, haemopoietic stem cells and immunity. *Haemopoietic Stem Cells, Ciba Foundation Symposium 13,* 285-297. Associated Scientific Publishers, Amsterdam.

Ogden, D. A. and H. S. Micklem. 1976. Ageing of haematopoietic cell populations. I. Fate of serially transferred bone marrow from young and old donors. *Transplantation:* submitted for publication.

Orgel, L. E. 1973. Ageing of clones of mammalian cells. *Nature 243:* 441-445.

Schofield, R. 1970. A comparative study of the repopulating potential of grafts from various haemopoietic sources: CFU repopulation. *Cell Tissue Kinet. 3:* 119-130.

Vogel, H., H. Niewisch and G. Matioli. 1968. The self renewal probability of haemopoietic stem cells. *J. Cell. Physiol. 72:* 221-228.

Wallis, V. J., E. Leuchars, S. Chwalinski and A.J.S. Davies. 1975. On the sparse seeding of bone marrow and thymus in radiation chimaeras. *Transplantation 19:* 2-11.

Hemopoietic Stem Cells in Tumor Bearing Hosts

P. K. LALA

Department of Anatomy, McGill University, Montreal, Quebec, Canada

It is a common experience to find significant changes in the level of hemic elements in human patients suffering from malignant diseases originating in tissues other than hemopoietic. Similarly, hemopoietic changes have also been observed in laboratory animals bearing experimental tumors (Delmonte *et al.*, '66; Milas and Tomlzanovic, '71; Hibberd and Metcalf, '71). A precise evaluation of the hemopoietic status of tumor-bearing host is relevant not only to the understanding of the host-tumor balance, but also towards optimizing a therapeutic protocol. Very often, injury to host lymphomyeloid tissues by the available anticancer agents stands as the most serious threat to therapy. A critical number of stem cells in host hemopoietic sites and their physiological state are important determinants of host survival. However, their role can hardly be considered in isolation from the differentiated elements, since the response of hemopoietic stem cells to the presence of a tumor at a distant site may constitute only a part of the events initiated in host hemopoietic organs. For this reason, we examined several parameters to evaluate the overall response of the host lympho-myeloid system to heterotopic tumor transplantation in mice: a) life history and properties of host leukocytes invading the tumor site, b) leukocyte dynamics in the host blood, bone marrow and spleen, and c) distribution and dynamics of host hemopoietic stem cells.

Various parts of this study used several tumor-host combinations: (i) Ehrlich ascites tumor (EAT) maintained in CF_1 strain mice by a weekly intraperitoneal passage of 10^6 tumor cells, (ii) EAT maintained in a similar fashion in CBA or CBA/HT_6 strain mice, and (iii) TA-3 (Stockholm) line tumor grown in A strain mice either as ascites tumors, using a similar protocol of transfer or as solid tumors by subcutaneous transplantation of 5×10^6 cells from the ascites form of the tumor in the medial aspect of left thigh. All animals used were young adult (12-16 week old) females. The strain of origin of the Ehrlich ascites tumor is unknown and it grows intraperitoneally in any mouse strain. TA-3 (St) tumor originated as a mammary carcinoma in A strain mice and is strain-specific. Animals bearing these various ascites tumors survived close to two weeks, although the total numbers of tumor cells during the plateau phase of tumor growth (second week) were different. These were approximately 750×10^6, 350×10^6 and 130×10^6 cells respectively for the EAT grown in CF_1 mice, the EAT in CBA or CBA/HT_6 mice and the TA-3 (St) ascites tumor in A strain mice. Latter strain mice carrying solid subcutaneous tumors survived for 5-6 weeks; tumor sizes nearly plateaued by the end of the third week and the maximal tumor diameters ranged from 3 to 4 cms. Experiments were confined during the first two weeks of tumor growth with special emphasis on the growing phase of the tumors.

LIFE HISTORY AND SURFACE PROPERTIES OF HOST LEUKOCYTES INVADING THE TUMOR SITE

DYNAMICS OF LEUKOCYTE ACCUMULATION WITHIN THE TUMOR

Leukocytes of all forms - lymphocytes, monocytes, macrophages and granulocytes - were found to accumulate at the developing tumor site both in the case of ascites as well as solid tumors. Although the absolute rates of increase in various cell types within the tumors differed from one tumor type to the other, the relative preponderance of various cells were qualitatively similar. For example, excepting the very early post-transplantation intervals, monocytes and macrophages together constituted the highest proportion amongst all leukocytes followed by the lymphocytes and then the granulocytes. In the plateau phase of growth of Ehrlich ascites tumors in CF_1 mice, the total leukocyte number was close to 55×10^6 (i.e. about 7% of the total ascites cell population including tumor cells) as compared to approximately 5×10^6 cells in the peritoneal space of a normal tumor free mouse. One hour following a single injection of ^{3}H-thymidine (^{3}HTdR), none of the leukocytes incorporated label (excepting a very minor labeling of some macrophages) at any stage of Ehrlich ascites tumor growth, indicating that the leukocyte accumulation was a result of migration from blood rather than local proliferation. In order to examine the post-mitotic age of these migrant cells (Lala, '74), a series of 12 injections of ^{3}HTdR (20 μCi every 8 hr) were given to mice prior to intraperitoneal ascites tumor transplantation (experimental group) or injection of isotonic saline (control group). Animals were then sacrificed at daily intervals to determine labeling indices in the blood and peritoneal leukocytes. Granulocyte labeling in the blood as well as peritoneal space was near 100% in both groups of animals at all intervals, as expected from the known granulocyte kinetics. Temporal changes in the labeling of lymphocytes (from 10% at 0 day to 22% at day 6) and monocytes (from 20% at 0 day to 57% at day 6) were identical in the blood and peritoneal space of normal animals indicating a free exchange of cells between these compartments. Labeling indices of these cell types increased more rapidly in the blood of tumor bearing hosts (to 40% for lymphocytes and to 80% for monocytes at 6 days) suggesting an accelerated turnover in the circulation. More strikingly, the labeling of these cells within the ascites tumor was even higher (viz. 65% for lymphocytes and 92% for monocytes at 6 days) indicating a selective migration and/or retention of newly formed cells within the developing tumors in contrast to a random migration into the normal peritoneal space. Furthermore, labeling of macrophages within the tumor was identical to that of monocytes indicating a rather short monocyte-macrophage transition.

Invasion by young monoclear cells does not seem to be a unique event peculiar to ascites tumors. In recent studies using closely similar protocol (Lala and Kaizer, unpublished), an identical phenomenon has also been observed for the lymphocytes and monocytes accumulating in the subcutaneous transplants of strain-specific TA-3 (St) tumors growing in A strain mice: at all intervals (5 to 14 days of tumor growth), the labeling indices in these cells were significantly higher than those in the circulation of the same hosts or normal control animals.

SOURCE OF NEWLY FORMED MONONUCLEAR CELLS APPEARING IN THE TUMOR

Blood-borne monocytes appearing in inflammatory exudates are definitely known to be marrow-derived (Van Furth, '70). However, bone marrow as well as the thymus are two well recognized primary sources of newly formed lymphocytes (Osmond, '72). In the present case, bone marrow was identified as the major source of these cells (Lala, '76) from a comparison of the temporal changes in the lymphocyte content of ascites tumors growing in sham-irradiated, whole-body irradiated or partially irradiated hosts. Whole body irradiation (900R) made the tumors virtually leukocyte free. Irradiation (1000R) of four limbs alone (to stop production in about half of the hemopoietic marrow) or whole body minus the four limbs (to stop production in about half of the marrow, in addition to the thymus and other lymphoid organs) prior to tumor transplantation led to similar patterns of lymphocyte depletion and recovery within the developing tumors. The depletion in the limb-shielded group was only slightly more than that in the limb-irradiated group. These results indicated that contribution of lymphocytes into the tumor from the extramyeloid organs was possibly much smaller than that from the bone marrow.

SURFACE MARKERS OF SMALL LYMPHOCYTES ACCUMULATING IN THE TUMOR

Two important surface markers are conventionally employed to classify mouse lymphocytes into two broad functional categories (Raff, '71): B cells showing immunoglobulin (IgM) on the cell surface and responsible for humoral immunity, and T cells having θ antigen on the surface and engaged in cellular immune response. In the mouse, these two cell types are known to be initially produced in the bone marrow and the thymus, respectively (Grieves *et al.*, '74; Osmond, '75). The presence of IgM or θ antigen was explored radioautographically on the surface of lymphocytes appearing within the Ehrlich ascites tumors grown in CBA/HT_6 mice, with a view to providing a link between their life history and functional potentials (S. Garnis and P.K. Lala, unpublished data). Surface IgM was detected with the technique of Osmond and Nossal ('74a). Cell suspensions were exposed to ^{125}I-labeled goat antimouse IgM for 30 minutes at 0°C at a wide range of antiglobulin concentrations (1, 3, 10 and 20 μgm/ml). Labeling index of small lymphocytes in the normal spleen reached a well defined plateau of approximately 46% at antiglobulin concentrations of 3 μgm/ml or higher. This reflected the incidence of mature B cells in this organ. In contrast, small lymphocyte labeling within the 7 day ascites tumor increased continuously in a linear fashion with increasing antiglobulin concentrations (from 12% at 1 μgm/ml to 36% at 20 μgm/ml) suggesting that these cells had a wide range of IgM density on the surface; it must be low on many cells, since they bound anti-IgM only at high concentrations. An identical phenomenon was also found in normal bone marrow small lymphocytes, as originally reported by Osmond and Nossal ('74a), who have also shown from double labeling studies (Osmond and Nossal, '74b; Osmond, this volume) that the surface IgM density in these cells increased with post-mitotic age and thus reflected their level of maturation. One may thus conclude that IgM bearing small lymphocytes appearing within the tumor constituted B cells at various stages of maturation.

The large proportion (Ca 64%) of the small lymphocytes within the 7 day tumor which did not show any detectable anti-IgM binding even at the highest concentration could theoretically represent one or both of the following cell categories: those bearing θ antigen on the surface or cells having neither surface IgM nor θ antigen. The latter category of "double negative" cells in the bone marrow were shown by Osmond and Nossal ('74b) to include very young cells of B cell lineage at the early post mitotic stage, which acquired surface IgM after further maturation for about 1.5 days (Osmond, this volume). To distinguish between the two categories, we devised a sandwich immunolabeling technique (P.K. Lala and S. Garnis, unpublished data) in which, prior to exposing cells to radioiodinated antimouse IgM (10 μgm/ml), they were either incubated at $0^{o}C$ for 30 min with compliment-inactivated normal mouse serum (control series) or anti θ serum (experimental series), followed by appropriate washing to remove unbound material. Preincubation with normal serum gave results identical to those obtained with the protocol for detection of surface IgM, where this step was omitted. Preincubation with anti θ serum detected θ+ve cells in addition to IgM+ve cells and thus the difference between the labeling indices in the control and the experimental series provided a measure of the incidence of θ+ve cells; cells remaining unlabeled in the experimental series gave a measure of the incidence of "double negative" cells. Results are summarized in Table 1.

TABLE 1

Incidence (percent) of small lymphocytes with IgM and θ surface markers (^{125}I anti IgM conc = 10 μgm/ml).

	IgM+ve	θ+ve	Double-ve
Normal marrow	47	7	46
Normal thymus	4	88	8
Normal spleen	46	40	14
7d EAT - ascites fluid	26	38	36
3-5d EAT - spleen	56	16-17	27-28

In the 7 day ascites tumor, about a third of the small lymphocytes had detectable θ antigen, about a quarter (at an anti-IgM conc. of 10 μgm/ml) to a third (at anti-IgM conc. of

20 μgm/ml) had detectable surface IgM and the rest had neither marker. Incidence of double negative cells was found to increase with increasing tumor age, e.g. from 23% at day 3 to 36% at day 7. A high incidence of this category of cells was also found within TA-3 (St) ascites tumors grown in A strain mice. Because of their tentative life history, they were considered as very young marrow derived elements of B cell lineage. We are presently examining this possibility by testing whether they can acquire surface IgM with time when cultivated *in vitro*.

Thus, it may be concluded that the tumor is invaded by a large number of newly formed lymphocytes. A majority of these are marrow derived, but many of them are still immature, and thus apparently not yet fully equipped to mediate humoral immune responses.

LEUKOCYTE DYNAMICS IN THE BLOOD, BONE MARROW AND THE SPLEEN

Absolute leukocyte numbers in these organs changed significantly after tumor transplanation. Despite quantitative differences, the temporal patterns of changes were qualitatively similar for the various tumor-host combinations. For example, following transplantation of ascites tumors, host blood as well as femoral marrow showed a rapid decline in leukocyte numbers (affecting lymphocytes most), reaching minimal levels at 3 days, followed by a recovery in the following 2-3 days. In most cases, this was followed by some overshoot. Since a transient lymphocyte depletion in these organs, although to a smaller degree, was also noted in saline injected control animals, this response can partly be explained by nonspecific stress. Blood hematocrit levels in all the ascites tumor bearing animals showed some decline during the second week; in the case of CBA or CBA/HT_6 mice bearing Ehrlich ascites tumors, some decline was also noted during the first week. These declines coincided with the appearance of red cells in the ascites fluid. All tumor transplanted animals, bearing either intraperitoneal or subcutaneous tumors, showed a steady rise in the splenic weight, paralleled by an identical increase in the nucleated cell content of the spleen, while little or no increase was seen in the saline injected controls. At one week after tumor transplantation, splenic weight or cellularity was approximately 2-fold in CBA (or CBA/HT_6) mice bearing Ehrlich ascites tumor, 2.5 fold in CF_1 mice bearing Ehrlich ascites tumor, 2.6 fold in A strain mice bearing TA-3 (St) tumor subcutaneously and 4.3 fold in A strain mice bearing TA-3 (St) tumor in ascites form. At 12 days these values ranged between 4 fold and 5 fold. During the first week, lymphoid cells accounted for most of the increase in cellularity. An examination of the spleens of A strain mice during the second week of subcutaneous growth of TA-3 tumors showed, in addition, a marked increase in the nucleated erythroid cells, predominantly mature elements, owing to local erythropoiesis in the spleen.

Possible mechanisms for increased cellularity were explored during the first 5 days of splenic enlargement in CBA/HT_6 mice bearing Ehrlich ascites tumors (Lala and Lind, '75), where lymphoid cells accounted for 90-97% of all nucleated cells. The increase was accounted in part by local proliferation primarily in large lymphoid and blast like cells, and to a significant extent by extraneous migration. The overall 1 hr ^{3}HTdR labeling index of nucleated cells increased from 0.7% at day 0 to 5% at day 5. If one assumed a DNA-synthesis time of 5 hours, local proliferation accounted for about a third of the increase in cell number between days 1 and 2, and four-fifths of the increase between days 4 and 5. The remaining increase represented a minimal estimate of influx from extra-splenic sites, since there must be some cellular efflux out of the spleen as well.

The life history of cells dividing in the spleen of ascites tumor bearing hosts was explored in two sets of experiments using the T_6 chromosome marker to trace cellular origin (Lala and Terrin, '74). Results indicated that a high proportion of these cells were recent immigrants from the bone marrow via circulation. When CBA mice parabiosed with CBA/HT_6 mice (having T_6 marker chromosomes in somatic cells) were injected with EAT cells intraperitoneally, the incidence of T_6 marker containing metaphases in their spleens increased rapidly to about 36% (compared to a maximal incidence of 12% seen in the saline injected controls) indicating a rapid influx of blood borne cells with proliferative potential into the spleen. Secondly, partial marrow chimeras were produced by local irradiation (1000R) of 4 limbs of CBA mice followed by a reconstitution of these marrow sites with CBA/HT_6 marrow cells given intravenously. Intraperitoneal inoculation of EAT cells in well established chimeras led to a fast rise in the incidence of T6 marker containing mitoses in the spleen (to 37% at 3 days compared to a maximum of 12% in the saline injected controls). Since possibly a maximum of 40-50% of the total body marrow was chromosomally marked, at least half of the cells dividing in the spleen, or their immediate precursors must have immigrated from marrow within the 3 days after tumor transplantation.

An increased incidence of newly formed lymphocytes was also observed in the spleen of A strain mice bearing subcutaneous TA-3 (St) tumors. When hosts were prelabeled repeatedly with ^{3}HTdR for several days, lymphocyte labeling indices in the spleens of hosts at various intervals after tumor transplantation were significantly higher than the respective values in saline injected control animals (P.K. Lala and L. Kaizer, unpublished data).

Surface properties of small lymphocytes in the spleen of CBA/HT$_6$ mice bearing Ehrlich ascites tumors were further characterized by immuno-radioautography as mentioned earlier (S. Garnis and P.K. Lala, unpublished). There was a reduction in the incidence of θ antigen bearing small lymphocytes and an increase in the "double negative" small lymphocytes (Table 1) quite early in tumor development. Later, there was also some decline in the proportion of IgM bearing cells; many of the IgM+ve cells were immature as judged from low IgM density on the surface. Thus, somewhat similar to the influx seen within the tumor, there is an accumulation of young marrow derived B cells within the host spleens.

DISTRIBUTION AND DYNAMICS OF HOST HEMOPOIETIC STEM CELLS

Spleen colony technique of Till and McCulloch ('61) was employed to measure the stem cell (CFU-S) content of the whole organ (femur or spleen) or per unit volume of blood in CBA/HT$_6$ mice at different days after intraperitoneal transplantation of ascites tumors (Lala and Terrin, '74; P.K. Lala and R. Scott, unpublished data). Data were expressed as ratios relative to the CFU numbers in the respective organs of saline injected, tumor-free control animals. Results are summarized in Table 2.

There was a fast decline in the CFU content per fermur in the first few days reaching a minimum at 3 days after tumor transplantation. A recovery followed, leading to a normal level by 10 days. The splenic CFU content nearly mirror-imaged that of the marrow at early intervals. It increased steadily reaching a maximum of 8 fold normal in EAT hosts and 6 fold normal in TA-3 (St) hosts on day 5. This was followed by some decline in the former and small fluctuations in the latter, but supranormal levels were always maintained. Since there was a rise in the blood CFU content during the declining phase of the marrow, the very early rise in the splenic CFU content is possibly a consequence of migration from the marrow, although some local CFU proliferation cannot be excluded even at very early intervals. This early response cannot be considered as a kinetic consequence of an increased demand for cell production following a large scale influx of newly formed mononuclear cells into the tumor; the

TABLE 2

Mean CFU-S contents of host femur, spleen and peripheral blood (experimental/control).

	Days after tumor transplantation							
	1	2	3	4	5	6	8	10
Femur								
EAT host	0.52	--	0.15	--	0.30	--	0.60	0.90
TA-3(St) host	0.50	--	0.41	--	0.53	--	0.84	1.0
Spleen								
EAT host	3.0	--	7.1	--	8.0	5.2	3.6	3.0
TA-3(St) host	1.3	--	4.0	--	6.0	5.0	5.5	--
Blood								
EAT host	2.2	3.2	2.4	2.5	--	--	--	--

latter is seen mostly after 3 days and onwards. It is possible that the nature of stimulus for stem cell migration is similar to that for the migration of young B cells from the marrow. Intraperitoneal inoculation of 10^6 sonicated ascites tumor cells or the purified plasma membrane fraction of 10^7 tumor cells also led to some 30-50% reduction in the femoral CFU content and some 2-3 fold amplification in the splenic CFU content between 2-3 days, suggesting that this response may be initiated by some plasma membrane associate components of the tumor.

Most of the splenic stem cells were found to be in cell cycle at 4-5 days of tumor development, as judged by their vulnerability to ^{3}HTdR suicide, comparable to the phenomenon observed during post-irradiation regeneration (McCulloch *et al.*, '74). Incubation of cell suspensions from these spleens *in vitro* with ^{3}HTdR (300-500 μCi/ml, 40 μCi/mM) prior to transfer into irradiated recipients reduced the CFU level to 30-40%, as compared to 76-90% values seen with normal control spleens.

We have found a high colony stimulating activity in the cell free ascites tumor fluid for *in vitro* growth of granulocytic or macrophage colonies (G. Keeb and P.K. Lala, unpublished data) on semisolid agar with normal bone marrow or spleen cells, using the technique of Bradley and Metcalf ('66). The precise cellular source of this activity (viz.

tumor cells or host derived cells within the tumor) remains to be determined. Since most of the early *in vivo* changes in the host spleens were indicative of lymphoid hyperplasia, it needs to be tested whether lymphoid colonies can be grown from these spleens, when provided with a feeder layer of tumor cells.

Stem cell dynamics in the host hemopoietic organs following transplantation of these fast growing ascites tumors is in many ways analogous to that found after treatment with leukomogenic viruses (Okunewick *et al.*, '72; Siedel, '73a,b; Iturizza and Siedel, '74), polycythemia-inducing viruses (Wendling *et al.*, '72; '74) or anemia inducing viruses (Wendling *et al.*, '74), or Bordella pertussis vaccine (Monette *et al.*, '72). Many other antigenic agents are also known to cause a rapid enhancement of splenic CFU content (Metcalf and Moore, '71). Thus the acute effects of tumor bearing on the hemopoietic stem cell dynamics, as seen in the present study, resemble antigen driven effects. However, all these effects may not be long sustained. For example, the ascites tumor bearing animals appear to recover from the early CFU depletion in the marrow. With somewhat slower growing or long established tumors, there may be a moderate elevation of the CFU level in the marrow in addition to that in the spleen. This has been observed in C3H mice bearing spontaneous mammary tumors (P.K. Lala and M. Terrin, unpublished) and mice bearing transplanted fibrosarcoma (Milas and Tomljanovic, '71). Somewhat similar response of *in vitro* colony forming cells has also been reported for C3H mice bearing a transplanted breast tumor (Hibberd and Metcalf, '71).

GENERAL REMARKS

Transplantation of several rapidly developing tumors in mice is found to produce significant effects on the dynamics of host hemopoietic cells, although none of these tumors, at least during the periods of study, were found to metastasize in the hemopoietic organs. There was a rapid influx of newly formed mononuclear cells into the tumor, largely from the host bone marrow. Young B lymphocytes at various stages of maturation represented a large component of this stream. These cells did not appear to proliferate within the tumor. An immigration of similar cells, or their precursors, including hemopoietic stem cells occurred from the marrow to the spleen, where at least some of them subsequently proliferated. The precise significance of these cell migration streams initiated by tumor transplantation remains to be determined. Migration of young B cells may reflect an initial host response ensuring an adequate supply of cells from which specific recruitment will take place leading to the antibody forming cells. It is obscure, however, why the lymphocytes do not show any local proliferation within the tumor. An increased hemopoietic stem cell content of the spleen, at least at the early stages, may be part of the same response directed towards lymphoid development.

ACKNOWLEDGEMENTS

This work was supported by the Medical Research Council of Canada and the U.S.P.H.S. grant CA1711-01 from the National Cancer Institute. Expert technical assistance by Mrs. D. Hodge and Miss D. Dixon is gratefully acknowledged. I am also indebted to several coworkers who participated in various areas of this study: M. Terrin, S. Garnis, C. Lind, L. Kaizer, R. Scott and G. Keeb; they have been acknowledged in appropriate places.

LITERATURE CITED

Bradley, T. R. and D. Metcalf. 1966. The growth of mouse marrow cells *in vitro*. *Aust. J. Exp. Biol. Med. Sci.* *44:* 287-299.

Delmonte, L., A. G. Liebelt and R. A. Liebelt. 1966. Granulopoiesis and thrombopoiesis in mice bearing transplanted mammary cancer. *Cancer Res. 26:* 149-159.

Grieves, M. F., J.J.T. Owen and M. C. Raff. 1974. *T and B lymphocytes. Origins, properties and roles in immune responses*. Excerpta Medica, Amsterdam.

Hibberd, A. D. and D. Metcalf. 1971. Proliferation of macrophage and granulocyte precursors in response to primary and transplanted tumors. In *Immunological Parameters of Host Tumor Relationships*. Vol. 1, D.W. Weiss, ed. Academic Press, New York. pp. 202-210.

Iturizza, R. G. and H. J. Seidel. 1974. Stem cell growth and production of colony stimulating factor in Rauscher virus infected CBA/J mice. *J. Natl. Cancer Inst. 53:* 487-492.

Lala, P. K. 1974. Dynamics of leukocyte migration into the mouse ascites tumor. *Cell Tissue Kinet. 7:* 293-304.

Lala, P. K. 1976. Effects of tumor bearing on the dynamics of host hemopoietic cells. In *Proc. N.C.I. Symp. "Cell Kinetics and Cancer Chemotherapy", Cancer Chemotherapy Reports Suppl*. In press.

Lala, P. K. and C. Lind. 1975. Cell kinetics in the spleen of tumor-bearing hosts. *Anat. Rec. 181:* 403.

Lala, P. K. and M. Terrin. 1974. Migration of cells with proliferative potential from bone marrow to the spleen of tumor bearing hosts. *Abstr. 3rd Ann. Meeting, Intl. Soc. Exp. Haemat.* p. 32.

McCulloch, E. A., T. W. Mak, G. B. Price and J. E. Till. 1974. Organization and communication in populations of normal and leukemic hemopoietic cells. *Biochimica et Biophysica acta 355:* 260-299.

Metcalf, D. and M.A.S. Moore. 1971. *Haemopoietic cells.* Amsterdam, North Holland. pp. 107-108.

Milas, L. and M. Tomljanovic. 1971. Spleen colony forming capacity of bone marrow from mice bearing fibrosarcoma. *Rev. Europ. Etudes Clin et Biol. 16:* 462-465.

Monette, F. C., B. S. Morse, D. Howard, E. Niskanen and F. Stohlman. 1972. Hemopoietic stem cell proliferation and migration after Bordella pertussis vaccine. *Cell Tissue Kinet. 5:* 121-129.

Okunewick, J. P., E. L. Phillips and P. Erhard. 1972. Increase in number of splenic transplantable colony-forming units in the SJL/J mice after infection with Rauscher leukemia virus. *J. Natl. Cancer Inst. 49:* 1101-1106.

Osmond, D. G. 1972. The origins, lifespans and circulation of lymphocytes. In *Proc. Sixth Annual Leucocyte Culture Conference.* M. R. Schwartz, ed. Academic Press, New York. pp. 3-32.

Osmond, D. G. 1975. Formation and maturation of bone marrow lymphocytes. *RES 17:* 99-114.

Osmond, D. G. and G.J.V. Nossal. 1974a. Differentiation of lymphocytes in mouse bone marrow. I. Quantitative radioautographic studies of antiglobulin binding by lymphocytes in bone marrow and lymphoid tissue. *Cell Immunol. 13:* 117-131.

Osmond, D. G. and G.J.V. Nossal. 1974b. Differentiation of lymphocytes in mouse bone marrow. II. Kinetics of maturation and renewal of antiglobulin-binding cells studied by double-labeling. *Cell Immunol. 13:* 132-145.

Raff, M. C. 1971. Surface antigenic markers for distinguishing T and B lymphocytes in mice. *Transplant. Rev. 6:* 52.

Seidel, H. J. 1973a. Hemopoietic stem cells in murine virus induced leukemia. I. Spleen colony assays in CBA mice after Rauscher virus infection. *Z. Krebsforsch. 79:* 123-134.

Seidel, H. J. 1973b. Hemopoietic stem cells in mice with virus-induced leukemia. II. Studies in C3H mice after Rauscher virus infection. *Z. Krebsforsch. 80:* 229-237.

Till, J. E. and E. A. McCulloch. 1961. A direct measurement of the radiation sensitivity of normal bone marrow cells. *Radiat. Res. 14:* 213-222.

Van Furth, R. 1970. The origin and turnover of promonocytes, monocytes and macrophages in normal mice. In *Mononuclear Phagocytes*. R. Van Furth, ed. Blackwell, Oxford. pp. 151-165.

Wendling, F., P. E. Tambourin and P. Jullien. 1972. Hemopoietic CFU in mice infected by the polycythemia-inducing Friend virus. I. Number of CFU and differentiation pattern in the spleen colonies. *Int. J. Cancer 9:* 554-566.

Wendling, F., P. Tambourin, P. Gallien-Lartigue and M. Charon. 1974. Comparative differentiation and numeration of CFUs from mice infected either by the anemia or polycythemia-inducing strains of Friend viruses. *Int. J. Cancer 13:* 454-462.

Teratocarcinoma: A Model of Stem Cell Differentiation

JOHN M. LEHMAN AND WENDELL C. SPEERS

Department of Pathology,
University of Colorado Medical Center,
Denver, Colorado

Numerous developmental systems have been defined that suggest a stem cell as the progenitor of various tissues and cells. Certain tumor systems have been described that contain stem cells and cells expressing varying degrees of differentiation. Three such models have given significant experimental data to support this fact. These are the murine teratocarcinoma, the squamous cell carcinoma of the rat, and primary mouse mammary tumors (Pierce, '67; Pierce and Wallace, '71; Wylie *et al.*, '73). All these tumors exemplify the phenomena of stem cell differentiation in a tumor.

Pierce ('67) has shown with the murine teratocarcinoma that a stem cell exists which is capable of proliferating and producing new stem cells and also "differentiating" into cells which are no longer capable of proliferation. The teratocarcinoma is a tumor which contains embryonal carcinoma, the stem cell, and numerous cells and tissues which are derived from all three embryonic germ layers (Pierce and Verney, '61; Pierce, '67). Stevens and Little ('54) described the murine teratocarcinoma which developed spontaneously in strain 129 mice from the germinal epithelium (primordial germ cell) (Stevens, '67; Pierce *et al.*, '67). Kleinsmith and Pierce ('64) showed following adaptation of these tumors to the ascitic form that cloned embryonal carcinoma was capable of producing a teratocarcinoma. These results demonstrated that the stem cell of this tumor was capable of producing more stem cells, and also phenotypically well differentiated tissues from all three embryonic germ layers. The differentiated cells in this tumor are not malignant and therefore may be truly end-stage differentiated cells (Pierce *et al.*, '60). The conclusion drawn from these experiments suggests that the teratocarcinoma may be a caricature of normal development (Pierce, '67). Therefore, this tumor might be utilized as a model system for the study of embryonic differentiation.

TISSUE CULTURE OF TERATOCARCINOMA

With these facts we adapted to cell culture the murine teratocarcinoma (Lehman *et al.*, 1974). It was obvious that this cell system *in vitro* was extremely complex, and had to be characterized to allow definition of the various cell types and their differentiation *in vitro*. Of prime importance was our concern whether the tissue culture system was analogous to the transplantable tumor, in respect to the various types of differentiated tissues. With light and electron microscopy and histochemistry, we analyzed the tissue culture line isolated from the teratocarcinoma that was maintained as a stationary culture for 4-6 weeks with frequent medium changes. These results are shown in Table 1.

TABLE 1

Tissue and cell types observed in the transplanted teratocarcinoma and teratocarcinoma cultivated in vitro

	In vivo	In vitro
Embryonal carcinoma	+	+
Parietal yolk sac	+	+
Neuroepithelium	+	+
Epithelium		
Squamous	+	+
Columnar	+	+
with cilia	+	0
Gland formation	+	+
Muscle		
Striated	+	+
Smooth	+	+
Cardiac	0	+
Cartilage	+	+
Bone	+	0
Bone marrow	+	0
Adipose tissue	+	+
Trophoblasts	+	+

Of those tissues identified *in vivo*, they were also consistently present *in vitro*, suggesting that the two systems were analogous.

Due to the complexity of the above cell system, we approached the problem in further experiments by isolating the stem cell, embryonal carcinoma, with the purpose of characterizing spontaneous differentiation in this system. This end was accomplished by selecting colonies of embryonal carcinoma which were preferentially removed from the culture dish with pancreatin. These cells have been cloned, and used in the studies described in this paper. They retain their potential to differentiate *in vitro*, and to produce a teratocarcinoma when injected into animals. The differentiated cell lines which have been isolated have low tumorigenicity; however when tumors are produced, they are composed of one cell type (parietal yolk sac, fibrosarcoma) not the multiplicity of cell types observed in a teratocarcinoma. These differentiated cell lines when cloned are unipotential, not multipotential. There is at present no one criteria to define the stem cell and the differentiated cells *in vitro*, rather a number of criteria are needed, of which the above are two. These markers include cytogenetics, morphology, growth characteristics, tumorigenicity and the response of the cells to various viruses (Table 2). The embryonal carcinoma cells exhibit a near-diploid chromosome number (39); however, differentiated clones which have been isolated and characterized are near-diploid or near-tetraploid (65-75). A unique difference between the embryonal carcinoma and differentiated cells is the response of the cell to DNA and RNA viruses (Swartzendruber and Lehman, '75; Lehman *et al.*, '75; Lehman *et al.*, '76). Polyoma and SV_{40} were capable of infecting the differentiated cells as assayed by the presence of the viral specific intranuclear T antigen. Polyoma will replicate in these cells and produce complete virus. However, SV_{40} will transcribe the early gene sequences (T antigen); but not the late functions, which are required for production of infectious viruses (Swartzendruber and Lehman, '75; Lehman *et al.*, '75). Transformation with SV_{40} has been observed with these cells. The response of the differentiated cells to SV_{40} and polyoma is similar to the response of mouse somatic cells to these viruses. When embryonal carcinoma cells are infected with these two viruses, no evidence of a virus function was detected (T antigen, V antigen, virus replication). Adsorption and uncoating progressed normally when measured by electron microscopy and labeled virus uptake (Swartzendruber and Lehman, in preparation). When

TABLE 2

Characteristics of embryonal carcinoma and differentiated cell lines

Characteristics	Embryonal carcinoma	Differentiated cells
Morphology	Multiple cell types	One cell type
Cytogenetics	Diploid	Diploid or tetraploid
Tumorigenicity	Highly tumorigenic (teratocarcinoma)	Low tumorigenicity (fibrosarcoma or neural or parietal yolk sac tumor)
Response to viruses		
SV_{40}	T antigen (-) Virus replication (-)	T antigen (+) Virus replication (-)
Polyoma	T antigen (-) Virus replication (-)	T antigen (+) Virus replication (+)
Mengo	Virus replication (+ ↓)	Virus replication (+)

infectious DNA was utilized to infect the embryonal carcinoma, no evidence of virus function was detected. Thus the embryonal carcinoma cells are refractory to infection with SV_{40} and polyoma. Mengovirus was used to compare the responses of embryonal carcinoma and differentiated cells to an RNA virus. Replication of Mengovirus was detected in both cell types; however, only 10% of the virus yield was produced in the embryonal carcinoma when compared to a differentiated cell (Lehman *et al.*, '76). These observations allowed us to distinguish between an embryonal carcinoma cell and a differentiated cell. Further studies are in progress to compare and define the response of these cells to virus infection, since this system offers a unique model to characterize the response of cells in different physiological and differentiation states to DNA and RNA viruses.

With a partial characterization of the embryonal carcinoma and the differentiated cells, we utilized certain compounds to direct or stimulate stem cell differentiation in this tumor. Several chemical compounds have been shown to affect the expression of differentiated cellular functions in tissue culture systems. 5-Bromodeoxyuridine (BrdU) will prevent or retard morphological and biochemical markers of differentiated tissues in experimental myogenesis (Bischoff and Holtzer, '70), and chondrogenesis (Abbott and Holtzer, '68) and adversely affect invertebrate embryonic development (Tenier and Brachet, '73).

Some preliminary results with the BrdU treatment of embryonal carcinoma cells have been reported (Speers and Lehman, in preparation). At concentrations of 1 to 20 μgm/ml of BrdU within 3-4 days, there was an increase in cells which were morphologically distinct from embryonal carcinoma. We will refer to these cells as BrdU-EC. Morphologically (light and electron microscopy), these cells resembled the differentiated cells that appear at the periphery of embryonal carcinoma colonies at 5-6 days after subcultivation; the treated cells were not obtained as a result of cell death or selection, since growth kinetics of the treated and untreated cultures were similar. If BrdU was removed after the change had occurred, the cells resembled BrdU-EC and did not revert to embryonal carcinoma. The response of BrdU-EC cells to SV_{40} and polyoma were studied, with the result that these cells responded to the two DNA viruses, as did differentiated cells of the teratocarcinoma. Polyoma intranuclear T, virus antigen and infectious virus was detected. Only the SV_{40}-T antigen was detected following infection of the BrdU-EC cells.

MATERIALS AND METHODS

The cell line utilized in these studies was preferentially selected as embryonal carcinoma, which retained the capability to differentiate *in vitro* (Lehman *et al.*, '74). At 4-6 days after subcultivation differentiated cells appeared around the periphery of the colony of embryonal carcinoma. These cells were characterized as differentiated by the criterion outlined in Table 2. The tissue culture medium utilized throughout these studies was Eagle's Minimal Essential Medium (GIBCO) supplemented with non-essential amino acids, 1X vitamins and 5% fetal calf serum (Microbiological Associates). Penicillin and streptomycin were added to a final concentration of 100 units/ml and 100 μg/ml, respectively. The cells were grown in Falcon 60 mm plastic petri dishes.

5-Bromodeoxyuridine, 5-iododeoxyuridine, thymidine, cytidine and ^{3}H-BrdU were dissolved in phosphate buffered saline and stored at -20°C as stock solutions after filter sterilization. Dilutions of the various compounds were made in growth medium and stored at 4°C in the dark and used within 7 days.

Autoradiography was utilized to follow the uptake and localization of ^{3}H-BrdU (specific activity 26.8 curies/mM). In these experiments 0.1 μc/ml of ^{3}H-BrdU was used. This technique has been reported previously (Lehman and Defendi, '70).

To measure the activity of alkaline phosphatase, cells were washed three times with PBS containing 6 mM glucose at 4°C, scraped from the plates, centrifuged, and resuspended in 1 ml distilled water followed by freezing and thawing three times. The samples were then assayed as described by Bowers *et al.* ('66). The reaction was carried out at 32°C and the change in absorbancy at 550 mμ was read in a micro sample spectrophotometer (Gilford 3400 System) at 10 second intervals. The average of six readings was recorded. Protein content of each sample was measured by standard methods (Lowry *et al.*, '51).

RESULTS AND DISCUSSION

After one day of BrdU treatment there was no appreciable difference between control and experimental cultures. However, at subsequent days it was obvious that the BrdU-treated cultures contained relatively and absolutely more BrdU-EC cells than the control cultures. An attempt was made to quantitate the relative number of BrdU-EC cells compared to embryonal carcinoma. This was done using the point counting method of point intersection (Henning and Meyer-Arendt, '63; Garner and Ball, '66). When the area of BrdU-cells to area of embryonal carcinoma cells was compared, a 4-5 fold increase in BrdU-EC cells was observed in the treated cultures at 4-5 days. This increase in surface area may possibly be a result of cell elongation and flattening. This has been described as an effect of BrdU on eukaryotic cells (Abbott and Holtzer, '68). As mentioned above, light and electron microscopy confirmed that there were morphological differences between the treated and untreated cultures.

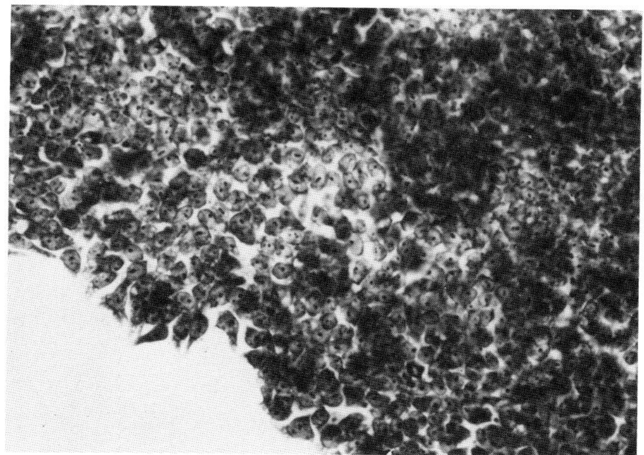

Fig. 1. Untreated culture at 4 days. Morphologically these cells resemble embryonal carcinoma. They are densely packed with multiple layers of cells. Magnification, 200X.

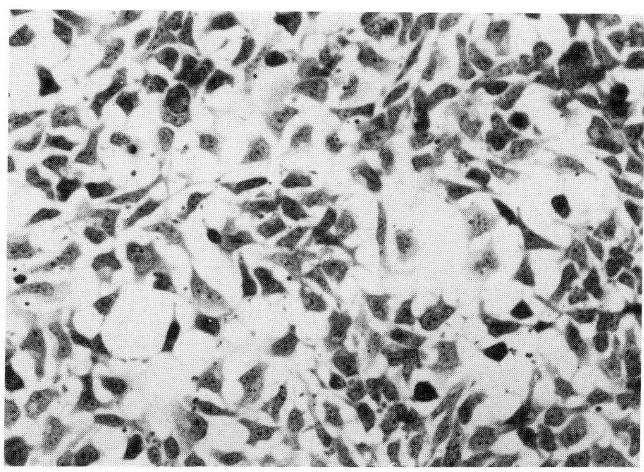

Fig. 2. Treated culture (10 μg/ml) at 4 days. The cells are less dense than the treated cultures, however, growth curves show similar cell numbers in treated and untreated cultures. These cells are flat and pleomorphic. Magnification, 200X.

The results of several experiments suggest that BrdU acts after incorporation into the cellular DNA in place of its analog, thymidine (Eidenoff *et al.*, '59). Autoradiography of cells treated with ^{3}H-BrdU showed almost exclusive nuclear localization which increased with time of exposure from 11 to 24 hours. Thymidine is known to inhibit BrdU effects. When thymidine was added in a 50-fold molar excess concurrently with the BrdU, there was no increase in BrdU-EC cells; however, there was no evidence of thymidine inhibition of division. BrdU added as a 24 hour pulse, followed by removal of BrdU with the addition of thymidine (50 μgm/ml), resulted in an increase of BrdU-EC cells (Table 3).

TABLE 3

Fixation of BrdU effect

Treatment	Presence of BrdU-EC cells
BrdU (20 μgm/ml) 24 hour pulse	Increased at 2-4 days
BrdU (20 μgm/ml) and thymidine (50 μgm/ml) 24 hour pulse	None
BrdU (20 μgm/ml) 24 hour pulse, then thymidine (50 μgm/ml)	Increased at 2-4 days
Thymidine (50 μgm/ml)	None
Cytidine (50 μgm/ml)	None
BrdU + cytidine	Increased at 2-4 days

Thus, thymidine added after the 24 hour pulse of BrdU could not reverse the morphological change of increased BrdU-EC cells (Table 3). Cytidine (50 μgm/ml) substituted for thymidine in these experiments showed no evidence of an inhibition of the BrdU effect. These results indicate that BrdU is incorporated into a stable thymidine resistant site within 24 hours after administration. This site is in the nucleus and is presumably the cellular DNA. Iododeoxyuridine (1-5 μgm/ml) also accelerated the appearance of BrdU-EC cells in the embryonal carcinoma cultures.

The specific activity of alkaline phosphatase was measured in the treated and untreated cultures. By day 2 an increase in activity in the treated cultures was observed, and continued to rise on days 3 and 4. The increase in activity of alkaline phosphatase paralleled the morphological change (Table 4).

TABLE 4

*Specific activity of alkaline phosphatase**

	Day 1	Day 2	Day 3	Day 4
Control cells	128	147	150	187
BrdU-treated cells (10 μgm/ml)	128	208	224	236

**International milliunits of activity per milligram of protein. Average of determinations on three separate cultures at each time point.*

This finding of increased alkaline phosphatase activity in the BrdU-treated cells is interesting since Bernstine *et al.* ('73) found decreased alkaline phosphatase activity in differentiated cell lines derived from teratocarcinoma cells. However, an increase in alkaline phosphatase activity has been reported in numerous mammalian cell lines after treatment with BrdU (Koyama and Ono, '73).

These results demonstrate that following treatment of embryonal carcinoma cells with BrdU, a morphological change is evident within 2-3 days. The ^{3}H-BrdU appears to localize in the nucleus and the BrdU effect is inhibited by thymidine, when both compounds are added simultaneously. When thymidine is added 24 hours after the BrdU treatment, the cells are not inhibited from progressing to the morphological change (BrdU-EC). This change was stable since removal of the BrdU did not reverse the morphological change. Previous studies showed that the BrdU-EC cells did not arise by selection or cell killing, since no difference in growth kinetics of the treated and untreated cultures were noted (Speers and Lehman, in preparation).

Morphologically these BrdU-EC cells resemble the spontaneously differentiated cells that appear around the periphery of embryonal carcinoma colonies. The obvious conclusion suggests that BrdU induced or accelerated this differentiation process. However, BrdU is known to block or arrest differentiation of normal differentiating cell systems *in vitro* (chondrogenesis, erythrogenesis, myogenesis). These results with the teratocarcinoma may suggest that BrdU may induced the opposite effect. The morphological data presented in this paper do not unequivocally prove this conclusion, since BrdU-treated cells change morphologically (Abbott and Holtzer, '68). Therefore, we have utilized other criterion that suggest this possibility.

The increased susceptibility of BrdU-EC cells to infection with polyoma or SV_{40} indicates that BrdU has changed the response of embryonal carcinoma cells to two DNA viruses. In previous studies, we showed that the embryonal carcinoma cells did not produce certain viral antigens or infectious virus, suggesting that these cells are refractory to infection with these viruses. However, the differentiated progeny of embryonal carcinoma are infected, and these cells respond as do mouse somatic cells (Swartzendruber and Lehman, '74; Lehman *et al.*, '76). The BrdU-treated cells also provide another system in which to study the response of differentiated cells to infection with polyoma and SV_{40}; and suggest that BrdU-EC cells are similar to the differentiated cells of the teratocarcinoma.

The alkaline phosphatase data demonstrated an increase in this enzyme following treatment of embryonal carcinoma cells with BrdU. Previous studies with embryoid bodies showed a decrease in the alkaline phosphatase as the embryonal carcinoma differentiates (Bernstine *et al.*, '73). Thus this biochemical marker which we utilized does not show a similar trend. These results may suggest that the BrdU increases the alkaline phosphatase level, as has been shown with other mammalian cell systems (Koyama and Ono, '73) or that possibly the BrdU is inducing specific differentiation of the stem cell, and that these cells have high alkaline phosphatase.

The results taken together certainly have not unequivocally demonstrated a BrdU effect which induces differentiation. However, these studies have suggested a trend which with further experimentation will conclusively prove or disprove this model. These preliminary results demonstrate the possible usefulness of the teratocarcinoma system in relation to stem cell differentiation. With more conclusive markers to identify the differentiated cells, this system may be a useful model for the study of stem cell differentiation.

ACKNOWLEDGEMENTS

The authors would like to extend their appreciation to Dr. G. Barry Pierce for his interest and helpful suggestions throughout these studies.

This work was supported by Grants CA-16030, CA-16823 and CA-13419 from the National Cancer Institute.

LITERATURE CITED

Abbott, J. and J. Holtzer. 1968. The loss of phenotypic traits by differentiated cells. V. The effect of 5-bromodeoxyuridine on cloned chondrocytes. *Proc. Nat. Acad. Sci., U.S. 59:* 1144-1151.

Bernstine, E. G., M. L. Hooper, S. Grandchamp and B. Ephrussi. 1973. Alkaline phosphatase activity in mouse teratoma. *Proc. Nat. Acad. Sci., U.S. 70:* 3899-3903.

Bischoff, R. and H. Holtzer. 1970. Inhibition of myoblast fusion after one round of DNA synthesis in 5-bromodeoxyuridine. *J. Cell Biol. 44:* 134-150.

Bowers, G. W. and R. B. McComb. 1966. A continuous spectrophotometric method for measuring the activity of serum alkaline phosphatase. *Clin. Chem. 12:* 70-89.

Eidenoff, M. L., L. Cheong and M. A. Rich. 1959. Incorporation of unnatural pyrimidine bases into deoxyribonucleic acid of mammalian cells. *Science 129:* 1550-1551.

Garner, A. and J. Ball. 1966. Quantitative observations on mineralized and unmineralized bone in chronic renal azotemia and intestinal malabsorption syndrome. *J. Path. Bact. 91:* 545-556.

Henning, A. and J. R. Meyer-Arendt. 1963. Microscopic volume and probability. *Lab. Invest. 12:* 460-464.

Kleinsmith, L. S. and G. B. Pierce. 1964. Multipotentiality of single embryonal carcinoma cells. *Cancer Res. 24:* 1544-1551.

Koyama, H. and T. Ono. 1973. Control of alkaline phosphatase activity in cultured mammalian cells: Induction by 5-bromodeoxyuridine, cyclic AMP and sodium butyrate. In: *Differentiation and Control of Malignancy of Tumor Cells.* W. Nakahara, T. Ono, T. Sugimura and H. Sugano, eds. University Park Press, pp. 325-342.

Lehman, J. M. and V. Defendi. 1970. Changes in deoxvribonucleic acid synthesis regulation in Chinese hamster cells infected with Simian virus 40. *J. Virol. 6:* 738.

Lehman, J. M., W. C. Speers, D. E. Swartzendruber and G. B. Pierce. 1974. Neoplastic differentiation: Characteristics of cell lines derived from a murine teratocarcinoma. *J. Cell Physiol. 84:* 13-28.

Lehman, J. M., W. C. Speers and D. E. Swartzendruber. 1975. Differentiation and virologic studies of teratocarcinoma *in vitro*. Excerpta Medica International Congress Series No. 349, Vol. 1, Cell Biology and Tumor Immunology, *Proceedings of the XI International Cancer Congress*, Florence, 1974, Excerpta Medica, Amsterdam, pp. 176-177.

Lehman, J. M., I. B. Klein and R. M. Hackenberg. 1976. The response of murine teratocarcinoma cells to infection with RNA and DNA viruses. In: *Roche Symposium on Teratomas and Differentiation*. D. Solter and M. Sherman, eds. In press.

Lowry, O. H., N. J. Rosebrough, A. L. Farr and R. J. Randall. 1951. Protein measurement with the folin phenol reagent. *J. Biol. Chem. 193:* 265-275.

Pierce, G. B., F. S. Dixon and E. L. Verney. 1960. Teratogenic and tissue forming potentials of the cell types comprising neoplastic embryonal bodies. *Lab. Invest. 9:* 583-602.

Pierce, G. B. and E. L. Verney. 1961. An *in vitro* and *in vivo* study of differentiation in teratocarcinomas. *Cancer 14:* 1017-1029.

Pierce, G. B. 1967. Teratocarcinoma: Model for a developmental concept of cancer. In: *Current Topics in Developmental Biology*. A.A. Moscona and A. Monroy, eds. Academic Press, New York, 2: 223-246.

Pierce, G. B., L. C. Stevens and P. K. Nakane. 1967. Ultrastructural analysis of early development of teratocarcinomas. *J. Nat. Cancer Inst. 39:* 755-773.

Pierce, G. B. and C. Wallace. 1971. Differentiation of malignant to benign cells. *Cancer Res. 31:* 127-134.

Stevens, L. C. and C. C. Little. 1954. Spontaneous testicular teratomas in an inbred strain of mice. *Proc. Nat. Acad. Sci. U.S. 40:* 1080-1087.

Stevens, L. C. 1967. Origin of testicular teratomas from primordial germ cells in mice. *J. Nat. Cancer Inst. 38:* 549-552.

Swartzendruber, D. E. and J. M. Lehman. 1975. Neoplastic differentiation: Interaction of Simian virus 40 and polyoma virus with murine teratocarcinoma cells *in vitro*. *J. Cell Physiol. 85:* 179-188.

Tenier, R. and J. Brachet. 1973. Studies on the effects of bromodeoxyuridine (BudR) on differentiation. *Differentiation 1:* 51-64.

Wylie, C. V., P. K. Nakane and G. B. Pierce. 1973. Degree of differentiation in nonproliferating cells of mammary carcinoma. *Differentiation 1:* 11-20.

Discussion, Session VI -
Stem Cells in Growth, Aging and Neoplasia

Dr. Moore's presentation

Dr. Lala enquired about the number of cell generations undergone by fetal and adult stem cells, respectively, during the formation of spleen colonies. Dr. Moore stated that stem cells from various tissue sources show marked differences in duration of proliferation under identical experimental conditions. Assuming Lajtha's figure of 6-8 hours for the cycle time of stem cells in spleen colonies to be representative of stem cells throughout the present colony transfer system it may be calculated that yolk sac stem cells can divide about 150 times (i.e. there are 150 population doublings) compared with 90 times for neonatal stem cells and 40 times for adult stem cells. These figures are approximate and subject to wide variation.

Dr. Miller raised the question of lymphoid development in fetal liver cultures, in the light of evidence suggesting that mouse fetal liver contains functional B cells by 14 days of gestation and functional B cell precursors as early as day 11. Dr. Moore commented that in the avian bursa of Fabricius cells having immunoglobulin in their cytoplasm, but not on their surface, have been shown by Kincade and Cooper to appear in early follicles 24-48 hours after stem cells enter the epithelial rudiment at 13-14 days. In mouse embryos the liver is colonized by stem cells at the 28 somite stage or approximately 10½ days gestation. Thus, it would be quite compatible with the avian evidence for cells synthesising immunoglobulin, but not bearing it on their surface, to appear in the liver by day 12, possibly by day 11. Incidentally, stem cells which colonize the bursa of Fabricius or the mouse embryo liver are assumed to be equivalent to CFU-S, although there is no direct evidence that this is so.

With regard to the predominance of erythropoiesis in the fetal liver, Dr. Wolf suggested that the local microenvironment as well as systemic factors may be important. His data agree with those of Dupont ('72) that the transfer of liver cells from 13 or 15 day old mouse fetuses into adult irradiated animals results in many more erythrocytic than granulocytic colonies in the recipient animals' spleens, the ratio being as high as 80 to 1 in Dr. Wolf's work. Thus, the CFU-S in the fetal liver appear to be largely committed to erythropoiesis, although toward the end of gestation the fetal liver cells tend to produce colony ratios typical of those produced by adult bone marrow cells. On the other hand, the transfer of spleen cells from anemic animals with stimulated erythropoiesis results mainly in granulocytic rather than erythrocytic colonies. These results suggest that systemic stimulating factors tend to cause stem cells to revert to a pluripotent form, while in the fetal liver the local microenvironment encourages an early erythroid commitment of stem cells. Dr. Moore emphasised that he has been totally unable to reproduce such data: studies of spleen colony morphology by histological sectioning show no significant difference in the distribution of erythrocytic, granulocytic, mixed and megakaryocytic colonies following transfer of cells from yolk sac, fetal liver or adult bone marrow, apart from an apparent shift to a category of more undifferentiated colonies using the early stem cell source. To reconcile the discrepancies between these two groups of apparently similar and straightforward experiments differences in animal strains and in the degree of anemia of the irradiated recipient animals were proposed.

In response to questions by Dr. Chui concerning the properties of yolk sac cells, Dr. Moore stated that, interestingly, injections of these cells do not cure the anemia of W/Wv mice, while the hemoglobulin produced by spleen colonies of yolk sac cell origin is of adult phenotype.

Dr. Micklem's presentation

Dr. Lala wanted to know the meaning of the term "quality of CFU" as used in the serial transfer experiments. Dr. Micklem explained that in this system CFU quality is an indirect concept, based on his observation that when large numbers of CFU are serially transplanted a decreasing proportion of them actually contribute to repopulation, the simplest explanation being that an increasing proportion of CFU become of a less efficient type. For blood cells the evidence is more direct; self-renewal is demonstrably more efficient in bone marrow derived colonies than in blood derived colonies.

Dr. Lala asked whether an evolution of new markers can be found in the residual host stem cells that eventually take over from transplanted stem cells. Dr. Micklem replied that such markers do occur frequently: they indicate a sequential development of host clones which appear, dominate and then disappear. In early work Ford attempted to follow these clones but Dr. Micklem himself has not done so.

Dr. Till suggested that the apparent reduction in the number of clones contributing to repopulation after serial transfer might be explained as the result of a birth-and-death process rather than a changing potential of the stem cell population with age. In the serial transfer system the clones multiply exponentially, while individual stem cells have a choice of either renewing themselves or disappearing by differentiating. Such a birth-and-death process can be shown to result in a skewed distribution, i.e. most of the progeny stem cells will be descended from very few of the initially transplanted stem cells. It resembles the problem of the distribution of family names in a human population: starting with a uniform distribution of family names the population undergoes a birth-and-death process which results, after many generations, in a skewed distribution in which large numbers of people have only a few of the family names. Dr. Micklem referred to his experiments using serial transfer of bone marrow cells from 2 separately identifiable syngeneic sources. Up to the third transfer the mitotic population of the bone marrow contains 50% of cells from each of the 2 donors, but in further recipients of this bone marrow the donor cell ratios are scattered widely, the cells of 1 donor type ranging from perhaps 5-10% to 90% in individual recipients. The transplanted CFU were assumed to show the same proportion of the 2 donor types as that of the mitotic populations. While this may be too simplistic an assumption Dr. Micklem wondered whether the birth-and-death phenomenon would manifest itself that suddenly. Dr. Till thought it could, but suggested that serial transfers of cells from individual donors to individual recipients, rather than pooled cells, would show the phenomenon most markedly.

Dr. Moens commented that populations are subject to at least 2 forces: large populations of surviving cells involve natural selection and the survival of the better adapted clones, while small numbers of cells may be influenced by "drift", a statistical variation due to small samples, as described by Wright and Fisher.

Dr. Lala's presentation

Dr. Moore commented that it would be ideal to compare the effects of progressing and regressing tumors on the host's hemopoietic and lymphoid system using spontaneous or syngeneic tumors, of which the C3H breast tumor could be the model of choice. The influence of immunomodulators on host cell migration would also be of interest. Dr. Lala agreed and reported preliminary data in C3H mice bearing mammary tumors which suggest that the numbers of CFU-S are increased 2-3 fold in the bone marrow as well as being markedly increased in the spleen. The decreased bone marrow CFU-S numbers noted in the preceding presentation might reflect differences between the two tumor systems or in the timing of the experiments; possibly, tumor-bearing animals may show an early fall in numbers of bone marrow CFU-S, undetected in spontaneous tumor systems, followed by a recovery and increase in CFU-S levels.

In reply to questions by Dr. Leblond, Dr. Lala noted that animals bearing ascites tumors die by the end of the second week and look sick by 7-8 days; they were examined during the first 5 days for cell migration studies and the first 10 days for assays of CFU-S. Animals live much longer when the same cells are grown as a solid tumor under the skin.

Concerning the mechanism of the hemopoietic changes induced by tumors Dr. Leblond asked if extracts of tumor cells were effective. Dr. Lala replied that with Dr. Preddie he has injected either sonicated tumor cells or their plasma membrane fractions and observed a 2-3 fold increase in number of CFU-S in the spleen after 2-3 days, whereas microsomal fractions were without effect.

Dr. Lehman's presentation

Dr. Leblond raised the possibility that biologically active substances such as erythropoietin or nerve growth factor might produce specific effects in the teratocarcinoma cultures. Dr. Lehman commented that Pierce has used biologically active substances in an *in vivo* model without inducing specific differentiation. Similar studies should be performed *in vitro* but some inductive agents, e.g. nerve growth factor, are not readily available.

Dr. Miller asked if the differentiation of cultured cells is expressed in other ways, such as the geographic distribution of various kinds of differentiated cells in the culture dish or the distribution of differentiated cell types observed over large numbers of cultures. Dr. Lehman replied that interesting looking distributions do occur but are difficult to quantitate. Striated muscle is found associated with cartilage, but only in prolonged cultures in which cells have piled up to form "tissues". The differentiating cells observed around the periphery of colonies are not identifiable morphologically -- they could be yolk sac, smooth muscle or neural cells. There is a need for suitable cell markers, particularly isozyme markers that could be used with histochemical or micro-acrylamide gel techniques. While the differentiation of a multiplicity of cell types has been demonstrated it has not yet been possible to show specific cell associations.

Dr. Cairnie enquired whether the teratocarcinoma grows better with BrdU than without it, noting that, while Holtzer finds a suppression of muscle cell differentiation by BrdU, Davidson has a cell line which both requires BrdU for the expression of differentiation and shows improved growth in the presence of BrdU. Dr. Lehman pointed out that the embryonal carcinoma cells themselves do not grow better with BrdU: they are apparently "pushed to differentiate" into morphologically changed cells which are then incapable of division. Attempts to produce growth of the differentiated cells have failed, apart from the transformation induced by SV40 and polyoma viruses.

Dr. Lala questioned whether "surface modulation" or some other term than "differentiation" should be used for the BrdU effect, in the absence of functional or biochemical evidence of differentiation. Dr. Lehman reiterated the need for multiple isozyme markers but referred to his use of alkaline phosphatase as a biochemical marker, the activity of alkaline phosphatase being higher in BrdU-treated cultures than controls. Although the literature is conflicting on this point a number of Japanese workers have also found that BrdU increases alkaline phosphatase activity in certain cell lines. Further, injection of BrdU-treated cells into animals results in a lower "take" of teratocarcinomas than the same number of untreated cells, consistent with a decreased number of stem cells and an increase in differentiated cells.

Dr. Oakberg remarked that the 129 animal is a pink-eyed chinchilla rather than an albino, a point which might be relevant in considering melanocytes and melanoblasts.

Dr. Till enquired about the possible properties of somatic cell hybrids formed by fusion of teratocarcinoma cells and cells having a differentiation marker. Dr. Lehman referred to one somatic cell fusion study by Ephrussi and Kahan in which embryonal carcinoma cells were fused with numerous L-cell lines but no differentiation markers could be shown in the tumors. A practical difficulty is to ensure that fusion occurs between the desired cell types, e.g. between differentiated cells or stem cells, respectively.

In reply to a question by Dr. Till concerning their karyotype, Dr. Lehman stated that cells of embryoid bodies, cultured for up to 3 years and composed primarily of embryonal carcinoma cells, have a near-diploid chromosome number of around 39. In contrast, those differentiated cell lines that are capable of proliferating are near-tetraploid. Measurements of DNA content per cell confirm the results of these chromosome counts and also indicate that as a proportion of cells become morphologically differentiated in the cultures a similar proportion of cells move from the diploid to the tetraploid peak of DNA content. The interpretation of this finding is uncertain: possibly those cells that are capable of proliferation have a hypo-tetraploid DNA content and chromosome number.

Dr. Lamerton asked if embryonal carcinoma cells can differentiate out completely, either in vitro or in vivo. Dr. Lehman answered that in some cultures apparently all the embryonal carcinoma cells do differentiate, so that if injected into an animal in doses of $40\text{-}50 \times 10^6$ cells the incidence of tumor take is almost 0. Moreover, in some cases the cultures appear to select for just one unipotential cell type.

Dr. Chui wanted to know if the sequence of differentiation in the cultures resembles that of embryogenesis in normal ontogeny and, in particular, whether a marker has been used to distinguish yolk sac, fetal or adult types of hemoglobin. Dr. Lehman commented that these tumors do not show much bone marrow *in vivo* and have only developed it once in culture. Martins and Evans suggest that the first cells to differentiate in embryoid body cultures are entodermal cells which may exert an inductive influence on the development of subsequent cell types.

Concluding Address

L. F. LAMERTON

*Institute of Cancer Research,
Sutton, Surrey, England*

Normally, the thought of having to prepare and give a final summary would have taken from me all enjoyment of a scientific meeting. This time, however, it has not, and I have enjoyed myself immensely. Scientifically it has been a most stimulating experience, with the added significance of being held in honour of Professor Leblond. Socially, the hospitality could not have been more generous, and this meeting has been a welcome opportunity for renewing old friendships and making new ones. I know I am speaking for everyone when I say how much we appreciate the work of the organisers, Drs. Cairnie, Lala, Clermont, Osmond and Till, and those behind the scenes - the secretaries and slide projectionists and helpers.

As regards summarising the meeting, of course I cannot; but I shall indulge instead in some generalisations, prejudices and personal impressions. Perhaps I can start off with the biggest generalisation of all, and ask why we came here to talk about stem cells. I think the answer has to be that we wanted to learn a little more about the mechanisms controlling the cell populations of the body, and how they can fail in diseases such as cancer. We all recognise that the behaviour of stem cells is by no means the only factor in the homeostasis of cell populations. It was Harvey Patt who once said that there is a lot of biology between the stem cell and the mature cell, and how right he was. At this meeting we have heard about a number of examples of control mechanisms that operate at the level of the maturing and mature cell, including probably the chalones.

In considering the nature of the forces acting on the stem cells, data presented on the gut and the skin gave further evidence that a normal rate of production of maturing cells may continue for some time after severe depletion of the early progenitor compartment. The same conclusion has emerged from other work on renewing tissues, for instance studies of the dynamics of growth of the rat incisor. This indicates that 'pull', as opposed to 'push', or in other words, a maturation stimulus, must play an important part in the cell population control of renewing tissues. On present evidence it would not be unreasonable to suggest that 'population pressure' may be of little significance in the control of stem cell populations, and that changes in proliferative rate of the stem cells are the secondary consequence of movement of cells out of the compartment. This being so, the question is what prevents the early progenitors from being run down. I don't think it is superfluous to remind ourselves that this question, together with that of the nature of the factors limiting continuous growth of stem cell populations, are the basic questions to which our experiments and discussions on stem cells are directed.

IDENTIFICATION OF STEM CELLS

We cannot get very far until we can identify the stem cells of the various tissues. In the case of the small intestine the elegant work that Professor Leblond and Dr. Cheng have done allows the crypt base columnar cells to be identified as the stem cells. The problem remains of how far the stem cell compartment extends up the crypt. In this context it is important to make the distinction, which Dr. Cairnie discussed, between 'functional' and 'potential' stem cells. The 'functional' stem cells - the cells which provide the permanent population of a renewing tissue - may be very small in number. However, in addition there are the 'potential' stem cells, which are those that have the capacity for repopulating the tissue under appropriate conditions. We have to be careful here. In the past one had tended to believe that certain types of maturation, for instance keratinisation of epithelial tissues or haemoglobin production in the erythroid series, represents an irreversible step towards loss of reproductive integrity. Nowadays, one can no longer be dogmatic. Recent work, particularly in the area of cell fusion, suggests that what have in the past been regarded as non-reproductive end-cells may, under the appropriate stimulus, be made to divide. Thus it becomes increasingly difficult to talk of any cells, save those that have lost their nuclei, as being irreversibly non-reproductive. In discussion of stem cells of tissues one should therefore restrict the concept of 'potential stem cells' to those with the capacity for long continued cell division under *physiological conditions*, and not take into account the capacity for division that may result from rather exotic laboratory manoeuvres. This necessarily introduces a certain vagueness, because we are undoubtedly unaware of a number of the stimuli for division that could be present under physiological conditions. The cell population measured in a clonogenic assay is generally taken to mean the total of functional and potential stem cells, but this will necessarily only be valid if the conditions of the clonogenic assay provide the maximum stimulus for proliferation.

Regarding the small intestine, there have been various suggestions made at this meeting concerning the size of the stem cell population, which has been equated with the clonogenic cell population. Radiobiological studies are giving information on this but do not yet allow a precise figure to be stated - the range of possible values is considerable, but it appears to be of the order of 50 to 100 cells, which is not inconsistent with the size of the cell population in the lower part of the crypt that is in exponential growth.

For the skin, the problem of identification of stem cells is now seen to be more difficult than it once appeared. The view held generally was that the monolayer of cells in contact with the basement membrane represents a homogeneous stem cell population. The work described and reviewed by Dr. Potten, based on histological and radiobiological studies, indicates that the population of clonogenic cells in the basal layer in the unstimulated skin may be only about 10%. The remaining cells are composed of obviously maturing cells, together with some of uncertain function. The hexagonal 'stacking' structure of cells in some types of stratified epithelium is fascinating, but I feel that its relevance to the problems of stem cell identification and control cannot be fully determined without parallel studies on epithelia not showing the structure, including skin stimulated into rapid division. Studies of the different types of cells in the lower levels of the epithelium during a period of increased proliferation, when the population of stem cells might be expected to increase, might also be of value for closer identification of the earliest progenitors.

The bone marrow, nowadays the most studied of all tissues from a dynamic point of view, is the one where there is most uncertainty about identifying the stem cell.

CHARACTERISTICS OF STEM CELLS

Professor Leblond discussed the characteristics of the stem cells of the small intestine and of other tissues where a reasonable identification of the population is possible. There certainly appears to be a degree of uniformity in the structure of stem cells, as shown by the electron microscope, with a close resemblance to embryonic cells. Other properties of the stem cells of the small intestine mentioned during the meeting concerned an apparently high sensitivity to tritiated thymidine and to actinomycin D. These indications need to be followed up.

A very interesting point that came out of the discussion of various tissues, and particularly the small intestine, was that the cells identified as stem cells often themselves showed evidence of commencement of maturation by the presence, for example, of specific cytoplasmic granules. The implication of this, from a conceptual point of view, is important. Perhaps we should not talk about a stem cell population as being distinct from the maturing population, but think in terms of a continuum of cells in various stages of maturation, with 'stop' points in the earlier stages, where maturation would be held up and the cells could continue to divide indefinitely. This would imply a heterogeneity of the stem cell population which has been suggested by various speakers, in particular with respect to the bone marrow.

The related problem is that of the factors that determine the limited proliferative capacity and life time of the later maturing cells, and here the type of work described by Dr. Altmann is beginning to give some interesting clues, in terms of limited stores of essential proteins, following cessation of production.

With regard to the proliferative characteristics of stem cell populations, it is probably a valid generalisation, on present evidence, to regard the stem cell population in unstimulated renewing tissues as turning over relatively slowly. The question is whether this indicates a relatively slow cycling of all the stem cells, or the presence of two populations of cells, one rapidly cycling and the other slowly cycling or 'resting' in the so-called G_O state. The indications from the small intestine is that the clonogenic cells include both the few slowly turning over cells at the base of the crypt and some part, at least, of the rapidly dividing population further up. In the skin I think it is true to say that there is evidence only of a slowly turning over population in the unstimulated state, and in the testis the stem cell is certainly slowly dividing. Regarding the bone marrow there is still much discussion. The extent to which the slowly turning-over populations contain true resting cells is very difficult to determine experimentally, as Dr. Blackett demonstrated. The finding by Hegazy and Fowler in unstimulated mouse skin, of a relatively narrow distribution of cycle times around a value of as long as 5 days, no longer makes it imperative to assume a resting state when intermitotic times of this order of duration are found. In view of these uncertainties, it would be much better if the term 'recruitment' of stem cells should, for the time being, be taken to imply either a shortening of a long cell cycle or a triggering out of a resting state.

From work presented on gut, by Dr. Altmann, and skin, by Dr. Fowler, it appears that the repopulation of depleted early progenitor compartments can be initiated without appreciable reduction in numbers of later maturing and mature cells. This implies that feed-back from the maturing cells is not a necessary condition for increased proliferation of stem cells, and that there is an internal size control, presumably related to some form of 'sensing' of the size of the stem cell population.

A concept is perhaps beginning to develop that the stem cell population is one that is highly shielded from the many types of stimulating and inhibiting factors that control the behaviour of the later maturing cells, and that the major external controlling factor concerns the withdrawal of cells on the maturation pathway.

THE MAINTENANCE OF STEM CELL NUMBERS

What then is the mechanism that prevents the number of cells at the earliest stage or stages of maturation from falling below a given level? The purely mathematically minded will see no problem. All that is needed is a small computer in the tissue, that organises the correct probabilities of stem cells to divide or to move out on the maturation pathway. I am afraid I would regard having to accept this as a last resort, and a confession of biological failure. Let us seek for biological mechanisms, and the area in which we have to seek is that of the microenvironment. The concept in the skin of the importance of contact with the basement membrane appeared at one time to be a very fruitful starting point. The recent evidence about the heterogeneity of the basal layer has brought difficulties, but I do not think we should forget the concept. The proliferating unit within a tissue is not an island, and the relationships with the stroma, with blood vessels and nerves, is a field where a great deal of work is required. It is very likely that the solution will require much work with disorganised tissues, and valuable leads are very likely to come from tumour systems in which the differentiation path can be varied, as in the mouse teratocarcinoma discussed by Dr. Lehman.

One aspect of the microenvironment concept is the interaction between cells, sometimes of very different types. Modern biology is full of surprises, and some very odd phenomena are now being recognised between cells - cells which need each other as well as those which hate each other. Reference at this meeting has been made to the possible 'managerial' function of the Paneth cells in the small intestine, but this area is at present one of hypothesis rather than of fact.

THE BONE MARROW

We have had a great deal of discussion on the bone marrow. I find it difficult to make up my mind whether the bone marrow is a good or bad model for looking at the generality of stem cell problems. It certainly has some unusual features, for instance in its possession of migrating stem cells which can increase greatly in number under certain conditions, as Dr. Lord pointed out. It is also a very complex system, and we are not at all sure what the earliest precursors look like and what their relationship with the various classes of later precursors is. On the other hand, with the availability of functional assays and other quantitative techniques, there is so much opportunity for study.

One area in which a great deal of information has been, and is, being obtained is the nature of the controls acting at the level of the maturing cell. Originally, with the discovery of erythropoietin, this was restricted to work on the red cell system, but the studies described by Dr. Moore showed how much progress was now being made on the granulocyte system. Both erythropoietin and the granulocyte colony stimulating factor (CSF) appear to exert a double action - increasing the proliferation rate of the cells on which they are acting and moving them on to a particular maturation pathway. A deeper knowledge of their mode of action should help a great deal in the understanding of fundamental mechanisms applicable to stem cell behaviour.

Another area in which work on the bone marrow is of particular significance to the subject of our meeting is the relevance of the microenvironment, especially of the stromal elements, in determining the pattern of maturation of stem cells. One cannot say that unanimity had been reached on this subject, but it is very important that the different points of view, as represented by Dr. Trentin and Dr. Till, should be resolved. One is tempted to say that some of the problem may lie in the present tendency for alphabetical haematology - CFU, ACU, ERA, etc. etc. Might I suggest that the terms applied to the haematological microenvironment problem, HIM (Haematopoietic Induction Microenvironment) and HER (Haematopoiesis Engendered at Random), be changed to the passionless IT (Inductive Territory)?

A third area in which bone marrow studies are providing data relevant to a basic problem of stem cell behaviour concerns the 'ageing' phenomenon, and the validity of Hayflick's concept of a limited number of divisions for somatic cells. A number of participants considered that a '50 doublings' limit was quite inconsistent with the known behaviour of renewing tissues. Dr. Micklem felt there was strong evidence of eventual failure of stem cell activity in repeated transplantations of bone marrow, and also for heterogeneity of CFU-S. There was nothing sacred about the limit of 50; it could be more. The situation remains unresolved.

Dr. Till reported his work on the search for surface markers of bone marrow stem cells. The primary aim of the work is to identify the stem cell population of bone marrow but, if the experiments are fruitful, it would provide a valuable tool in a much wider field of study.

THE TESTIS

The seminiferous epithelium is certainly the most fully worked out of the renewal tissues of the body from the point of view of the identification of the stem cell population, the spatial relationship of stem cells and the various types of maturing cell and the timing of the process, as the presentations of Dr. Clermont and Dr. Oakberg showed, though there are still important areas where there is not full agreement, concerning the existence and nature of the 'reserve' stem cells, and the irreversibility of the earlier maturation stages.

There were a number of interesting aspects of the relationships between stem cells and maturing cells in the testis discussed at this meeting, particularly concerning the connections between cells, and the syncytial structure present during maturation. As with bone marrow, we have to ask the question whether the testis is a special case and if, as we gain more knowledge of the dynamics of the various renewal systems of the body, we shall become more impressed by their differences rather than their similarities.

However, so far as the earliest progenitor cells of the

testis are concerned, the basic problem is common to other tissues. In describing the seminiferous epithelium of the monkey, Dr. Clermont said that the real mystery was why, on the average, each division of the stem cell gave rise to one stem cell and one maturing cell. The cell pattern observed precluded asymmetric or differential division, which is the same conclusion that has been reached for most other renewing tissues. How far is this a question of microenvironment, of intrinsic programming, or of messages between cells of which we at present know very little? This remains the fundamental question.

We have not solved many of the problems of stem cells at this meeting, but I feel that we are beginning to ask some of the right questions, and there is no doubt that progress now depends on closer integration of the different approaches and techniques that have been presented here. This, I am sure, is the thought that Professor Leblond would wish to leave with you.

May I now assume the role of Conference Chairman which, because of the skill and energy of the co-Chairmen, has been a sinecure. I would like to close the meeting by expressing again the thanks of all participants, and particularly those who have come from distant parts, to you, Professor Leblond, and all your colleagues, and to the various Canadian agencies which generously provided funds, for the privilege of participating in this Conference.

Subject Index

M